Y0-ACH-909

UTILITY MANAGEMENT FOR WATER AND WASTEWATER OPERATORS

UTILITY MANAGEMENT FOR WATER AND WASTEWATER OPERATORS

Frederick Bloetscher

Copyright © 1979, 1995, 2003, 2011 American Water Works Association.
All rights reserved.
Printed in the United States of America.

No part of this publication may be reproduced or transmitted in any form or by any means, electronic or mechanical, including photocopying, recording, or any information or retrieval system, except in the form of brief excerpts or quotations for review purposes, without the written permission of the publisher.

Project Manager: Melissa Valentine
Editor: Bill Cobban
Production Editor/Cover Design: Cheryl Armstrong
Production Services: TIPS Technical Publishing, Inc.

Disclaimer

Many of the photographs and illustrative drawings that appear in this book have been furnished through the courtesy of various product distributors and manufacturers. Any mention of trade names, commercial products, or services does not constitute endorsement or recommendation for use by the American Water Works Association or the US Environmental Protection Agency. In no event will AWWA be liable for direct, indirect, special, incidental, or consequential damages arising out of the use of information presented in this book. In particular, AWWA will not be responsible for any costs, including, but not limited to, those incurred as a result of lost revenue. In no event shall AWWA's liability exceed the amount paid for the purchase of this book.

Library of Congress Cataloging-in-Publication Data
Bloetscher, Frederick.
Utility management for water and wastewater operators / by Frederick Bloetscher.
p. cm.
Includes bibliographical references and index.
ISBN 978-1-58321-823-5
1. Waterworks. 2. Water utilities—Management. 3. Water-supply. 4. Sewerage. I. Title.
TD485.B55 2011
363.6'1068—dc22 2011005847

6666 West Quincy Avenue
Denver, CO 80235-3098
303.794.7711
www.awwa.org

Table of Contents

Appendixes on Enclosed CD

About the Author

Frederick Bloetscher, PhD, PE, is currently an assistant professor at Florida Atlantic University in Boca Raton, Fla., and president of Public Utility Management and Planning Services, Inc. (PUMPS), a firm he started in 2000 to provide local governments with help in the evaluation of utility systems, needs assessments, condition assessments, strategic planning, capital improvement planning, grant and loan acquisition, inter-local agreement recommendations, bond document preparation, consultant coordination, permitting, and implementation of capital improvement construction. He has over 25 years experience in local government and municipal water and sewer provision, including four years as a town manager in North Carolina. He holds a PhD in Civil and Environmental Engineering from the University of Miami, a Master of Public Administration from the University of North Carolina, and a BS in Civil Engineering from the University of Cincinnati.

Mr. Bloetscher is the past chair for the Water Resource and Sustainability Division Trustees for the American Water Works Association, past chair of the Technical and Educational Council Education Committee, and past chair of the AWWA Groundwater Committee. He currently chairs two education transfer committees and continues to serve as a trustee on the Water Resource and Sustainability Trustees. He has extensively published on water and sewer issues, and coordinates the technical program for the AWWA Florida Section annual conference. He has helped develop training materials for AWWA, including training materials related to this book. He can be contacted at P.O. Box 221890, Hollywood, FL 33022-1890 or h2o_man@bellsouth.net.

Acknowledgments

I wish to acknowledge the contributions to and support of this project by the following: Mark G. Lawson, longtime friend and co-worker, whose firm, Bryant, Miller and Olive, P.A., specializes in utility finance, impact fees, and assessments, who provided help with the legal aspects of rates, fees, and borrowing as well as counsel for many years on a variety of issues; A. John Vogt, professor at the School of Government at the University of North Carolina at Chapel Hill for his thoughts and wisdom; Whitfield R. Van Cott, former utilities director for the city of Hollywood, Fla.; George Jaskulsky, former ESS manager, city of Hollywood; Don Bayler, city of Pompano Beach; Bob Boyce (ret.), Lisa Blouin, Gail Kasha, Kim Dematteo, and Gail Thrasher, city of Hollywood; Dominic F. Orlando, public services director, Leo Williams, Jose Urtecho, and Jim Baker of the city of Dania Beach, Fla.; Fernando Vazquez and Michael Sheridan, former public services directors for the city of Dania Beach, Fla.; Patricia Varney, finance director, city of Miami Gardens, Fla.; Dan Colabella and Bruce Taylor, town of Davie, Fla.; Brian Shields, former DEES manager, city of Lauderhill, Fla.; H. Peter Schalt, Collier County (Fla.) Water/Sewer District; Brian Wheeler and Deb Beatty at the Toho Water Authority; Karl Kennedy, Calvin, Giordano and Associates; Jeanine D. Plummer, associate professor, Worcester Polytechnic Institute; William Jarocki, formerly the environmental finance center manager, Boise State University; David A. Chin and James D. Englehardt, professors, University of Miami; P.D. Scarlatos, professor, and Daniel Meeroff, associate professor, Florida Atlantic University; Ana Maria Gonzalez, Hazen and Sawyer, P.C.; John R. Proni, formerly at National Oceanographic and Atmospheric Administration and now at Florida International University; Terri Messner for help with dictation and stenography; Fabiola Roll, Bill Lauer, Gay Porter DeNileon, Melissa Valentine, and Molly Beach at AWWA in Denver; Bill Cobban, editor; Albert Muniz, Hazen and Sawyer, P.C., and John Stubbart in Lanai, for help with proofreading, review comments to AWWA, their input, graphics, and commentary; Mom, Dad, and Aragorn; and my fiancée Cheryl Fox, for her love, thoughts, and support.

DEDICATION

This book is dedicated to the hardworking engineers and operators in the water and wastewater industries, and especially to my fiancée Cheryl Fox.

The structure of the utility usually involves a governing body or board of directors. The governing board is tasked with ensuring that the appropriate policies are in place to provide fairness to all customers. The less political the board, the more it can focus on long-term goals and service reliability. Most local officials want to develop water and sanitary sewage systems that will meet the water and sewerage needs of the areas served by the utility, yet ensure that existing and future utility systems are constructed, operated, and managed at a reasonable or fair cost to the users (without outside subsidies), and to develop a system that is compatible with the area's future growth. Meeting this objective involves understanding the utility's operating environment, performing planning activities, and making financial decisions that will provide appropriate funding to achieve the utility's needs.

Executive managment reports to this board. Management's task is to ensure that the appropriate tools and resources, including personnel, are available to effectively provide service. It usually benefits the utility if the management staff possesses relevant expertise, education, and technical skills, such as engineering. Legal services should also be available to the management staff; many of today's issues are both technical and legal in nature. Permits are a typical example. Personnel with technical skills to demonstrate permit compliance and provide training are important and should report directly to management. The functions to ensure adequate personnel (human resources), finances (budget and accounting), and materials (purchasing) as divisions should report directly to management and should be held accountable to provide the assets and tools necessary to operate the utility system.

Figure 1-1 shows one example of how a utility might be organized, but there are endless variations. As a result, except for implementation of impact fees or operations that consistently violate regulations, there is no "one way" to provide service, meet the utility goals, or establish fees. The proof is that in the United States, for the most part, publicly owned water and sewer systems comply with all regulations on a consistent basis, but the backlog of infrastructure needs means that longer-term compliance may be an issue.

One objective of this book is to outline each of the areas on the organization chart while providing utility and local officials, both technical and nontechnical, with a basic understanding of what water and sewer systems consist of; what is required for proper operations, management, supervision, maintenance, and infrastructure investment needs; and appropriate fiscal standards. The guidelines used herein are principles that have been shown to be effective and understandable to the public, an important issue for elected and municipal management officials. To this end, there are four objectives for this document.

The first objective is to provide the reader with a basic understanding of the environment in which the utility operates. To accomplish this objective, a basic introduction to the components of a treatment facility that are necessary to treat water to appropriate standards, distribute the finished potable water to customers, and then retrieve the water for treatment after usage in the wastewater system is included. Much of the utility's costs are derived from the construction or maintenance of piping and treatment facilities. Ongoing upgrades are necessary to meet changing regulations affecting the water and wastewater industries. Because both public health and the health of the ecosystem are involved, the number of regulations with which utilities must comply is significant. Because utilities are often point sources for pollution, they have been, and remain, obvious targets for regulatory action. Understanding how to meet regulatory challenges and the focus of regulatory agencies is therefore imperative. Capital improvements and ongoing operations are

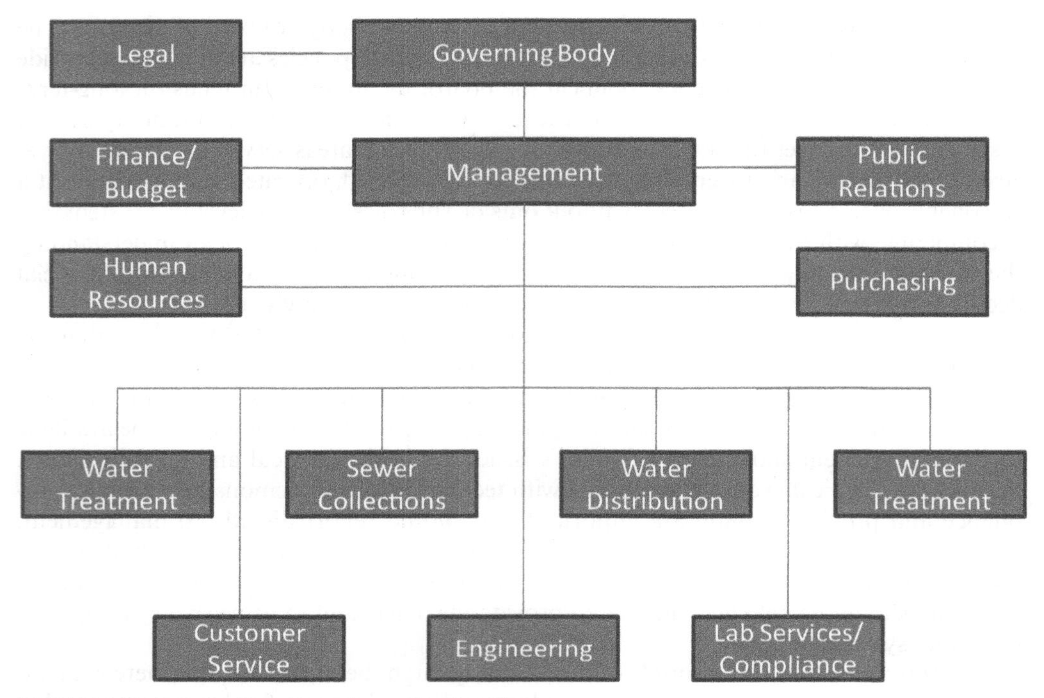

FIGURE 1-1 Organization chart for an ideal utility

intertwined with the ability of a utility to meet regulatory requirements. After the events of September 11, 2001, meeting security needs also requires ongoing maintenance and monitoring of the utility system, often without direct public knowledge. The second objective concerns the development, maintenance, and expansion of the utility system, from current day-to-day operations to long-range planning. Governing boards, whether local elected officials or appointed district board members, and the management staff must balance the needs and goals of the utility to meet the expectations of the public (i.e., to provide safe drinking water in sufficient quantities, and remove and treat wastewater so that it creates no adverse health effects). Because public perception of water quality and the quality of operations is often very different than reality, efforts are needed to preserve the public trust in the utility system's ability to deliver adequate supplies of safe drinking water. The bottled water and home faucet filter industries depend on negative public perception of local utilities to sell expensive home treatment units to residents in markets where perfectly safe potable water supplies are present at a small fraction of the cost.

Implicit in the operations and maintenance of the utility is the need to understand employees, including hiring, training, supervision, and matching skill sets with organizational needs. Because many utility managers come from the ranks of engineers or front-line operators, management and supervision are often areas where they receive training. "Management" encompasses not only managing people, but equipment, assets, finances and budgets, and ultimately infrastructure, which incorporates parts of all other aspects. The future loss of operators and engineers with experience in the utility industry as a result of retirement has been identified by AWWA and other entities as a serious concern for the industry.

Having a "vision" of the utility in the future is helpful. Visioning can only be developed by a public planning process that reaches some community consensus. Long-range planning is needed because current customers are not the only ones making demands for water—developers will also ask for water and wastewater guarantees, and adjacent communities may want to participate in bulk agreements to acquire water and/or wastewater service from the utility. The financial impacts of these added demands must be evaluated carefully. Regulatory and customer-driven demands may stretch limited available funds, requiring prioritization or reallocation of project monies.

The third objective is to provide enough background for interested parties to understand the proper financial basis for utility operations. This includes an outline of how enterprise funds work and why, policies to encourage self-sufficiency and effective operations, and development of appropriate fees. Rather than the general funds of most local governments, water and sewer utilities should rely almost exclusively on user fees. Fees can be developed to encourage growth, conservation, or other local objectives. Impact fees are an issue in many states. Significant litigation has occurred in states such as Florida that form the basis in case law for imposition of impact fees throughout the nation. Impact fees are designed to have growth pay for itself, as opposed to forcing existing residents to pay for growth. While this is an important goal, there are cases where balancing the needs of the local economy and the costs to current residents must be weighed. If large capital projects are anticipated, borrowing funds will be required. Bonds, loans, and grants are available to most municipal systems. Grants are severely limited (usually to small, economically depressed areas). Borrowing will require securing a payback, so policies that provide the lender with confidence are important for securing favorable borrowing rates.

The final goal of this document is the introduction of a series of emerging issues for water and wastewater systems. These issues include water conservation and the impacts that conservation may have on the financial stability of the system. Water and wastewater utilities are in business to sell water service; restricting service may have revenue and efficiency implications. Regionalization is a potential solution to some operating problems—especially for utilities with limited treatment capacity or disposal options. Other issues, such as pathogens in water, the presence of pharmaceutically active substances (the next frontier for regulations), and watershed protection are also discussed. These emerging issues are all technical in nature, however, and this book is not intended to provide detailed technical discussions.

The intended audience for this book is municipal management staff who desire to become more familiar with utility operations and new managers of utility operations. It is also an introductory text for those going into the field. Water and wastewater system functions are linked by regulations, operations, capital, and finances. To this end, this document is broken into the following chapters:

Chapter 2—Regulations Affecting Water and Wastewater Operations: provides a background on the water and wastewater regulations that impact daily and long-term operation of the utility system, as well as some historical perspective on the reason the rules have been implemented and what might be expected in the future.

Chapter 3—Water and Wastewater Operations: outlines what happens in the field and why, including a brief introductory discussion of various treatment processes that might be employed and the intended constituents to be removed by each. (Note: this section is not intended to be comprehensive—a good water and wastewater design handbook will provide additional detail.) Numerous photos of the processes are included to help the reader understand what the processes look like and envision the operations and maintenance that might be required.

Chapter 4—Planning for Operations and Maintenance of Assets: informs the reader about the personnel, equipment, and maintenance needs of the system to ensure it operates continuously as expected by the public. Topics cover asset management, including creation of an asset summary and methods to value it with time, depreciate it with time, and compare choices.

Chapter 5—Employment Rules and Managing Employees: outlines the expectations of supervisors, supervisory concepts, and goals.

Chapter 6—Supervision of Operations: includes numerous topics related to supervision.

Chapter 7—Management of the Utility: focuses on the expectations of the public, governing bodies, and regulatory agencies, and how the utility management team should strive to meet the expectations and convey their message.

Chapter 8—Utility Planning: examines a variety of plans, including emergency response planning.

Chapter 9—Developing the Capital Improvement Plan: includes information on how to determine which capital projects should be pursued.

Chapter 10—Capital Construction: focuses on the solicitation and hiring of consultants, contractors, and other entities, and construction management.

Chapter 11—Capital Funding Mechanisms: includes explanations of borrowing options and municipal leasing.

Chapter 12—Cost of Service Delivery: focuses on noncapital expenses and includes a comparison of issues involved with similarly located but differently sized utilities.

Chapter 13—Methods for the Establishment of Rates, Fees, and Charges: explains methods for the development and application of periodic charges and impact fees.

Chapter 14—Financial Policies and Evaluation of Financial Health of the Utility: describes various policies and vital measurements of the operation.

Chapter 15—Customer Service and Public Relations: discusses numerous topics in these important areas, as well as water conservation strategies.

Chapter 16—Emerging Issues: reviews issues that the author anticipates will affect utilities in the near future.

Chapter 17—Public Responsibility: discusses the responsibilities of operating a utility.

The appendixes include examples of Florida policies and ordinances that may provide some guidance to municipal utility systems. Several papers published by the author on emerging issues are also included for those who want more detail than is provided in chapter 16.

Readers of this book may focus on topics of particular interest to them, and links between the subjects are provided. Case studies are provided throughout the document to illustrate applications. Examples are provided to illustrate specific points. Diagrams and photos of utility components are included throughout as a means to facilitate understanding. The old saying "a picture is worth a thousand words" will ring true, it is hoped, as many officials may have infrequently visited the water treatment plant or observed the piping systems only during adverse conditions.

CHAPTER 2

Regulations Affecting Water and Wastewater Operations

The goal of water utilities is to provide its customers high-quality water that does not pose a public health concern, and to provide it in the sufficient quantities needed. Public health is the major issue that utilities face each day, and has been for as long as public water systems have been in existence. Adequate, safe water supplies are a part of the requirements for a stable society, so governments have attempted to ensure stability for society through public water works. The Roman aqueducts are an example where the Roman government constructed facilities to transport large quantities of water great distances to maintain public health in Rome. Wastewater was also recognized as potentially causing disease, so streams and rivers were also used to remove wastes from cities, although it was not until the 19th century that the direct connection between wastewater and disease was made.

Rivers have long served as both water supplies and wastewater disposal sites, yet the regulatory focus for many years was on water quality of drinking water. The history of drinking water regulations in the United States goes back to the 1890s. Little concern was focused on the disposal of industrial and domestic wastewater, and much of it was piped directly to water bodies as had been done historically. Dilution was thought to be the solution for wastes, so there was little concern until fires occurred on polluted water bodies. River fires are unusual in the United States, although not as unusual as in Europe where there has been a longer history of pollution, especially in Eastern Europe.

Figures 2-1 and 2-2 show the 1936 and 1952 Cuyahoga River fires. In both figures it is apparent that the river is literally on fire as a result of contamination from industrial and other urban contributions (this was not an infrequent event from 1930 to 1969). Interestingly enough, most people don't remember the 1969 fire, let alone the earlier ones. The 1969 fire was a relatively minimal one that only damaged one trestle of a railroad bridge.

Public outcry over the conditions of the nation's waterways, which are generally the nation's drinking water supplies, and finding volatile organic compounds in groundwater in Louisiana at the same time, indicated the need to regulate contaminants, and motivated Congress to approve legislation to address the failing quality of the nation's waters. The result was the approval of three major acts that affect water supply utilities. The National Environmental Policy Act was approved in 1970, followed by the Clean Water Act in 1972, and the Safe Drinking Water Act in 1974. Each remains in place. All have been reauthorized. Each affects a different aspect of the water system.

FIGURE 2-1 Cuyahoga River burning, 1936
Source: Cleveland Public Library.

FIGURE 2-2 Cuyahoga River burning, 1969
Source: Cleveland Memory Project, Cleveland State University.

NATIONAL ENVIRONMENTAL POLICY ACT

As mentioned, Congress approved the National Environmental Policy Act (NEPA) in 1970. It required reporting of environmental impacts and set up internal federal mechanisms to respond to them. NEPA requires federal agencies to consider the environmental impacts of their proposed actions in their decision-making processes and to provide reasonable alternatives to those actions in cases where they indicate an adverse environmental impact.

Title I of NEPA contains a Declaration of National Environmental Policy, which requires the federal government to use "all practicable means to create and maintain conditions under which man and nature can exist in productive harmony" (USEPA, 2009a). This act spawned the environmental impact statements that required large projects and those obtaining federal funds to consider the environmental impacts of the project. The NEPA process consists of three levels of analysis depending on whether or not an undertaking could significantly affect the environment. These three levels include categorical exclusion determination, preparation of an environmental assessment/finding of no significant impact (EA/FONSI), and preparation of an environmental impact statement (EIS). The act did not require the project sponsor to do anything, only to report the impacts. However, reporting was not enough. A mechanism was needed to respond to the impacts or to mitigate adverse environmental impacts, which led to the Clean Water Act, Public Law 92-500, in 1972.

CLEAN WATER ACT

The Clean Water Act, Public Law 92-500, was signed into law by President Richard M. Nixon in 1972. Section 101(a)(2) declares the intention of the act to be to achieve water quality sufficient for the protection and propagation of fish, shellfish, and wildlife, as well as for recreation in and on the water, by July 1, 1983. The preamble for the Clean Water Act is as follows (source: Public Law 92-500):

> The objective of this act is to restore and maintain the chemical, physical and biological integrity of the Nation's waters…

Note that cost was not a consideration in the act. Congress further stated that it is the nation's policy that the discharge of pollutants in toxic amounts be prohibited. As a result, this act affects water and wastewater utilities in four ways: surface discharges of treated wastewater to freshwaters, ocean discharges by wastewater plants, the disposal of concentrated process waters from water plants (such as concentrate from membrane facilities), and disposal of residuals (sludge). Implicit in the act also is that stormwater and agricultural runoff issues may affect potable water supplies and are subject to regulation, although neither was an initial focus of the act.

Legislation was first directed to wastewater because discharging it to a stream made it the source water for downstream communities, and the visual condition of the nation's waterways was obvious to most people. Hence, the Clean Water Act focused on trying to get waters clean at the point of pollution, which would meet the dual objectives of reducing pollution in drinking water and providing an ecological and public health benefit. If wastewater could be cleaned up before it went into the rivers, then this might reduce the amount of treatment that would be needed for drinking water. The intent was good and the focus was primarily on wastewater operations and setting the concentrations of contaminants allowed to be discharged from wastewater treatment plants.

At the same time, a variety of other issues were addressed, such as the attempt to reuse wastewater for beneficial uses like irrigation, dealing with industrial pretreatment so that metals and other contaminants that would disrupt the wastewater treatment process would not be discharged to the sewer system, and the idea that stormwater might contribute to problems. There was also an interest in separating sanitary and storm sewers, which were commonly together in the North and Midwest at that time.

Surface Discharges—Fresh Water

Discharges into surface waters, where surface waters remove the waste from the point of generation, is one of the oldest methods of disposing of waste. Reduction of the waste occurs downstream because of dilution and natural degradation processes (due to bacteria in the water). Given sufficient treatment prior to discharge, these mutual processes work to reduce the waste to relatively minimal levels. Failure to treat the discharge adequately will overload the natural attenuation ability of the water body, resulting in noticeable pollution.

In a surface discharge, the physical discharge point is usually a pipe on the side of a stream (for small systems) or a pipe into the bottom of the water body with one or more holes or diffusers (for larger systems). Each diffuser in the pipe has a percentage of the total flow going through it (see Figure 2-3). The pipe connects the effluent wet well or chlorination chamber to the discharge point in the water body.

The most important potential environmental effect of wastewater effluent discharge into a surface water body is a decrease in dissolved oxygen levels in the surface water downstream of the discharge. This is due to consumption of oxygen by the oxidation of organic materials in the wastewater. The process is defined by the Streeter-Phelps equation (Masters, 1998). For this reason, secondary treatment is generally required in the United States prior to discharge of wastewater effluent to surface water. (For more information on what this entails, see chapter 3.)

Other potential environmental impacts of surface discharge arise from the precipitation of metals and other heavy compounds on the bottom of the receiving water body, downstream of discharge. Heavy metal accumulation among benthic populations and the possible recycling of these metals through the food chain is a subject of current research. Industrial processes like mining are notorious for leaching metals. Figures 2-4 and 2-5 show an old mining area where the mines still exist, but have not been active for over 70 years, yet the water leaching is bright red in color.

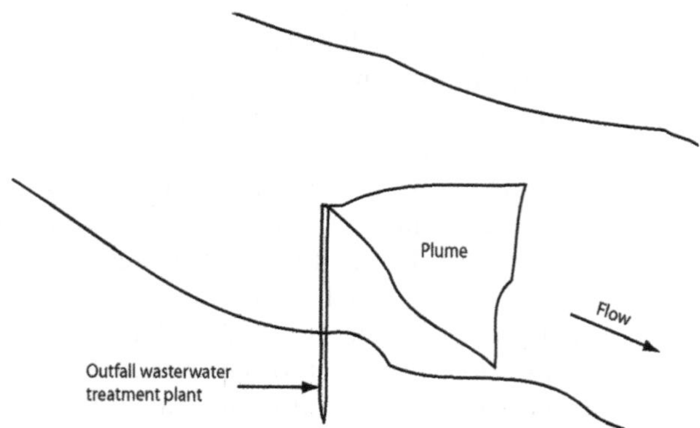

FIGURE 2-3 Typical means to discharge wastewater to surface streams

FIGURE 2-4 Typical example of a historical mining area

FIGURE 2-5 The dark water in this photo is a deep red color that comes from the leaching of metals for tailing in the mining area in Figure 2-4

The antidegradation policy of the act requires that any discharges to a water body be such that the water body's designated most beneficial present and future uses are not diminished. Water quality standards for discharges are established by the US Environmental Protection Agency (USEPA) based on the designated classification of the water body. The classifications range from Class I potable waters to Class V navigational waterways in ports.

In the 1980s, the USEPA delegated enforcement of the surface water programs to most states. The regulations state that no facility may discharge effluent that does not meet the applicable secondary treatment, basic disinfection, and pH levels prior to discharge into receiving surface waters. Both minimum and general criteria were established. To comply with the "minimum criteria," all surface waters must, at all places and at all times, be free from

- Domestic, industrial, agricultural, or other artificially induced discharges that alone or in combination with other substances or in combination with other components of discharges may create adverse impacts to aquatic life;
- Putrescent deposits or other nuisances;
- Floating debris, scum, oil, or other matter in such amounts as to form nuisances;
- Color, odor, taste, turbidity, or other conditions in such degree as to create a nuisance;
- Acute toxins;
- Concentrations that are carcinogenic, mutagenic, or teratogenic to human beings or to significant, locally occurring wildlife or aquatic species; and
- Dangers to the public health, safety, or welfare.

Mixing zones are often permitted in water bodies with significant flow volumes. Mixing zones supply initial dilution zones within which the discharge standards must be met because the surface water quality criteria apply to all surface waters outside of zones of mixing.

It should be noted that near-shore marine discharges and estuarine discharges are held to standards more stringent than most freshwater discharges due to the lack of motion of the water and the potential for nutrients to encourage algal blooms. Near-shore marine and estuarine discharges generally are required to meet advanced wastewater treatment (AWT) standards, which include reduction of phosphorus and nitrogen nutrients.

Ocean Outfall Disposal

Ocean outfall discharge of wastewater effluent also has a long history throughout the world. Mechanisms of public health protection achieved by outfalls include dilution, advection away from shore, and natural environmental degradation processes. Secondary treatment is usually required prior to discharge, but the use of ocean discharges is limited to coastal states. An ocean discharge has two primary differences separating it from other surface water discharges—higher density of the receiving water (due principally to salinity) that causes the discharge plume to rise toward the surface, and much greater volume of the receiving water. The result is that the ocean outfall dilution is immediate and considerable.

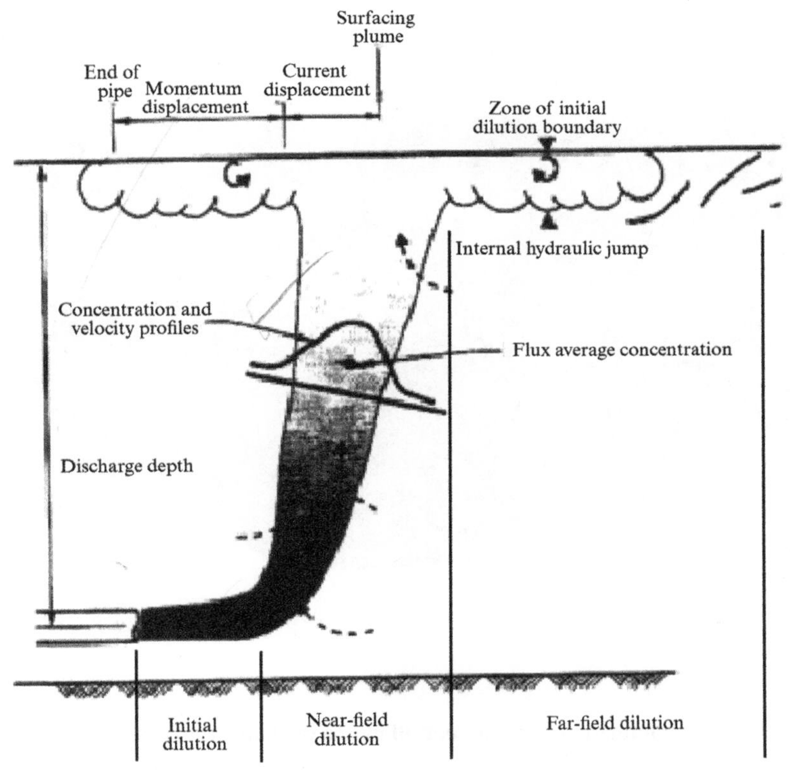

FIGURE 2-6 Outfall plume
Source: Hazen and Sawyer, 1994.

Buoyancy of the plume, marine currents, and turbulence result in three distinct phases of dilution (see Figure 2-6):

1. The initial plume dilution takes place from the time the effluent leaves the outfall until it begins to make a turn to the surface, which is a very short distance. The initial dilution phase is a rapid process, taking a few seconds.
2. Near-field dilution occurs as the plume rises to the surface. The generally freshwater effluent creates a turbulent, rising plume in the salt water because of buoyant forces exceeding the horizontal velocity forces. Once at the surface, the vertical momentum of the plume is translated to horizontal momentum, and the plume is radially dispersed at the surface. Near-field dilution takes place within the water column, the boil, and areas adjacent to the boil. This process takes a few minutes.
3. Far-field dilution results from the interaction of the mixing plume and surface convective processes. After the initial mixing processes, the subsequent dilution will be dominated by oceanic turbulence. The effects of buoyant spreading are much less in the far-field than in the near-field dilution.

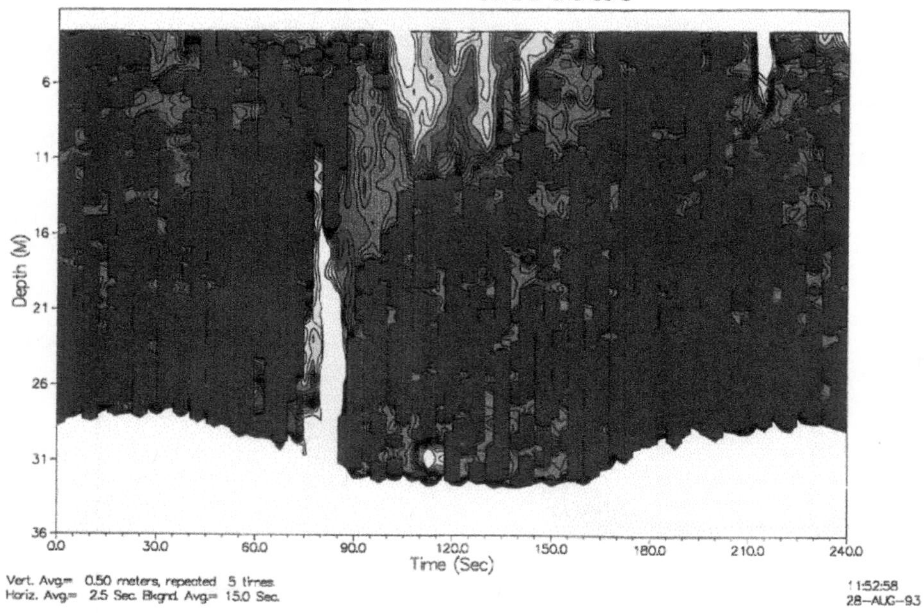

FIGURE 2-7 Acoustic photograph of ocean outfall plume in southeast Florida
Source: SEFLOE/NOAA, Hazen and Sawyer, 1994

FIGURE 2-8 Boil at the ocean

Figure 2-7 shows an acoustic photograph of a typical plume rise in southeastern Florida that mimics the theory illustrated in Figure 2-6 (these plumes rise to the surface—100 ft [30.5 m] while those in deeper waters may rise only 200 ft [61 m]). Figure 2-8 shows a photograph of the actual boil at the surface. What cannot be seen is any measurable difference in water quality or color. Waters within three miles of the coast are considered waters of the state. In all but three coastal states, the federal government has delegated responsibility for regulation of the open ocean outfall program to the states. Most of the surface discharge standards apply to ocean outfalls as well, but there are differences caused by wave and current regimes, as well as density.

All domestic wastewater treatment plants discharging to open ocean waters are required, at a minimum, to provide secondary-level treatment. Typical parameters for the effluent prior to discharge are not more than 30 mg/L chemical-biological oxygen demand (CBOD5) and 30 mg/L total suspended solids (TSS), or 85 percent removal of these pollutants from the wastewater influent, whichever is more stringent. The Southeast Florida Ocean Outfall Experiments (SEFLOE), conducted between 1986 and 1994, were designed to gather scientific information relevant to the future of ocean outfalls. SEFLOE characterized the initial dilution, dispersion, and decay rates of the effluent wastes in surrounding waters (Hazen and Sawyer, 1994). SEFLOE remains the defining research on ocean impacts in the southeastern United States.

Ocean outfalls must be designed to ensure structural integrity to minimize potential damage from natural occurrences (e.g., wave action and storms) and human activities (e.g., anchorage of boats). Ocean action during storms has the most significant potential impact to the integrity of the outfall, so there is a need to anchor boats properly. Outfalls must be designed, with respect to depth and location, to minimize adverse effects on public health and environmental quality.

SAFE DRINKING WATER ACT

The Clean Water Act did not address all problems, especially groundwater pollution found in a quarter of wells during the 1960s, so tap regulations came shortly thereafter in the form of the Safe Drinking Water Act (SDWA). The SDWA was used as guidance for drinking water regulations in the United States, Canada, and the international community. The SDWA built upon earlier public health laws in the United States that had been in existence since the 1890s.

Historical records indicate that the regulations or standards required for water systems were generally not applied until late in the 19th century. Up until that point, the basic philosophy was that water should be free of any organisms that cause disease and high quantities of minerals and other substances that could produce adverse physiological effects. Even the ancient civilizations realized that certain waters were healthful while others were not, the latter being water that contained some unseen contaminant that caused death or disease. With the realization that numerous epidemics, typically typhoid and cholera, were caused by waterborne diseases, people came to realize that the quality of drinking water could not be discerned simply by looking at it, tasting it, or smelling it. As a result, more stringent quality criteria were required to protect the public health.

The federal authority to establish drinking water standards for water systems originated in 1893 as a part of the Interstate Quarantine Act. This act authorized the US Public Health Service Director to "make and enforce such regulations as in his judgment are necessary to prevent the introduction, transmission or spread of communicable diseases from foreign countries into the States or possessions or from one state or possession to any other state or possession." The initial efforts to legislate improvements to drinking water quality focused on disinfection based on results found in Great Britain in the 19th century. Permanent disinfection was first started in Jersey City, N.J., in 1908, and adopted by over 1,000 utilities by 1914. The result was a reduction in typhoid deaths from over 110/100,000 persons to under 20/100,000 within 10 years (USEPA, 2009c). Because of the significant change it made in the amount of bacteria in the water, disinfection became a standard for the water industry very shortly thereafter in 1914.

The first formal set of standards was adopted in 1914. It focused on limitations on the number of fecal coliforms (chosen as an indicator microorganism) that were permissible within the water system. The initial federal public health rule required that fecal coliform counts be below 2 per 100 milliliters (mL). By the 1920s, chlorination had become routine in many water systems because of its ability to eliminate waterborne diseases. In 1925, there was a reduction in the limit of coliforms to 1 per milliliter due to increasing use of chlorine and the ease with which most large utility systems were able to meet regulations.

In the 1940s, metallic contamination, primarily in the form of lead and chromium, was identified as creating problems with brain development in children. In 1942, a federal advisory committee presented a series of drinking water regulations that established the number of water samples to be taken each month, the procedures for examining those samples, and the maximum permissible concentrations for lead, fluoride, arsenic, and selenium (as well as heavy metals that had deleterious physiological effects). These regulations became the primary drinking water standards. In 1946, a minor addition added hexavalent chromium and several other metals to the drinking water standards.

While the 1942 standards were primarily oriented toward lead paint, the fact that lead and chromium were in water supplies demanded that these issues be addressed. By the early 1960s, over 19,000 water systems had been identified. The Public Health Service found that only 60 percent of the systems surveyed met all drinking water standards and that small systems had the most deficiencies (USEPA, 2009b). The public was becoming aware of the chemical and industrial wastes that were polluting many of the surficial water sources for the water systems. In addition, concerns about radioactive pollutants arose. In 1962, maximum concentrations for a series of substances such as cyanide, nitrate, and silver, as well as some language on radioactivity were added to the regulations. The 1962 US Public Health Service standards set the stage for the SDWA in 1974.

The Safe Drinking Water Act was enacted by Congress in 1974. The passage of the SDWA mirrored a change in attitude of Americans toward the public drinking water supplies that was spurred by the detection of organic contaminants in drinking water in many areas of the country. As mentioned earlier, a 1969 USEPA survey indicated that nearly a quarter of 1,000 water systems sampled contained substances that resulted from industrial processes, organic solvents, and pesticides.

Public concern was raised especially because the contamination was found in groundwater, whereas previously contamination had been thought to be limited to surficial water sources. The SDWA and its associated amendments are focused on protecting public health from various contaminants in potable water supplies. Whether they address surface waters, groundwaters, or operation and treatment, the SDWA has basic requirements that must be met. The law had two sets of standards: primary and secondary. The primary drinking water standards were metals and other contaminants that mostly had been taken from the 1962 public health requirements. Secondary standards were set to encourage people to use water. These are aesthetic standards and focus on the fact that people will tend to drink water that looks cleaner. Thus in the 1970s, the idea was to try to make the water as appetizing as possible. Removal of biological activity remained an important part of water quality requirements, as it had been since the 1890s.

In the SDWA, not only were there fecal coliform requirements, but heterotrophic plate counts (HPC) were required to be below 500 colony-forming units per 100 mL. Monitoring requirements were developed to indicate compliance with the regulations, with the caveat that the standards could change over time as improvements were made to treatment equipment, analytical techniques, and instruments. The SDWA required that these water quality standards apply to all public drinking water systems serving at least 25 persons.

Congress authorized the federal government to establish health-based national drinking water regulations by setting maximum permissible levels of a significant number of pollutants in drinking water. The initial regulatory framework adopted the concept that if the concentration of constituents was below a certain level of contaminants, then the water would be safe. With metals, scientists learned that above a certain level, they were toxic, but below that they did not seem to have significant effects and, therefore, the standards were based on this understanding (with appropriate factors for uncertainty).

The original method for setting drinking water standards was termed *reference dose.* The concept was to conduct a dose–response assessment to determine at what point there would likely be an acute response of the organism tested. A series of doses was tested (generally in log scales) to determine the lowest observable effect level (LOEL) and point where there was no observable effect level (NOEL, see Figure 2-10). Depending on the organism, the dose was divided by 10, 100, or 1,000 to set the reference dose that became the standard. (For the divisor, in order for the number to be something other than 1,000, human subjects needed to be involved, which was rare, hence the standards were set by dividing the NOAEL result by 1,000.) This method for setting standards continued through the 1960s and 1970s.

Ensuing decades saw the efforts of the USEPA, established in 1972, focus first on metals and synthetic organic chemicals (SOCs) that occurred from industrial contamination of surface water supplies. USEPA reported that over 1,000 SOCs had been found nationwide in drinking water samples, although the potential risk for the majority of the population was deemed to be limited from an acute perspective. However, some of the compounds being found in water supplies were carcinogenic. Carcinogens are substances that have a potential to cause cancer, and many synthetic organic compounds, volatile organic compounds, and even some metals appeared to have longer-term carcinogenic potential. The existing method for setting water quality standards proved inadequate for suspected carcinogens that had both toxic effects as determined from the reference dose method, but also longer-term effects that appeared to create cancer.

In the early 1980s, USEPA's focus turned to volatile organic chemicals (VOCs) commonly used as solvents and found in groundwater systems throughout the United States. While the majority of VOCs were present at very low levels, pollution had occurred in these water supplies previously thought to be pristine (AWWA, 1987). In the mid-1980s, USEPA examined the legal use of pesticides, since numerous pesticides had been found in both surficial and groundwater supplies. Efforts are still ongoing to survey and assess the extent of pesticide contamination in water supplies.

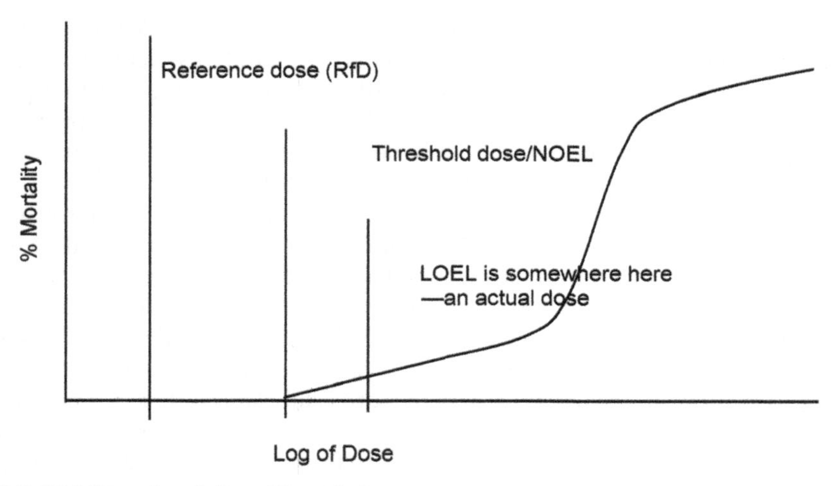

FIGURE 2-9 Risk-based cost-benefit analysis
Note: Threshold—NOEL—no observable effect level. LOEL—lowest observable effect level. RfD—NOEL/safety factors (USEPA limits [1970s]).

Throughout the period, USEPA continued to focus on microbiological outbreaks and the use of disinfectant techniques. From 1972 to 1981, there were 335 reported outbreaks of waterborne disease involving 78,000 people. Viruses contributed to 11 waterborne outbreaks involving 5,000 cases. Still, while these levels are historically lower and continue to decline, concerns over *Giardia* and *Cryptosporidium,* bacteriological contaminants, and viruses persist.

For the most recent 10-year period recorded (1997–2006), 137 waterborne disease outbreaks were reported to the Centers for Disease Control (CDC), with a total of 8,498 illnesses and 17 deaths (see Table 2-1; Barwick et al., 2000; Lee et al., 2002; Blackburn et al., 2004; Liang et al., 2006; and Yoder et al., 2008). Of the outbreaks with a known cause (101), 17 were attributed to chemical or toxin poisoning and 84 to pathogens. Bacteria were the most commonly implicated pathogen. Table 2-2 provides the breakdown of chemical and pathogenic agents, showing the highest number of outbreaks due to *Legionella, Giardia, Campylobacter,* norovirus, and *E. coli* O157:H7.

In 1986, Congress approved a comprehensive amendment to the Safe Drinking Water Act requiring USEPA to set regulations for nine contaminants by June 1987, 40 by 1988, and 73 remaining standards by June 1989. From that point forward, USEPA was to set standards and monitoring requirements for at least 25 conditional contaminants every three years. The 1986 amendments required USEPA to specify procedures by which utilities can comply with the maximum contaminant levels (MCLs) with the use of the best available technology (BAT) and techniques for those systems that cannot comply. A priority list of contaminants that may have adverse impacts on public health and are known to, or are anticipated to, occur in public drinking supplies was also to be compiled. Possible sources of these contaminants included industrial and chemical production and use sites, landfills, septic tanks, and runoff areas. USEPA was also given the authority to allow states to acquire primacy, i.e., the responsibility for enforcement of the SDWA, upon compliance with specific criteria.

TABLE 2-1 Waterborne disease outbreaks reported to the CDC, 1997–2006

Compiled from Barwick et al., 2000; Lee et al., 2002; Blackburn et al., 2004; Liang et al., 2006; and Yoder et al., 2008)

Years	Number of:			Outbreaks for which agent identified	Etiologic Agent				
	Outbreaks	Illnesses	Deaths		Chemical or toxin poisoning	Bacteria	Virus	Parasite	Mixed
1997–1998	17	2,038	0	12 (71%)	2	4	0	6	0
1999–2000	39	2,068	2	22 (56%)	2	9	4	7	0
2001–2002	31	1,020	7	24 (77%)	5	9	5	5	0
2003–2004	30	2,760	4	25 (83%)	8	13	1	1	2
2005–2006	20	612	4	18 (90%)	0	12	3	2	1
Totals	**137**	**8,498**	**17**	**101**	**17**	**47**	**13**	**21**	3

TABLE 2-2 Agents identified in waterborne disease outbreaks reported to the CDC, 1997–2006

Type	Agent	Number of outbreaks
Bacteria	*Campylobacter* (including *C. jejuni* and *C. lari*)	13
	Escherichia coli O157:H7	10
	Helicobacter canadensis	1
	Legionella (including *L. anisa, L. micdadei,* and *L. pneunophilia*)	27
	Salmonella (including *S. bareilly* and *S. typhimurium*)	3
	Shigella (including *S. sonnei*)	2
Parasites	*Cryptosporidium* (including *C. parvum*)	5
	Giardia intestinalis	14
	Naegleria fowleri	1
Viruses	Hepatitis A	1
	Norovirus	11
	Small round structured virus	1
Unidentified	None	31
Chemical	Bromate and other disinfection by-products	1
	Cleaning product	1
	Copper	5
	Gasoline by-products	1
	Nitrate	1
	Sodium hydroxide	3

USEPA was directed to promulgate these drinking water standards to protect consumers from adverse health affects. USEPA assessed the potential for harm from a given pollutant, evaluated the level at which effects may occur, and then determined how to attain maximum reduction of the harmful agents. When insufficient data are available, USEPA is mandated to include significant safety factors to provide a conservative standard. As a result, federal drinking water standards were designed to judge contaminant safety or the point at which risk to the consumer is minimized. All water systems are required to submit laboratory analysis to the regulatory agencies to demonstrate compliance with the regulations. The analyses currently include over 130 regulated substances. Submittals are made periodically, on a basis ranging from every three years to monthly, depending on the potential risk for the substance as judged on the USEPA criteria.

It was soon determined that even small concentrations of some substances might eventually lead to unpredictable effects. Since risk-based decision-making was in vogue in the 1980s, Congress proposed a risk assessment solution to address the risk and uncertainty associated with low dosages of carcinogens suspected of causing long-term effects. As a part of the 1986 revisions to the SDWA, USEPA developed risk-based methods to deal with carcinogens. Since no one really knows why cancer occurs, those substances that cause cancer are assumed to have some risk or exposure threshold. The concept can yield useful answers, but in most cases there are little data, or data that do not provide useful or definitive answers.

Risk assessment based on extrapolation of dose–response for chronic toxicity was believed to be the solution to assess 90,000 (and counting) chemical constituents for which there is minimal health effects information. This remains in place today. Most of the chemicals lacked data on their toxicity. The result was an assumption that there were no longer any thresholds and that minimal amounts might pose some risk to persons who were susceptible. The analyses create cumulative density functions or "S-curves," as shown in Figure 2-10, that measure dose versus effect. The doses are typically in logs, as are the effects. Hence, as noted in Figure 2-11, there are no zero-risk alternatives.

Because there are no zero-risk alternatives, and because there are finite limits to the amount of resources that can be used to either define risk or maximize risk reduction, the concept of "acceptable risk" has been developed. From the USEPA perspective, an acceptable risk is the 1 in 10,000 annual or 1:1 million lifetime chance that an impact will

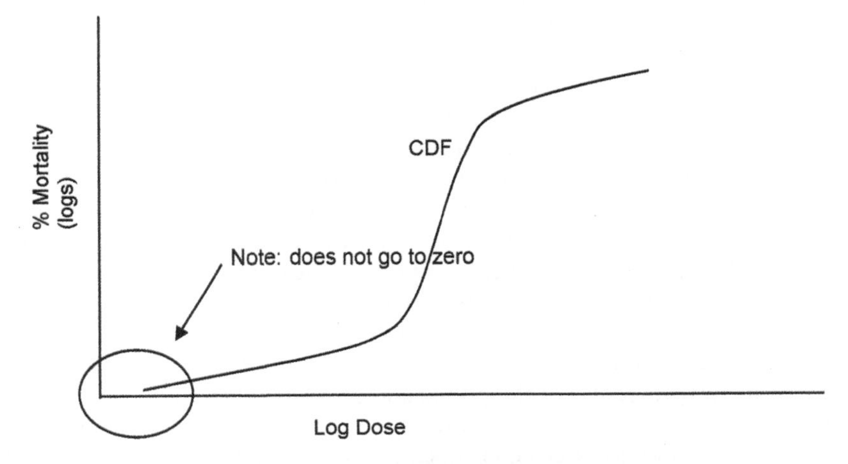

FIGURE 2-10 Typical dose-response curve (Highlighted area enlarged in Figure 2-11.)
Note: CDF=cumulative distribution function

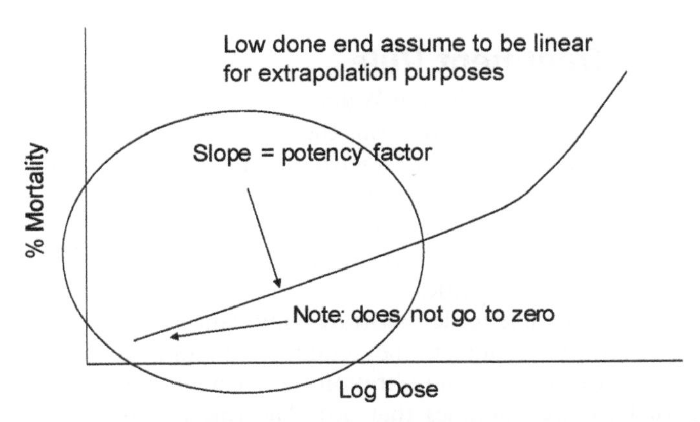

FIGURE 2-11 Typical dose-response curve. The area on the far left is assumed to have a constant slope called the *potency factor*, which can be looked up on the IRIS database.

occur to the general public as a result of an activity. Acute responses (immediate impacts), e.g., microbial water quality regulations, may be written around the probability that less than 1 in 10,000 will contract a disease from drinking water in a given year. For carcinogens, the tendency is to use the 1:1 million lifetime (chronic) risk more frequently than acute exposures, since the effects may not manifest themselves for many years. Regulations are typically written around these "acceptable risks." The results were to combine risk with cost–benefit analysis. The risk cost–benefit analysis remains highly controversial because it basically tries to put a price on public health and human lives.

In environmental impact assessments, risk–cost assumptions may be used. Figure 2-12 shows an example of such an analysis whereby the acceptable risk is defined, and the associated cost of compliance is calculated. As risk decreases, the cost rises. In some political circles, much of the focus is on whether risk or cost should be the limiting factor in determining acceptable risk. Obviously if cost is the limit, the risk would likely be much higher. Political conflicts over resources are one of the problems encountered with the acceptable-risk concept. The public, of course, wants the risk minimized, whereas industries lobbying Congress are more concerned about initial costs.

Limited data and conflicting data create confusion and uncertainty in the risk assessment process. From a regulation standpoint, conservatism is beneficial; from a conservative political viewpoint, such conservatism in setting a regulation level may create significant costs. Public distrust of science and the political process, together with a lack of understanding of risk applications, further muddies the issue. However, when weighing local decisions on expenditures, it should be noted that neither the preface to the Clean Water Act nor the 1996 amendments to the Safe Drinking Water Act mention cost as a part of the laws, only protection of the public and ecosystem health.

The resulting slopes on the cumulative density function (CDF) at the low dose level are called *potency factors* and, for a given chemical, may be found on USEPA's IRIS Web site. The problem with the risk-based method is that public perception can be significantly different from reality. People readily grasp events where consequences are high but the likelihood is very rare; they have difficulty grasping the high-incidence, low-risk consequences.

Surface Water Treatment Rule

Under the auspices of the Safe Drinking Water Act, a series of separate regulations were promulgated that affect utilities. The first was the Surface Water Treatment Rule (SWTR). The SWTR was published in the *Federal Register* by the USEPA on June 29, 1989. Significant revisions occurred in August 1996, when the Safe Drinking Water Act was reauthorized by Congress. The intent of the rule is to require that systems that use surface waters for water supplies filter their water (except in limited circumstances) and apply disinfection. The rule focuses on larger utilities.

Those systems that are able to demonstrate compliance with the source water quality criteria, meet the inactivation (contact time) requirements, and maintain an effective watershed control program are able to avoid filtration. Very few utilities can meet this test (Seattle and New York City are examples that do). The rule focuses on the inactivation of viruses and removal of *Giardia lamblia* cysts by removing turbidity and suspended solids that interfere with the disinfection process. The inactivation requirement for *Giardia lamblia* is 3-logs and viruses is 4-logs. Significant amounts of suspended solids require additional chlorine be used in areas where the organisms might hide and avoid disinfection,

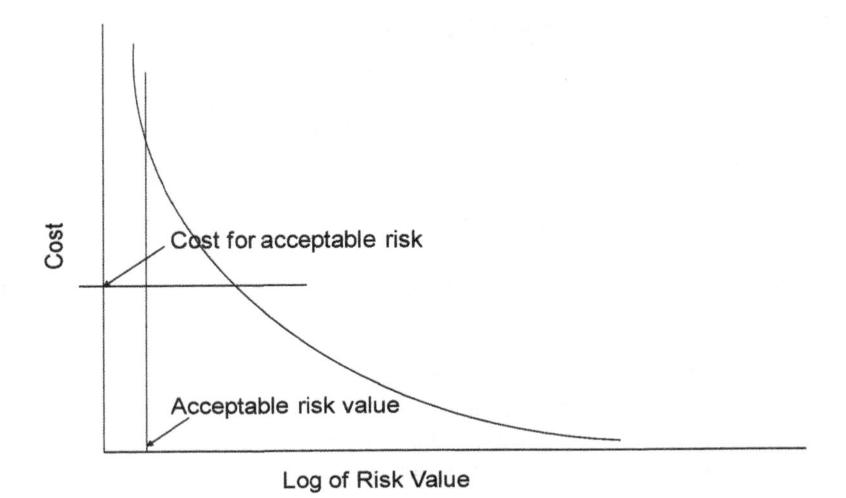

FIGURE 2-12 Risk-cost analysis

which may create conflicts with the portion of the SDWA's Disinfectant/Disinfection By-Products (D/DBP) Rule having to do with DBPs. The rule established maximum contaminant levels (MCLs) for turbidity, heterotrophic plate count for bacteria, *Giardia lamblia* cysts, *Legionella spp.*, and enteric viruses, along with publishing a set of approved testing methodologies for determining compliance (these can be found in 40 CFR Part 141). In addition, a source considered groundwater under the direct influence of surface water is also required to be filtered and disinfected. Appendix M of the SWTR Guidance Manual—AFT Protocol stressed the use of particle counting for determining if appropriate filtration was occurring and to serve as a surrogate for *Giardia* and *Cryptosporidium*. It was limited to engineered systems (pilot or full scale).

In addition to surface water systems, the rule affects groundwater sources that may be directly connected to a surface water. The concern is that if there is minimal filtration that occurs between the surface water body and the water withdrawn from wells, then contaminants, especially microbiological constituents, may contaminate the water source. Water sources that meet this criterion are considered "groundwater under the direct influence of surface water (GUDISW)." "Ground water under the direct influence of surface water is defined as: any water beneath the surface of the ground with (i) significant occurrence of insects or other macroorganisms, algae, organic debris, or large diameter pathogens such as *Giardia lamblia*, *Cryptosporidium parvum* or (ii) significant and relatively rapid shifts in water characteristics such as turbidity, temperature, conductivity, or pH which closely correlate to climatological or surface water conditions" (USEPA, 2008). If an aquifer is determined to be groundwater under the direct influence of surface water, and therefore vulnerable to contamination by disease-causing organisms found in surface water, the well water must be treated under the same requirements as a surface water system, meaning mandatory disinfection and filtration.

Long Term 1 Enhanced Surface Water Treatment Rule

On Feb. 13, 2002, USEPA finalized the Long Term 1 Enhanced Surface Water Treatment Rule (LT1ESWTR). The purposes of the LT1ESWTR are to increase monitoring and control of microbial pathogens, specifically the protozoan *Cryptosporidium*. *Cryptosporidium* came to the regulatory forefront as a result of the incident in 1993 in Milwaukee when over 400,000 people were affected by water contaminated by *Cryptosporidium*. In addition, several other minor outbreaks of *Cryptosporidium* in Nevada, Oregon, and Georgia, and a finding by the Science Advisory Board in 1990 that identified drinking water contamination, specifically disease-causing microbiological contaminants (such as bacteria, protozoa, and viruses), as the greatest remaining health risk management challenge for drinking water suppliers and as one of the most important environmental risks. The LT1ESWTR expanded the application of the Surface Water Treatment Rule to surface water small systems serving populations less than 10,000, which need to meet the more stringent filtration requirements. The potential for compliance issues with the disinfection by-products rules arose again, as it did under the SWTR. Optimization of disinfection practices and alternatives such as eliminating prechlorination disinfection to reduce disinfection by-products are required to be evaluated to ensure that microbial protection is not jeopardized if significant changes are made to the disinfection practices.

The LT1ESWTR amends the existing Surface Water Treatment Rule, strengthening microbial protection for surface water systems, including provisions specifically addressing *Cryptosporidium,* in addition to requirements that water systems must continue to meet existing requirements for *Giardia lamblia* and viruses. Specifically, the rule includes

- Maximum contaminant level goal (MCLG) of zero for *Cryptosporidium*;
- 2-log *Cryptosporidium* removal requirements for systems that filter;
- Strengthened combined filter effluent turbidity performance standards;
- Continuous turbidity monitoring at all individual filters, with combined filter turbidity monitoring every four hours;
- Disinfection profiling and benchmarking provisions to demonstrate that their disinfection by-product concentrations for total trihalomethanes (TTHM) are less than 0.064 mg/L and haloacetic acids (HAA5) are less than 0.048 mg/L;
- Subjection of systems using groundwater under the direct influence of surface water to the new rules dealing with *Crypdosporidium*;
- Inclusion of *Cryptosporidium* in the watershed control requirements for unfiltered public water systems;
- Requirements for covers on new finished water reservoirs; and
- Sanitary surveys, conducted by states, for all surface water systems, regardless of size.

The 2-log removal (99 percent) of *Cryptosporidium* can be accomplished via conventional filtration and disinfection or strengthened combined filter effluent (CFE) turbidity performance requirements.

Long Term 2 Enhanced Surface Water Treatment Rule (LT2ESWTR)

USEPA proposed the LT2ESWTR on Aug. 11, 2003, and finalized it Jan. 5, 2006, as a result of a consensus among the members of the Stage 2 M-DBP Federal Advisory Committee. The committee consisted of organizational members representing USEPA, state and local public health and regulatory agencies, local elected officials, tribes, drinking water suppliers, chemical and equipment manufacturers, and public interest groups (USEPA, 2005). USEPA published the agreement in a Jan. 5, 2006, *Federal Register* notice.

Like the SWTR and LT1ESWTR, the LT2ESWTR applies to all public water systems that use surface water or groundwater under the direct influence of surface water. The stated purpose of the LT2ESWTR is to reduce disease incidence associated with *Cryptosporidium* and other disease-causing microorganisms in drinking water.

The current regulations require filtered water systems to reduce source water *Cryptosporidium* levels by 99 percent (2-log), but recent data reviewed by USEPA on *Cryptosporidium* indicated that additional treatment is necessary for certain higher-risk systems (USEPA, 2005). As a result, this rule will supplement existing regulations by targeting additional *Cryptosporidium* treatment requirements to higher-risk systems. Systems initially monitor *Cryptosporidium* in their water sources for a period of two years to determine treatment requirements. To reduce monitoring costs, small filtered water systems will be permitted to monitor for *E. coli*, and will monitor for *Cryptosporidium* only if their *E. coli* results exceed specified concentration levels.

The rule applies to all systems that use surface water or groundwater under the direct influence of surface water (USEPA, 2008). The rule also contains provisions to reduce risks from uncovered finished water storage facilities and to ensure that systems maintain microbial protection as they take steps to reduce the formation of disinfection by-products. The largest systems (serving at least 100,000 people) began monitoring in October 2006 and the smallest systems (serving fewer than 10,000 people) began monitoring in October 2008.

Filtered water systems will be classified in one of four treatment categories (bins) based on their monitoring results. Most systems are expected to be classified in the lowest bin and as a result will not face additional requirements. Systems classified in higher bins must provide additional water treatment to further reduce *Cryptosporidium* levels by 90 to 99.7 percent (1.0- to 2.5-log), depending on the bin. Systems will select from different treatment and management options in a "microbial toolbox" to meet their additional treatment requirements. All unfiltered water systems must provide at least 99 or 99.9 percent (2- or 3-log) inactivation of *Cryptosporidium*, depending on the results of their monitoring (USEPA, 2005). After completing monitoring and determining their treatment bin, systems usually have three years to comply with any additional treatment requirements. Systems must conduct a second round of monitoring six years after completing the initial round to determine if source water conditions have changed significantly.

Other requirements include (USEPA, 2005):

Uncovered finished water reservoirs. Systems that store treated water in open reservoirs must either cover the reservoir or treat the reservoir discharge to inactivate 4-log virus, 3-log *Giardia lamblia*, and 2-log *Cryptosporidium*. These requirements are necessary to protect against the contamination of water that occurs in open reservoirs.

Disinfection benchmarking. Systems must review their current level of microbial treatment before making a significant change in their disinfection practice. This review will assist systems in maintaining protection against microbial pathogens as they take steps to reduce the formation of disinfection by-products under the Stage 2 Disinfectants and Disinfection By-Products Rule, which USEPA is finalizing along with the LT2ESWTR.

The LT2ESWTR allowed for a demonstration of performance (DOP) for riverbank filtration as a pretreatment alternative for systems that filter but have high *Cryptosporidium* concentrations in the raw water. In addition, under existing regulations (SWTR alternative treatment provision), any state or primacy agent can grant bank filtration credit for *Giardia* or *Cryptosporidium* removal. Thus a system may avoid construction of a filtration plant based on site-specific data (which creates some confusion).

Ground Water Rule

The Ground Water Rule (GWR) was approved in 1992 and updated on Oct. 11, 2006, with publication in the *Federal Register* on Nov. 8, 2006. The rule also applies to any groundwater system or water system that mixes surface water and groundwater if the groundwater is added directly to the distribution system and provided to consumers without treatment. It does not apply to groundwater sources that are GWUDISW.

The GWR was developed after evaluation of data on waterborne illness outbreaks indicated that there is a subset of groundwater systems (GWS) that are susceptible to waterborne viral and bacterial pathogens and indicators of fecal contamination in their wells. USEPA took a targeted risk-based approach to require groundwater systems that are identified as being at the greatest risk of contamination to take action to protect public health. The GWR requires that groundwater systems disinfect water upon withdrawal unless the utility can demonstrate that the water meets the requirements for "natural disinfection" or if the system qualifies for a variance. Ostensibly, this rule was passed in 1992 to deal with unchlorinated well systems. However, the most recent update to the GWR, in conjunction with regulations aimed at improving water quality from surface systems, was to "evaluate groundwater for fecal contamination, and, if source is at risk, then the utility must disinfect in order to provide 4.0-log inactivation of viruses."

Periodic sanitary surveys of systems require the evaluation of eight critical elements of a public water system and the identification of significant deficiencies (e.g., a well located near a leaking septic system). The eight elements are

- Water source;
- Water treatment systems and method;
- Distribution system;
- Finished water storage;
- Pumps, pump facilities, and controls;
- Monitoring, reporting, and data verification;
- System management and operation; and
- Operator compliance with state requirements.

Sanitary surveys have been conducted for years by local and state health departments, but the GWR requires that the sanitary surveys focus specifically on these eight areas of the treatment and distribution operations. With the passage of the GWR, the focus of the sanitary survey conducted by the state is to look at critical components of the water system. There are two monitoring provisions:

Triggered monitoring for systems that do not already provide treatment that achieves at least 99.99 percent (4-log) inactivation or removal of viruses, and

Assessment monitoring when a system (that does not already treat drinking water to remove 99.99 percent [4-log] of viruses) identifies a positive sample during its Total Coliform Rule monitoring.

USEPA recommends that the following risk factors be considered by states in targeting high-risk systems (USEPA, 2008):

- High population density combined with on-site wastewater treatment systems;
- Aquifers with restricted geographic extent, such as barrier island sand aquifers;
- Sensitive aquifers (e.g., karst, fractured bedrock, and gravel);
- Shallow unconfined aquifers;
- Aquifers with thin or absent soil cover; and
- Wells previously identified as having been fecally contaminated.

The sanitary survey provisions in this rule build on existing state programs established under the 1989 Total Coliform Rule and the Interim Enhanced Surface Water Treatment Rule and give states the authority to define both outstanding performance and significant deficiencies. For those systems that already treat drinking water to reliably achieve at least 99.99 percent (4-log) inactivation or removal of viruses, the rule still requires regular compliance monitoring to ensure that the treatment technology installed is reliably removing contaminants.

The rule indicates that the states must complete the initial sanitary surveys by Dec. 31, 2012, for most community water systems (CWSs) and by Dec. 31, 2014, for CWSs with outstanding performance. If a groundwater system sanitary survey shows a significant deficiency or a fecal-positive groundwater source sample (either by the initial triggered sample, or positive repeat sample, as determined by the state), the groundwater system must implement one or more of the following corrective action options (USEPA, 2008):

- Correct all significant deficiencies (e.g., repairs to well pads and sanitary seals, repairs to piping tanks and treatment equipment, control of cross-connections);
- Provide an alternate source of water (e.g., new well, connection to another public water system);
- Eliminate the source of contamination (e.g., remove point sources, relocate pipelines and waste disposal, redirect drainage or runoff, provide or fix existing fencing or housing of the wellhead); or
- Provide treatment that reliably achieves at least 4-log treatment of viruses (using inactivation, removal, or a state-approved combination of 4-log virus inactivation and removal).

Disinfectant/Disinfection By-Products Rule

The use of chlorine invokes the D/DBP, which regulates trihalomethanes and other carcinogenic organic compounds that can be produced from chlorine disinfection. These regulations apply to all drinking water regardless of source water. When water is disinfected with chlorine, and organics are present in the raw water, certain disinfection by-products may be formed in the finished water. Some of these disinfection by-products are trihalomethanes. Since trihalomethanes are thought to be carcinogenic, reducing levels is desirable, without diminishing disinfection capability. If the source water is groundwater, the USEPA Groundwater Rule may require that the water be disinfected upon withdrawal unless the water meets the requirements for "natural disinfection" or if the system qualifies for a variance. The major problem this rule poses is that to achieve the disinfection desired for raw water, the amount of chlorine by-products is significant, thus creating a conflict between violating fecal coliform and disinfection by-product standards.

Underground Injection Control (UIC) Program Regulations

The SDWA requires USEPA to protect underground sources of drinking water. USEPA's permitting authority to govern underground injection programs results from rules promulgated in 1981 pursuant to the Safe Drinking Water Act under 40 CFR 144 and 146. These regulations focus on design, construction, and operation of injection wells and on monitoring the impact of the waste. Figure 2-13 shows the construction of a typical well. Figure 2-14 shows how the well works. The regulations were aimed at regulating disposal of waste via underground injection, especially the injection of hazardous materials, which is common in some parts of the United States. As a part of the delegation of federal programs to the states during the Reagan administration, 40 states now administer their own programs for all or a portion of the underground injection control (UIC) program.

Upward migration is the concern. The regulations restrict injection where the water quality of an aquifer may lend itself to be a future source of drinking water (i.e., waters with less than 10,000 mg/L total dissolved solids [TDS]). To accomplish this, injection well systems are required to monitor the potential for degradation of the quality of other aquifers adjacent to the injection zone that may be used for other purposes. The regulations require abandonment of injection wells where the injected wastes can be shown to be migrating into any potential underground source of drinking water (USDW).

The federal regulations include an extensive set of definitions concerning injection wells. The federal regulations segregate injection wells into the classes listed below.

Class I injection wells. Used by generators of hazardous waste or owners and operators of hazardous waste management facilities that inject those hazardous wastes beneath the surface. The waste is required to be injected beneath the lowermost formation within a quarter mile of a well bore for an underground drinking water source. This requirement includes other industrial and municipal disposal wells that inject fluids beneath the lowermost formation containing potable drinking water supplies.

Class II wells. Used to inject fluids that are brought to the surface in connection with conventional oil and natural gas production, or enhanced recovery of oil and natural gas, and the storage of hydrocarbons.

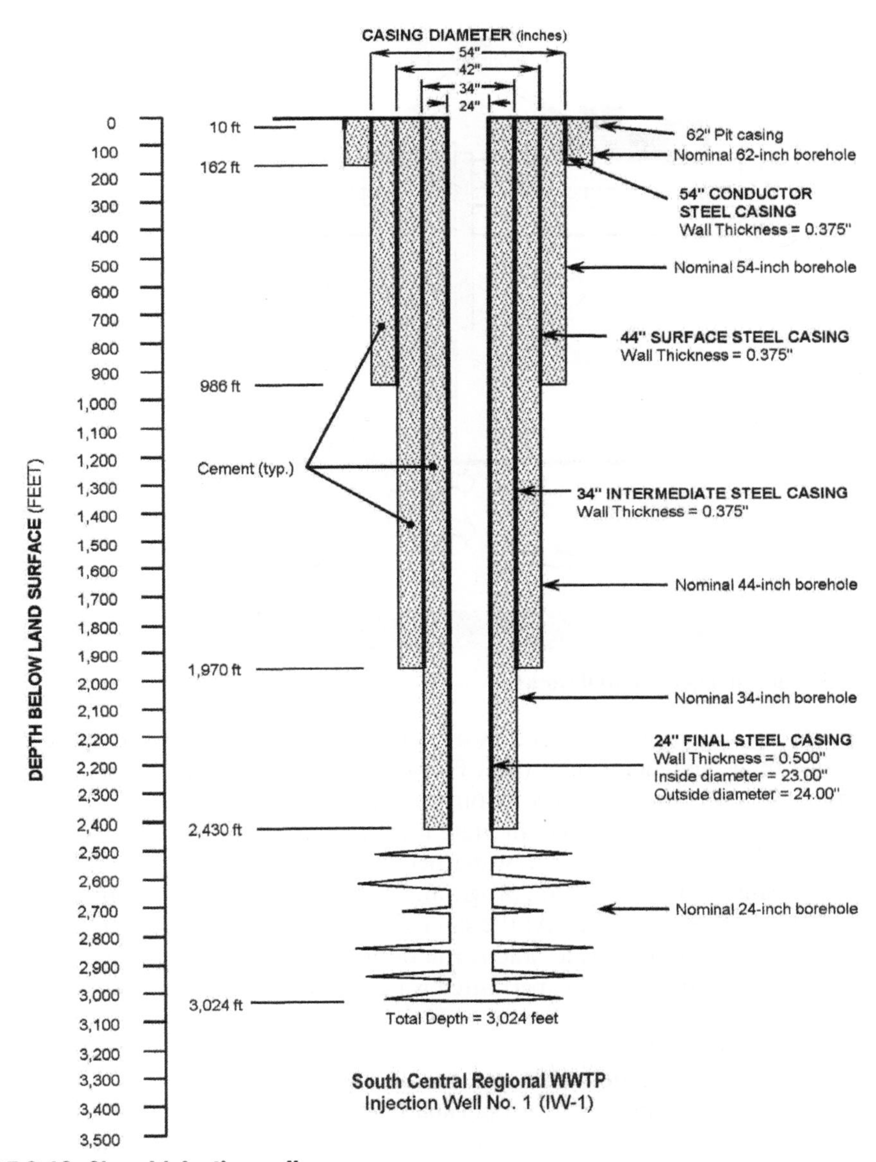

FIGURE 2-13 Class I injection well
Courtesy: Hazen and Sawyer, PC.

Class III wells. Used for the extraction of minerals.

Class IV wells. Used by generators of hazardous radioactive waste, which inject water below the lowermost drinking water zone.

Class V wells. Wells that are not included in class I, II, III, or IV. (Aquifer storage and recovery wells are classified as class V wells.)

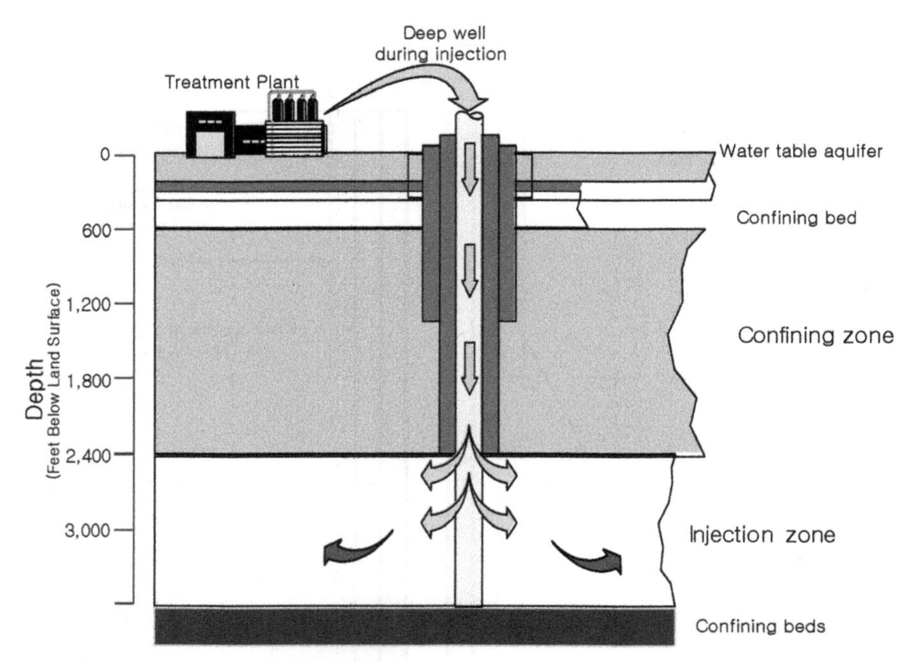

FIGURE 2-14 How an injection well works

The state regulations generally follow one of three paths: rules that are more stringent and extensive than the federal rules (e.g., Florida), those that are minimal, or those that focus mostly on water rights issues as opposed to injection concerns (western states). Some states add the caveat that the injectate must remain in the injection zone, and that there is no unapproved interchange of water between aquifers. Where state UIC programs exist, they typically mirror the federal program. Ongoing issues require continued input by USEPA to address the differences. At the same time, state injection programs are likely to be affected by a series of other rules and regulations that may or may not have federal references, such as rules for water use, permitted water consumption, reclaimed water, wastewater treatment quality, and/or indirect potable reuse.

Class I injection wells and class V systems are the two areas that affect utilities. Municipal wastewater injection wells must meet the technical and permitting requirements of the federal and state (where applicable) UIC programs for class I injection wells, including construction and operation permits.

Another injection program involves aquifer storage and recovery wells used to manage water supplies. ASR programs are regulated as class V wells. SDWA regulations govern the water being stored and recovered, as opposed to the UIC regulations, which protect the native and surrounding waters in the aquifer. The applicability of different SDWA regulations depends largely on the source of the stored water. A further discussion of this concept is included in appendix F.

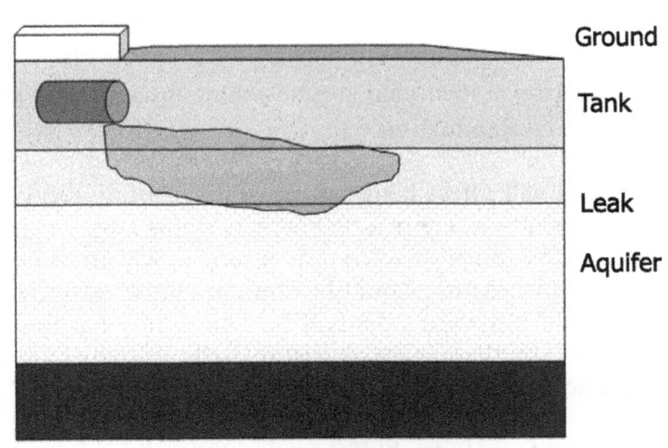

FIGURE 2-15 An example of a surface problem that could affect groundwater

Watershed/Wellhead Protection

The SWTR addressed protection of the watershed. Watershed protection regulations can be broadly defined as a means to reduce the threat to water supplies from contaminants in runoff or as a result of surface activities. Watershed protection is a requirement of the SDWA amendments of 1986 (Section 1428), and states are required to submit a plan to implement source protection. Utilities should be aware of the impact of surface activities on their water supply and make additional efforts where needed to protect it. Figure 2-15 presents an example of a surface problem that could affect downstream surface water supplies (or groundwater). Improperly abandoned wells are in violation of proper watershed protection strategies, as are active wells where the slabs are in poor condition and may allow surface inflow.

Water table aquifers can be contaminated easily by spills from surface activities. A common problem has been leakage from gas tanks and dry cleaners, which have contaminated hundreds of sites across the United States. Under the Superfund program, the USEPA prioritized cleanup sites. Because of their small volume, gas stations and dry cleaners have not always been high on the priority list. Some states have made them a higher priority but the number of sites and limitations in funding hurt cleanup efforts. Once cleanup is needed, remediation is a long process. A better practice is to locate well fields where contamination is unlikely, then protect the well field from surficial impacts. This is the focus of wellhead protection programs, which are a part of the Safe Drinking Water Act amendments.

Improperly abandoned wells should be properly abandoned as soon as possible. As noted earlier, in the 1960s, USEPA found that nearly a quarter of well fields were impacted by surface contaminants. Improved chemical analysis and more complete sampling of well water has revealed a large number of public water supply wells contaminated by careless use and disposal of synthetic chemicals. States and provinces are now implementing regulations to protect underground sources of water. Some actions being taken include:

- New requirements for installation and testing of underground storage tanks;
- Increased regulation for handling, using, and transporting toxic chemicals to reduce the possibility of spills;

- Greatly increased regulation of landfills and other waste disposal sites;
- Closer control of the use of pesticides and agricultural chemicals;
- Sampling and monitoring of identified groundwater contamination locations; and
- Actions to remove the contamination.

All of these programs will affect land use, possibly affecting local constituents and creating difficulty for the utility where private property rights laws are an issue. However, modeling the well fields with sophisticated computer models will allow the water system to develop areas where certain activities should be limited. Figure 2-16 is an example of the well field protection areas in Broward County, Fla. The county has four tiers for regulation based on travel times from the surface to existing wells. Activities are limited based on these zones (zone 4 has no requirements):

Zone 1: Provides for up to a 10-day buffer around the well field. No hazardous chemicals (regulated substances) are permitted within zone 1.

Zone 2: Provides up to a 30-day buffer. Businesses are required to be licensed and test the groundwater at their facilities for regulated substances they store or use on-site.

Zone 3: Provides up to a 210-day buffer. Businesses are required to be licensed and secondary containment is mandated for their stored regulated substances.

Zone 4: Area with no significant restrictions.

In conjunction with wellhead protection efforts, water systems should identify any groundwater sources they are using that may be directly affected by surface water. The concern is that if there is minimal filtration that occurs between the surface and the water withdrawn for wells, contaminants, especially microbiological constituents, may contaminate the water source (Bloetscher, 2009).

OTHER NORTH AMERICAN RULES

Canada has negotiated a set of guidelines with the provinces to ensure that drinking water meets water quality criteria outlined in the *Guidelines for Canadian Drinking Water Quality*. Each province is responsible for the formulation of guidelines within the province, but the provinces will typically adopt the Canadian guidelines with only minor modifications, just as the states do in the United States. Alberta and Quebec have made parts of the *Guidelines for Canadian Drinking Water Quality* into laws (which makes them regulations rather than guidelines). In these provinces, it can be a criminal offense to distribute water that does not meet regulations. Most industrialized countries in the world regulate drinking water in this manner.

In Mexico, water quality and use are federal properties. Water use has been historically tied to the principle that water resources are the property of the nation, and are therefore a free, constitutional right of all citizens. This is similar to some eastern states in the United States. Mexican water quality laws have been approved in the past 15 years. Laws on the management of water supplies and the use of cost–benefit analyses in the application of regulatory standards have also been approved.

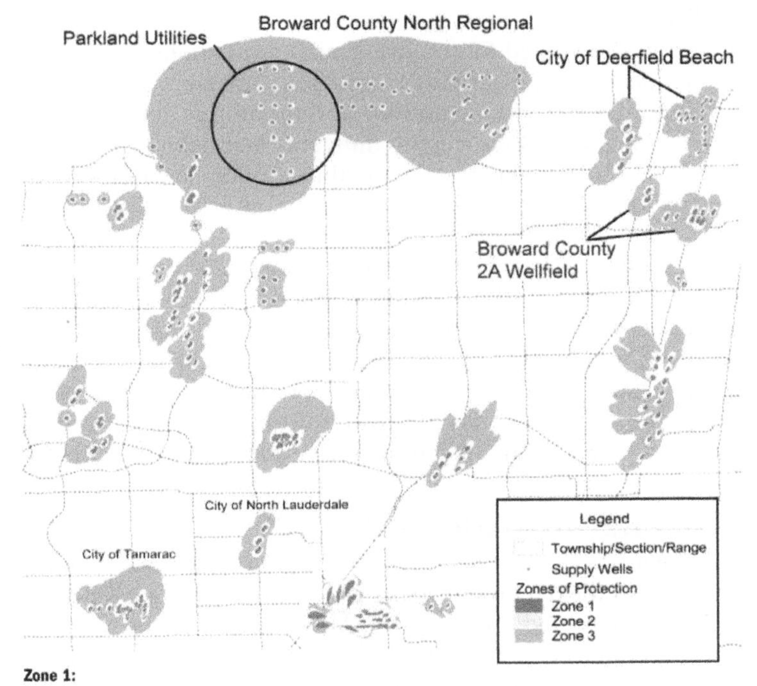

Zone 1:

Provides up to a 10-day buffer around the wellfield. No hazardous chemicals (regulated substances) are permitted within Zone 1.

Zone 2:

Provides up to a 30-day buffer. Businesses are required to be licensed and test the groundwater at their facility for regulated substances they store or use on site.

Zone 3:

Provides up to a 210-day buffer. Businesses are required to be licensed and secondary containment is mandated for their stored regulated substances.

Source: Broward County 2001.

FIGURE 2-16 Broward County wellhead protection zones
Courtesy of Broward County, Fla.

FUTURE WATER QUALITY REGULATIONS

The future of regulations is going to be focused on more technology-based solutions to remove contaminants at very low levels. Because detection methods have improved, there is an improved ability to detect very small quantities of contaminants that might be leading to effects. As a result, there will be more emphasis placed on limiting certain types of discharges and, where discharges must occur, minimizing contamination. This is just a furtherance of the Clean Water Act. In addition, stormwater and agriculture are starting to receive significant attention because the water quality of many streams has not improved, possibly as the result of stormwater or agricultural operations. Hazardous materials will be regulated and tracked more closely and pharmaceutically active substances will likely receive further attention. Chapter 16 includes a discussion about future regulatory directions.

CLEAN AIR ACT

The Clean Air Act in the United States, like similar regulations in other countries, also affects utilities, specifically with regard to chlorine gas. As a part of the amendments in the early 1990s to the Clean Water Act, all utilities storing chlorine gas were required to create a risk management plan, and to take steps to mitigate potential off-site impacts from chlorine releases. These include looking at air quality around the plant in the event of a release, figuring out what the worst release would be, and then developing containment methodologies.

Facility safeguards to accomplish this containment came from the risk management plan as noted. The rules also require plant operators to be licensed, that there be ongoing training of plant operators, and that chlorine systems be converted from pressure systems to vacuum systems to minimize the likelihood of chlorine leaks. It should be noted that despite the focus on chlorine, there has been only one operator death associated with chlorine gas use in the United States and no off-site deaths or impacts caused by chlorine releases. The water industry has an outstanding record regarding use of chlorine gas.

DEALING WITH REGULATORY AGENCIES

Having a series of regulations to follow, regardless how voluminous they may be, is helpful, but does not address the day-to-day issues of operating a utility system. Utility operations are variable by nature and damage to facilities, facility failures, and/or unexpected growth can impact the ability of the utility to comply with the regulations. As a result, understanding the rules is not the only necessity, so is understanding the regulatory perspective (Bloetscher, 2009).

Regulatory agencies serve many distinct groups, as shown in Figure 2-17. Utilities are but one of many groups, including environmental interests, agricultural interests, elected officials, lobbyists, and others. These groups tend to have different objectives that are at times mutually exclusive or incompatible. The result is that the agency personnel react to competing pressures, sometimes creating inconsistent direction. In part this is an attempt by the regulatory agencies to balance the competing demands.

Because of criticism often leveled at regulatory agencies when one group's objectives are not met, the agencies tend to be risk-averse. Requests that are identified in the rules explicitly and complied with in every detail are approved. Where judgment is required, the easiest course of action is denial unless the evidence to make a different decision is incontrovertible and extensive. This is a very hard goal to achieve for utilities requesting permits where little is consistent, unless enough redundancy is present that no potential failures of equipment can occur.

Therefore, for utilities to progress through the regulatory maze, several steps must be taken. First, the agency needs to cultivate a relationship with a reasonable person at the decision-making level of the agency. This relationship needs to be supportive of the agency; combatants rarely get what they want quickly. The relationships need to extend before and beyond times when utility requests are considered—this is a real relationship to foster support and trust, not a relationship to get a permit when desired.

Next, as a part of its ongoing efforts, the utility needs to determine what the goals of the agency are in permitting matters so that the utility can design its efforts to demonstrate how the goals will be met. Once the goals are known, the appropriate data can be

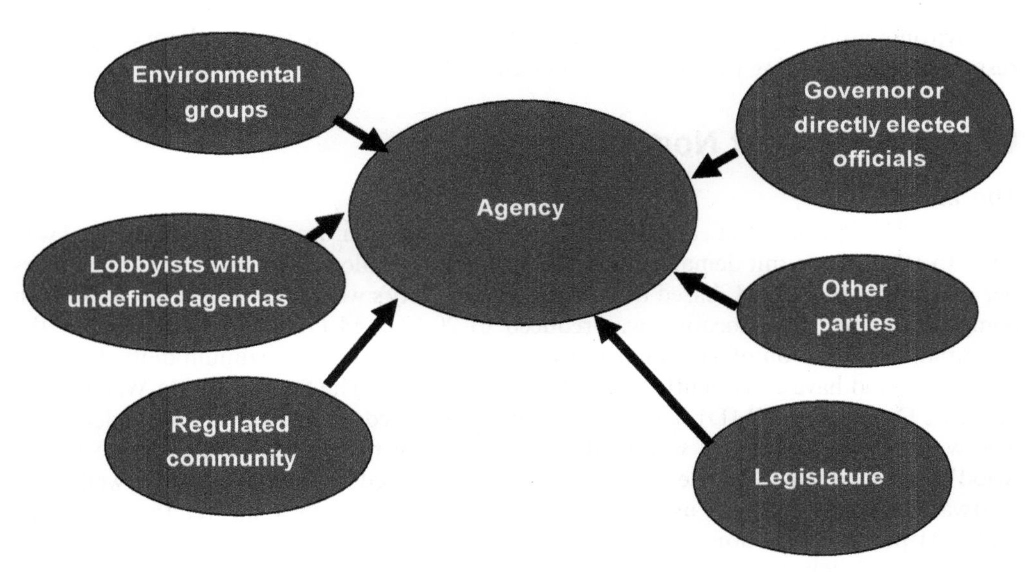

FIGURE 2-17 Groups providing input to regulatory agencies
Source: Bloetscher, 2009.

submitted. Too often copious volumes of data are supplied with permit applications, none of which addresses the agency needs or goals. How goals of the agency are met should be reflected in any submittal or request by the utility. Some policies may be disadvantageous to the utility, but the time to deal with those issues is before they are permit conditions or restrictions, not when the permit is needed (Bloetscher, 2009).

Not only do permit requests illustrate this problem, but so do responses to consent orders. Consent orders are negatively perceived by elected officials, government managers, and utility directors, mostly as a result of publicized operations failures that result in consent orders with large fines. For some consent orders, this reputation is deserved; usually these are utilities that have not taken appropriate action to resolve a problem.

Regulatory agencies view consent orders very differently than elected officials or utility directors, as evidenced by the fact that the fines are often reduced or restructured into desired improvements to the utility system. Regulatory agencies view consent orders as agreements that can be used for a variety of purposes, including quick modifications to existing permits. Consent orders can be very useful in dealing with temporary conditions, construction problems, early operation of some part of a capital project, or upgrades in usage due to special circumstances affecting neighboring jurisdictions. None of these issues would have fines or penalties involved. Some utilities have actually requested consent orders to resolve political issues with other jurisdictions. The key is knowing what the agency goals are and cultivating relationships that permit the agency to trust that the utility will address problem areas in a timely manner.

When permitting a project or requesting modifications to existing permits, it is important that the utility understand its goals and the steps it needs to take to meet those goals, as well as identify the goals of the agency. Permits are agreements, just like consent orders. The permit allows the utility to achieve its goal of performing some activity under certain guidelines (such as discharging wastewater effluent to a stream). All permits require a balance of environmental, regulatory, utility, and other interests, and the regulatory agency is responsible for ensuring this balance (Bloetscher, 2009).

Example—Permit Negotiation

The issue

The city of Hollywood, Fla., applied for a renewal of its water use permit for its coastal well field. In 1989 the permit demands were for 28.8 million gallons per day (mgd) (1,065 liters per second [L/s]). In 1994, based on modeling of the city's well field using a density-driven computer model, the allocation was reduced to 20.67 mgd (765 L/s) as a result of the potential for migration of salt water toward the wells. The city's maximum daily demands over the period have consistently been 28 mgd (1,040 L/s). The South Florida Water Management District (SFWMD) has goals and policies to reduce coastal well field demands. The westward movement of well fields is one measure to accomplish this. Since Hollywood's well field was near the coast, the district proposed moving the coastal well fields westward, and for the city to use the county's regional well supply. However, the 20.67-mgd (765-L/s) permit restriction on withdrawals had a number of consequences for the city, including the following:

- The city's current lime softening process (capacity 24.8 mgd [918 L/s]) cannot treat the county's raw water source;
- High iron content in the county's raw water will adversely impact membrane life of the city's membrane system (projected to need replacement two years early); and
- The cost of water in a bulk agreement whereby the city sold potable water to the county was being contested by the county; meanwhile the county's raw water costs continued to rise, which created a fiscal shortfall for the city.

As a result, the city began discussion in 1994 about modifying these permit limits in the next water use permit (1999 was the anticipated date). The city staff participated with the district staff on several projects and committees and built relationships within the organization. They also planned activities that would help meet the district's goals to reduce coastal water use, while permitting consideration of an increase in water use in the city.

What the city did

Starting in 1994, the city of Hollywood oriented many of its efforts toward improving its long-term water supply and treatment capability to meet future regulatory demands, customer needs, and the goals and objectives of the SFWMD. These efforts included the following activities:

- Initiated the first reclaimed water system in the county to meet effluent disposal needs. The target recipients were those who competed with the city for raw water supplies—4 mgd (148 L/s) were saved in the Biscayne aquifer.
- Completed the county's first integrated water resource plan, including analysis of all water use options (submitted to SFWMD in 1998).

- Expanded/improved water treatment capability by installing the largest membrane water treatment facility in Florida at the time. This is for treatment of the county water sources as well as to meet water supply and regulatory demands.
- Installed four brackish Floridan Aquifer wells and a reverse osmosis component to its membrane water treatment facility. The city was able to reduce its Biscayne aquifer water use by about 10 percent during the 2001 drought season.
- Moved the bulk of its water supply wells over a mile west as a part of the 1999 water use permit modification. The south well field pumping demand was reduced from 20.67 mgd (765 L/s) to 12.5 mgd (463 L/s), with 12 mgd (444 L/s) from the new wells.
- Identified a weir structure at a golf course that met the need identified by the SFWMD for salinity control in the southern part of the county. Modeling indicated that a weir structure that could hold water levels at 2 ft NGVD (national geodetic vertical datum, 1929) would minimize restrictions in Hollywood.
- Facilitated initial discussions between the Florida Department of Transportation (FDOT) and the SFWMD to move water from the C-9 basin to the C-10 basin to reduce impacts on Biscayne Bay and increase flows to Hollywood.
- Entered into agreements with the county to supply water service to all of District 3. This has eliminated 4 mgd (148 L/s) of eastern water withdrawals over the past ten years.
- Investigated, in partnership with the SFWMD, an innovative pilot project to use reclaimed water to retard salt-water intrusion into Hollywood.

The western movement of the city's wells, when modeled for the water use permit support, indicated that an average day withdrawal of 24.8 mgd (918 L/s) was viable from the city's wells as a result of the reduced demand from competing users and meeting SFWMD's goals to retain water and find other sources during drought periods. The city requested this allocation in its renewal application. In addition the city brought these activities to the attention of SFWMD staff to demonstrate how the city has attempted to comply with SFWMD directives, such as water conservation, conservation rate structures, restriction activities, and cooperation with surrounding agencies.

Not unexpectedly, there was an objection to this request. However, because of the city's efforts, the executive director for SFWMD deemed that the city had more than met the SFWMD's objective to reduce coastal water use and that their request was justifiable technically (via modeling) and reasonable. He directed staff to recommend the city's 24.8-mgd (918-L/s) request for approval. Without the effort the city had expended in the preceding seven years, this would not have happened.

Rulemaking Efforts

Rulemaking and modifications to existing regulations happen constantly. Many rulemaking decisions are made in the initial drafting process, so it is important for appropriate utility staff members to be involved in the rulemaking process—the earlier the better, before unyielding positions are taken. This is an important opportunity that utilities often miss. The proactive approach is respected by regulatory agencies and most of the agencies' "clients." With a public health issue like water and sewer, it is very difficult to react to a proposed regulation from a defensive posture.

Rule Compliance

Most utility systems comply with the water quality standards they are required to meet. However, simply meeting the water quality standards is sometimes not good enough. The secondary standards are aesthetic, not health related. Likewise, most states have rules that require a tidy workplace and can impose fines for not having things cleaned up. A sloppy workplace might be thought of as indicative of poor operations or unacceptable water quality.

WATER RIGHTS AND ALLOCATIONS BY RULE

A water right is a legal right to use, store, divert, or regulate water for some purpose. Across the United States, water rights methodologies and regulations differ, although in most cases the regulatory body is the state or some direct subdivision thereof. Federal laws affect military installations, national parks, and tribal lands. Surface and groundwater are treated somewhat differently and their uses may be different (e.g., you cannot go boating or fishing in groundwater).

Water rights are usually codified in some manner. In the eastern states, surface water allocations are typically administered under a riparian system of water law, and statewide programs for water use permitting are typically not used. Riparian rights have evolved where water is plentiful. Typically in riparian states, the water is owned by the state, not individuals. However individuals are permitted to use those waters on or adjacent to their property, although some restrictions may exist. Cech (2002) noted that there are two principles upon which riparian rights have developed. The first, the reasonable use principle, indicates that the riparian property owner can use as much water as needed as long as it does not interfere with the reasonable use by other riparian owners. This means that upstream riparian users cannot obstruct the natural water course or divert water so downstream users are deprived of the benefit. Likewise downstream users cannot obstruct the water course and flood upstream properties.

The second principle is correlative rights. This principle indicates that riparian users must share the total flow. A portion is allocated to each user based on the amount of waterfront each owner has. There is no priority for water use, and the concept assumes equal access and a reasonable minimum amount of water to all users. The concept seems reasonable until one tries to determine what happens during drought periods, or how utilities that serve large areas can serve their customers without making large land acquisitions (usually they can't). Conflicts can and do result from the correlative rights principle, and environmental effects are often ignored.

As a result of existing and potential lawsuits over water uses, many eastern states are moving toward a system of regulated riparian rights in which permits are issued based on a set of regulations. Permits define the withdrawals and uses that each user has for a particular water body and provide the users with some degree of confidence that their needs will be met (important for utilities). Issues involved in the issuance of permits are need, characteristics of the water body, infiltration rates (groundwater), modeling of the water body, and climate. There may be reporting requirements for water use above a specified amount, which may be voluntary and unenforceable.

Groundwater withdrawals in the eastern states may be affected by two doctrines:

Reasonable use. In the eastern states, in conjunction with the concept of reasonable use for surface waters, the principle is applied to groundwater. A property owner can use the water under his/her site, but cannot adversely affect the rights of the neighboring properties to do the same unless the use is reasonable (as opposed to wasteful). The concept suffers from the same concerns as for surface waters, i.e., what if all the uses are reasonable? The situation is notable in eastern North Carolina, where the Black Creek and Middendorf aquifers' use was reasonable for all users, but the aquifer has now become mined, and as a result, efforts are being made to remove users from the aquifer to attempt to restore the water. A further extension of the concept is where the groundwater withdrawal affects the surface (e.g., it dries up a stream). Such withdrawals are not considered reasonable under any circumstances.

Withdrawal permits. In some eastern states, groundwater withdrawals may be regulated within specific aquifers. States where water withdrawals from designated aquifers are regulated include Georgia, South Carolina, North Carolina, Virginia, Maryland, Delaware, Florida, and New Jersey.

The scenario is vastly different in the 18 western states that are governed by the Prior Appropriation Doctrine of water law. The Prior Appropriation Doctrine establishes the right of a water user to use a specific amount of the state's water resources for a specific "beneficial use," as stated in the water right by rule or statute. The water right holder has a priority date determined by a court or agency. The water users with older or more "senior" water rights have the right to use their water first. If the demand for water is greater than the available supply, then the senior water rights holders can "place a call" on the water source (typically a river). This may result in junior water rights holders being prevented from using their water if doing so would impact senior water rights holders. Hence a senior water rights holder can divert an entire water body from downstream users, but they are restricted from changing the way they withdraw, divert, or use their water so that those changes do not impact other existing water rights holders (including more junior water rights holders). Use of water or damage to senior water rights require compensation to the senior water rights holder(s), or are not permitted at all unless those rights can be secured. The doctrine applies to both groundwater and surface water.

ENVIRONMENTAL LAW AND SUSTAINABILITY

Legal issues have become an increased part of the utility industry during the past 20 years, and will continue to be an ever-present factor in utility decision making. As a result, a brief discussion of the sources and issues of law that affect utilities follows (Buck, 1996).

The basis of law in the United States stems from five sources:

1. Common law
2. Statutes and ordinances
3. Rules and regulations
4. The Constitution
5. Judicial interpretation

Common law is the oldest doctrine, going back to the 13th century in England. Common law is based on the concept of precedent, i.e., what happened the last time the issue was raised. Written laws do not significantly factor into common law. Certain water issues in eastern states were resolved this way in the beginning of the United States.

Statutes and ordinances are approved by bodies of elected officials. Congress, state legislatures, county commissions, and local entities all approve statutes and ordinances that have the force of law. Usually these statutes and ordinances are fairly specific and many are limited to things such as operations of the governmental entity (budget documents, for example). Governing bodies approve ordinances for utilities with regard to wastes that can be discharged to sewers; rates, fees, and charges for utility service; mandatory connection to available water and sewer system; and developer requirements. Many local agencies approve resolutions that enact regulations to mimic the force of law, such as for utility extension policies, standard materials specifications, and the use of reserves and other funds.

Rules and regulations are technically administrative measures developed by staff for operations. Rules and regulations cover a wide variety of issues, many of which may be internal. Most are approved by the governing body. To have the force of law, two tests typically must be met: the procedural due process and substantive due process. Procedural due process means that the governing body followed established rules for notification of the public that a rule or regulation was going to be published (such as new rates). Substantive due process determines whether or not the agency has the right or standing to approve the rule (e.g., the governing body of a utility would appear to have the right to establish rates). In general, the enabling legislation to form the governing body stipulates the boundaries of its actions.

The Constitution is a major factor in environmental and eminent domain cases. The Constitution protects the rights of the citizenry from government. The use of governmental powers to take land and personal property rights are Constitutional issues. The Constitution also includes the concept of Federalism, which is actually defined as an intergovernmental system of state and local governments. As a result, the majority of ordinances, laws, rules, and regulations are created at the state or local level, not the federal level. During the 1980s, many of the national programs for water and wastewater were delegated to the states under this doctrine. As a result, most utilities will deal almost exclusively with state and local regulators, not the USEPA.

Judicial decisions, termed *case law*, are the largest body of law. Case law is used to determine what decision a judge (and jury perhaps) might make if an issue was brought before them. Normally issues that go to court have not been tested before, and the court case is intended to resolve some confusion or unaddressed aspect of a statute, ordinance, rule, or regulation. Judicial precedents in case law are typically upheld by the courts.

Issues of importance to utilities that may involve lawyers include:

- Permitting (which relates to sustainability and off-site impacts),
- Eminent domain,
- Worker safety and public health (see operations discussion),
- Construction and contracts (see chapter 10), and
- Borrowing (discussed in chapters 9 and 11).

Virtually all of these issues involve local, state, or case law components. However, to raise a legal challenge, one must have *standing*. Standing is a right under common law indicating that the party has received actual impacts or injuries as a result of the rule or action. In federal cases, to have standing, one also must have a stake in the outcome. The case must also be brought to the appropriate venue—a state, local, or federal court that will hear the case.

Permitting and eminent domain are common areas where legal challenges arise. Property rights are rooted in English common law, supplemented by constitutional rights (Fifth Amendment). Eminent domain is the right of governmental entities to take private property for beneficial public use. Governments use eminent domain where private property owners are unwilling to negotiate a fair price for acquisition of property. For example, a pipeline has a certain route. The utility may have to buy property rights (usually easements) through the properties of a number of private owners in order to provide services to a new neighborhood. Most private property owners recognize the need to help their utility provide service, but, at times, a reasonable price cannot be agreed on. As the pipeline cannot be constructed without securing all property on the pipe route, and moving the pipeline would compound the acquisition problems, the right of eminent domain is used to conclude remaining property acquisitions.

To take property, the governmental entity must demonstrate a public benefit. A well site, pipeline easement, lift station site, or treatment plant site could all be valid public purposes. That another site would not be appropriate may also be a requirement to meet this public interest test. However, the Fifth Amendment guarantees that if government takes private property, the private property owner must be justly compensated. Significant bodies of law exist in every state regarding eminent domain. There are lawyers who specialize in eminent domain cases.

Permitting is another area of potential conflict. Permitting issues are usually site- or utility-specific, and commonly involve wastewater discharge requirements, permitting raw water supplies, or development requirements (or limitations). Utilities attempting to increase wastewater effluent discharges to surface water may receive permits that have much more stringent treatment requirements. Increases in wastewater volumes and more stringent limitations require additional capital facilities. Negotiated settlements and administrative hearings will generally conclude these cases. Environmental groups may challenge the impacts of discharges on surface water species or uses of the water body.

Examples of conflict over water withdrawal issues (as discussed previously) are common. In the absence of willing parties to conclude water rights or withdrawal rights, these cases often go to court. Issues such as salmon fishing in the West or wetland impacts to stream diversions that may affect endangered species or reduce wetland area may be difficult to resolve. Also increasing the potential areas of conflict is ongoing research studying how groundwater withdrawals may affect surface water flows. A significant amount of thought is evolving in this area with regard to the sustainability of water sources. In this case, *sustainability* should perhaps be called *regeneration*.

Unfortunately the current literature continues to fail to address the bigger-picture issues. Water supply decisions affect not only the current and proposed water users, but may have far-reaching economic affects. One of the conflicts in the Klamath Basin on the Oregon/California border involves the impact of water diversions to agriculture in the upper basin at the potential expense of downstream salmon fishing and eco-tourism throughout the basin, especially during droughts. Each group can demonstrate (or project) economic benefits to the region resulting from their use of the water (in terms of revenue and jobs). The economic value of each sector is in the hundreds of millions of dollars each year, and some believe that the eco-tourism/salmon fishing may be four or five times the benefit of agriculture. In many years, these uses can coexist. However, when agricultural users are prioritized for water over the other uses, as in 2000 and 2001, there may be longer-term and perhaps permanent effects on the ecosystem that would reduce the revenues of the eco-tourism/salmon fishing sector. The reverse is not particularly true, although some compensation would be expected by farmers to mitigate damage claims. Clearly the sustainability issues from the macro level should be discussed and evaluated when new allocations are considered.

Barten and Ernst (2004) discuss the multiple-barrier approach to sustainability with respect to source water protection efforts outlined in the Clean Water and Safe Drinking Water acts, and the improvement in water quality when the effects of nonpoint sources are factored into upstream water bodies. They cite a Trust for Public Lands/American Water Works Association study of 29 water suppliers that indicates that treatment costs are inversely related to the proportion of the watershed that is protected by wetlands and forests. Where open space contained large tracts of agricultural lands, the potential for pesticides, herbicides, and manure runoff to contaminate downstream drinking water supplies is significant. As a result, in contrast to their historical position, many utilities may find themselves having much more in common with environmental supporters than developers and agricultural interests. Conservation of land and protection of upstream habitat through easements, purchase, or other measures may reduce the risks for public health failures of a drinking water supply as well as protect endangered habitat or species if a use arrangement can be developed. New York City and Seattle are examples of water systems that have acquired the watersheds for their systems, which both benefits the wildlife of the area and reduces treatment requirements. The National Academy Press published a report in 1994 that points out that environmental resources may be severely undervalued in sustainability discussions, as much of the intrinsic value may be undefined at present (NAP, 1994).

REFERENCES

AWWA. 1987. *New Dimensions in Safe Drinking Water: An Overview of the 1986 SDWA Amendments and Proposed Primary Drinking Water Regulations*. Denver, Colo.: American Water Works Association.

Barten, P.K., and C.E. Ernst. 2004. Land Conservation and Watershed Management for Source Protection. *Jour. AWWA*, 96(4).

Barwick, R.S. et al. 2000. Surveillance for Waterborne-Disease Outbreaks—United States, 1997–1998. *Morbidity and Mortality Weekly Report,* 49(SS04):1–35.

Blackburn, B.G. et al. 2004. Surveillance for Waterborne-Disease Outbreaks Associated with Drinking Water—United States, 2001–2002. *Morbidity and Mortality Weekly Report,* 53(SS08):23–45.

Bloetscher, F. 2009. *Water Basics for Decision Makers: What Local Officials Need to Know about Water and Wastewater Systems*. Denver, Colo.: American Water Works Association.

Buck, S.J. 1996. *Understanding Environmental Administration and Law*. Washington, D.C.: Island Press.

Cech, T.V. 2002. *Principles of Water Resources: History, Development, Management and Policy*. Somerset, N.J.: Wiley Publishers.

Code of Federal Regulations, Title 40, Volume 15, Parts 136 to 149. Revised as of July 1, 2000 (40 CFR 146). *Federal Register.* Washington, D.C.: U.S. Government Printing Office.

Hazen and Sawyer. 1994. *Southeast Florida Ocean Outfall Experiment II Report*. Hollywood, Fla.

Lee, S.H. et al. 2002. Surveillance for Waterborne-Disease Outbreaks—United States, 1999–2000. *Morbidity and Mortality Weekly Report*, 51(SS08):1–28.

Liang, J.L. et al. 2006. Surveillance for Waterborne Disease and Outbreaks Associated with Drinking Water and Water Not Intended for Drinking—United States, 2003–2004. *Morbidity and Mortality Weekly Report,* 55(SS12):31–58.

Masters, G.M. 1998. *Introduction to Environmental Engineering*. Englewood Cliffs, N.J.: Prentice Hall.

National Academy Press (NAP). 1994. *Assigning Economic Value to Natural Resources.* Washington, D.C.: National Academy Press.

USEPA. 1981. *Underground Injection Control Rules*, 40 CFR 144 and 146. Washington, D.C.: USEPA.

———. 1999. *The Class V Underground Injection Control Study, Volume 20—Saltwater Intrusion Barrier Wells.*

———. 2000. *Ground Water Rule,* www.epa.gov/ogwdwo/standard/occur.html. Accessed 7/26/07.

———. 2002a. *Safe Drinking Water Act*, http://www.epa.gov/safewater/sdwa/sdwa.html. Accessed 7/26/07.

———. 2002b. *Primary Drinking Water Standards*, http://www.epa.gov/safewater/mcl.html. Accessed 7/26/07.

———. 2005. LT2 Rule, http://www.epa.gov/safewater/disinfection/lt2/regulations.html#rule. Accessed 8/26/2009.

———. 2008. Ground Water Rule. http://www.epa.gov/fedrgstr/EPA-WATER/2006/November/Day-08/w8763.pdf. Accessed 8/26/2009.

———. 2009a. National Environmental Policy Act of 1969 as amended, PL, 91–190, 42 USC 4321–4347, http://ceq.hss.doe.gov/nepa/regs/neap/nepaeqia.htm. Accessed 10/16/2010.

———. 2009b. Factoids: Drinking Water and Ground Water Statistics for 2009. Office of Water (4601M), EPA 816-K-09-004, November 2009. Available on-line at: www.epa.gov/safewater/data.

———. 2009c. From Risk to Rule: How EPA Develops Risk-Based Drinking Water Regulation (PowerPoint training presentation). http://water.epa.gov/learn/training/dwatraining/upload.dwaNPDWR_risktotuletraining.pdf.

Yoder, J. et al. 2008. Surveillance for Waterborne Disease and Outbreaks Associated with Drinking Water and Water Not Intended for Drinking—United States, 2005–2006. *Morbidity and Mortality Weekly Report*, 57(SS09):39–62.

CHAPTER 3

Water and Wastewater Operations

In chapter 2, the discussion focused on the regulatory requirements regarding water and wastewater utilities. In this chapter, the components of water and wastewater systems will be discussed. Understanding these facilities will provide a framework for demonstrating some options in dealing with risks in relation to utility operations and provide a background for the concepts of maintenance and operations management, staffing, capital needs, and asset management. This chapter will introduce treatment and piping systems in both words and illustrations.

Beyond the extensive rules that must be complied with, water and wastewater systems can be highly complex and are expected to operate 24 hours per day, seven days per week, without interruption. For most utilities and their management staff, the highest-risk categories usually involve service disruptions. If the systems do not operate properly, the public knows about it quickly through disruptions in service, which may become significant public relations problems. As a result, minimizing risks requires redundancy, reliable equipment, and skilled personnel to identify and react to potential problems before those problems manifest themselves as public health risks or property damage. Unfortunately, the infrastructure is composed largely of hidden components that are easily forgotten when budget decisions are being made and, as a result, may lack long-term funding (Bloetscher, 2009).

Water and sewer (the latter generally referred to as *wastewater*) will be kept separate as most regulations require. Figure 3-1 shows the basic components of a utility system that provides all services. This model will be used in the following paragraphs.

RAW WATER SUPPLIES

Raw water supplies are necessary for any water system. There are two options for raw water supplies—groundwater and surface water. Groundwater is used more often for smaller systems, while surface water is the choice of most large systems. Each has its advantages and disadvantages, but in many locales, the options are limited to one or the other. Different rules are applied for treatment of the water supply based on risk of contamination as discussed in chapter 2. The choice of a given community, if multiple options are available, usually comes down to the amount of treatment needed—less is better, because rates are lower. The concept of the sustainability of the supply is an issue that is a more recent consideration.

Groundwater Supplies

Groundwater supplies are generated by constructing wells into an aquifer (a water-bearing layer of rock beneath the ground) and pumping water out of the aquifer. Several important issues arise with aquifer systems, including the safe yield of the aquifer, water quality,

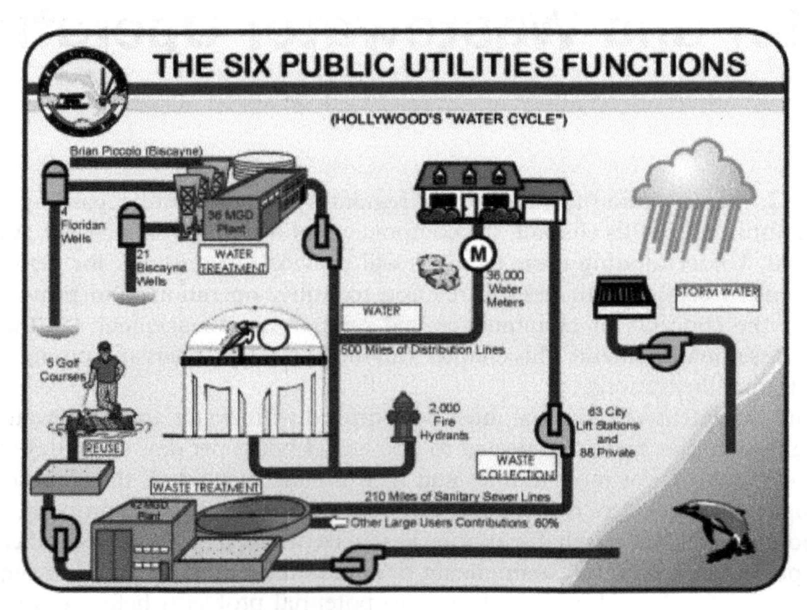

FIGURE 3-1 Example of a utility system diagram

and potential contamination. Understanding these issues requires field investigations to confirm site-specific characteristics, usually by a registered hydrogeologist. Finding a location favorable to the utility and its customers is also important.

For many years, groundwater was assumed to be pristine, unaffected by surface activities and therefore safe to drink without treatment. While some groundwater sources are safe to drink without significant treatment, the public perceptions of purity have created expectations of consistently reliable groundwater sources. As noted in chapter 2, this is not the case. Contaminants are present in some water supplies in sufficient quantities to render the water supplies unsuitable for drinking water. The contaminants included minerals, solvents, and other hydrocarbons. In addition, bacteria exist naturally in most aquifer systems; almost any aquifer with an organic content will have bacteriological activity to some degree, and some of these bacteria will be pathogenic to humans. Aquifers that have lowered groundwater levels may have higher natural constituents, such as fluoride and salts.

The evaluation of regional groundwater conditions and the potential for development of the resource should be based on the following factors:

- The quantity and quality of water required;
- The availability of water resources, i.e., the safe yield of the potential source basins;
- The cost to develop the water supply;
- The long-term land uses affecting quality and quantity of groundwater recharge;
- The nature of existing or likely future pollution sources within the recharge area;
- The effectiveness of regulatory controls to limit pollution sources;
- Competing water users; and
- The conveyance and treatment costs for the water supply.

High-quality groundwater minimizes treatment needs. Highly transmissive aquifers that have adequate recharge will provide sustainable water supplies for many years provided potential contaminants are kept out through well field/source water protection programs.

Well field protection efforts begin with identification of any groundwater sources that may be directly affected by surface water. The concern is that if there is minimal filtration that occurs between the surface and the water withdrawn for wells, contaminants, especially microbiological constituents, may contaminate the water source. In addition, many layers of rock that were previously thought to be highly confining may contain fractures or intrusions (such as inappropriately constructed or abandoned wells) that create hydraulic pathways for contaminants to move into underlying aquifers. As a result, the preferred location of wells is usually in undeveloped areas. This helps in avoiding potential pollution sources (e.g., gas stations) in developed areas that make well location/source water protection efforts more difficult.

The protection of public water supply wells through wellhead protection programs is considered an important component of comprehensive state groundwater protection programs. Many states have promulgated minimal rules that delegate specific implementation of the program to regional water management districts and/or counties. This may be appropriate from a technical perspective since geological and hydrogeological controls on the aquifer system are typically localized.

Developing a wellhead protection program involves various diverse perspectives and interests of the community, and requires cooperative efforts on all governmental levels and between units of government because of the potential of movement of water across jurisdictional boundaries. The regulatory process should be responsive to local needs and allow local autonomy, which is important in many governmental jurisdictions and in areas where multijurisdictional cooperation is needed. From a local perspective, protection of drinking water sources is the first step in safeguarding public water supplies, reducing the need to use expensive alternative treatment techniques.

With regard to groundwater, implicit in these guidelines is the need for properly constructed wellheads, sanitary seals, and the location of wellheads above flood stages to ensure the wells are not contaminated due to weather events. Protection of wells by locating them away from streams, septic tanks, and lift stations that might leak is also important.

Operations

Groundwater supplies are developed using wells (see Figure 3-2). Wells have these basic components: the borehole and casing (see Figure 3-3), the column pipe and pump (Figure 3-4), and the well screens. A well type usually refers to the method of well construction: dug, bored, driven, cable tooled, or drilled. A sixth type of well, which is not named for its method of construction, but rather a description of its configuration, is the radial collector well. Each type of well has advantages or disadvantages, including characteristics such as ease of construction, storage, capacity, ability to penetrate various formations, and ease of safeguarding against contamination. A hydrogeologist or engineer should be employed to guide the utility in the design and configuration of wells.

Figure 3-5 shows the basic aquifer terms. Water table aquifers are those that have no overlying formations that confine them from surface activities, thus they are termed *unconfined*. These aquifers are the most prone to pollution contamination. In contrast, if water completely fills an aquifer that is overlain by dense rock or clay, the water in the aquifer is called *confined*. Wells drilled into confined aquifers where the water level rises

FIGURE 3-2 Photograph of a well

FIGURE 3-3 Photograph of a well casing

toward the surface are referred to as *artesian wells.* If the water level in an artesian well rises above the land surface, the well is a termed a *flowing artesian well.*

The well design procedure starts with some expectations of potential water yield from an aquifer, based in part on the type of aquifer and the initial investigations by a (registered) hydrogeologist. The quantity and quality of groundwater depend on factors such as aquifer depth, rainfall, and geology. Typically the utility requires a specific water yield, so the engineer and hydrogeologist must bring this requirement together with treatment and other issues to design a cost-effective and efficient well configuration.

FIGURE 3-4 A submersible well pump

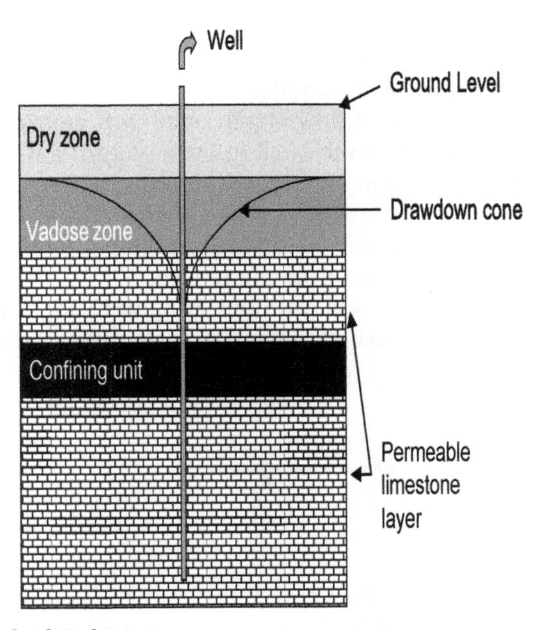

FIGURE 3-5 Aquifer and related terms

In general, wells completed in unconsolidated formations, such as sands and gravels, are equipped with screens to prevent sand from entering the well during pumping. Screens allow the maximum amount of water from the aquifer to enter the well with a minimum of silt and sand. Screens are sometimes installed in fractured formations to prevent the collapse of equipment into the borehole, thereby trapping equipment in the borehole.

Drawdown is the change in water level in the aquifer as a result of pumping. Figure 3-6 shows a cross section of a confined aquifer. The basic equation governing groundwater flow is Darcy's law:

$$Q = -KA\frac{\Delta h}{\Delta L} \tag{3–1}$$

where the amount of flow (Q) is related to how much change in head (water level) there is between two points (L), the thickness of the aquifer (b) or cross-sectional areas (A), a concept called *hydraulic conductivity* (K), change in vertical head (Δh), and distance between two wells (ΔL). Equation 3–2 outlines the relationship for a confined aquifer. Figures 3-7 and 3-8 show two wells (or monitoring wells) adjacent to a well where the flow volume (or drawdown) is desired. The equation shows that there is a relationship between the radius from the water supply well and the drawdown in the adjacent wells:

$$Q = \frac{2\pi Kb(h_2 - h_1)}{\ln\frac{r_2}{r_1}} \tag{3–2}$$

Where

Q = rate of flow
K = hydraulic conductivity
b = the thickness of the aquifer
h_1 and h_2 = differences in drawdown levels at monitoring wells 1 and 2
r_1 and r_2 = the distances from the well to the monitoring wells 1 and 2
ln = natural logarithm

The combination of the hydraulic conductivity (K) and aquifer thickness (b) is also termed *transmissivity*, which is defined from pump tests.

Figures 3-9 and 3-10 show one well (or monitoring well) adjacent to a well where the flow volume (or drawdown) is desired. Equation 3–3 outlines the relationship for an unconfined aquifer:

$$Q = \frac{2\pi K(h_2^2 - h_1^2)}{\ln\frac{r_2}{r_1}} \tag{3–3}$$

where Q is the rate of flow, b is the thickness of the aquifer, K is hydraulic conductivity, h_1 and h_2 are differences in drawdown levels at monitoring well 2 and the well in question 1, and r_1 and r_2 are the distances from the well to the monitoring well 2 and the radius of the well in question 1. Because the aquifer level will continually change in an unconfined aquifer as a result of pumping, the aquifer thickness is not used, only the water level (or head).

More exotic are horizontal wells that are used across the Midwest under rivers and lakes to allow surface waters to filter through the underlying soil into a horizontal screen. The Long Term 2 Enhanced Surface Water Treatment Rule (described in chapter 2) has a

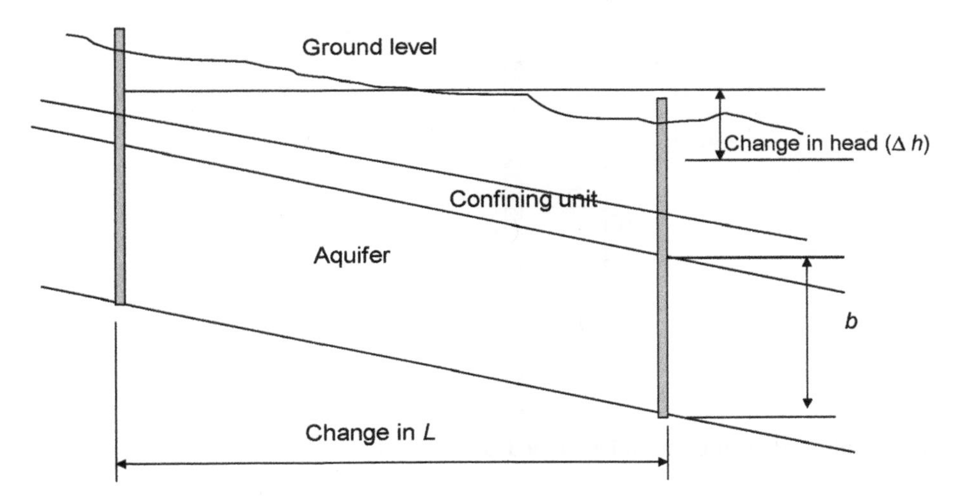

FIGURE 3-6 Confined aquifer and related terms

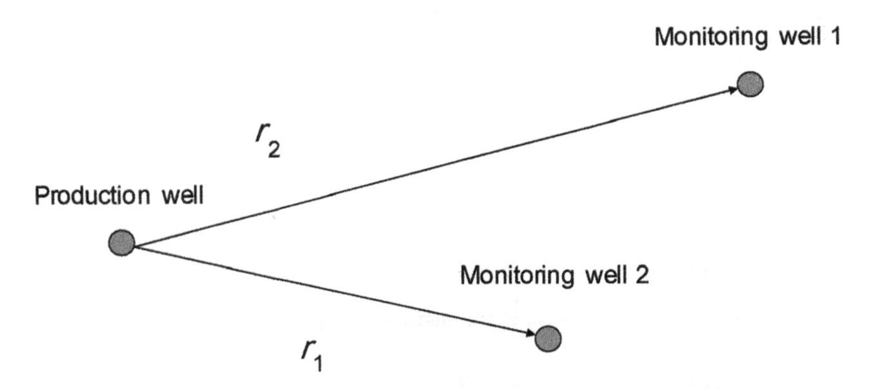

FIGURE 3-7 Confined aquifer and related terms as defined in Equation 3-2 (r_1 and r_2 are the distances from the well to monitoring wells 1 and 2.)

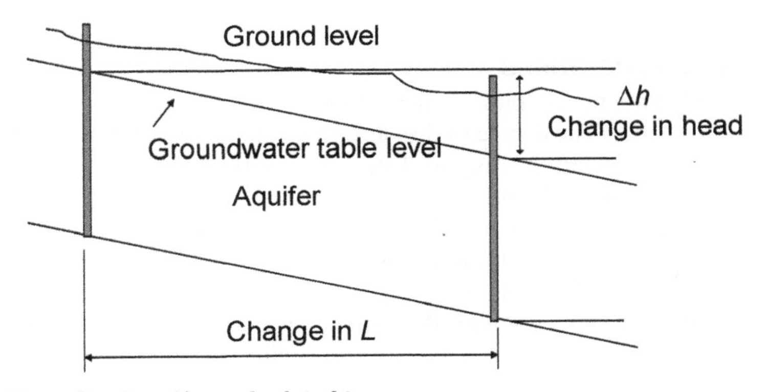

FIGURE 3-8 Unconfined aquifer and related terms

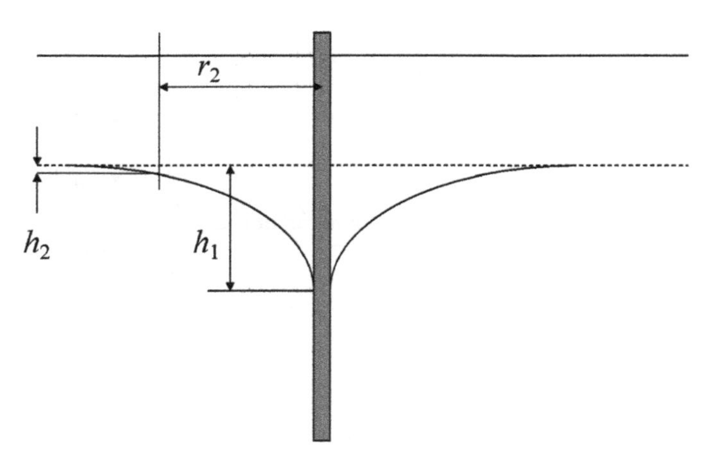

FIGURE 3-9 Illustration of terms in Equation 3-3.

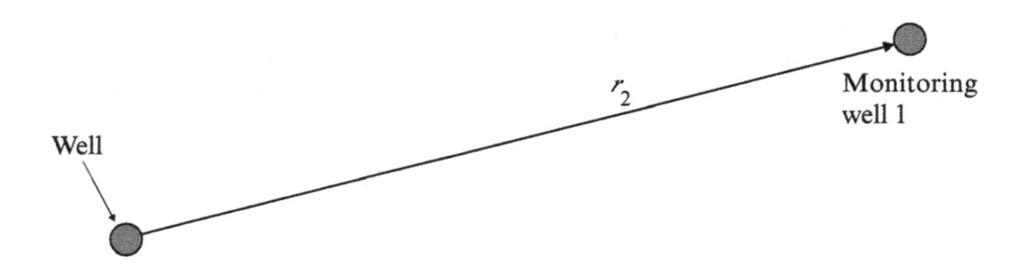

r_2 minimum = pipe diameter

FIGURE 3-10 Confined aquifer terms (r_2 is the distance from the well to monitoring well 1.)

section on this concept, called *riverbank filtration*. Riverbank filtration can allow utilities to avoid filtration if certain criteria are met, or to reduce the contact time (CT) values in their plant. Figure 3-11 shows the top of a riverbank filtration well in Louisville, Ky., where significant effort has been made to investigate this technology.

Water quality considerations

Groundwater supplies are never "pure" water—there are always some constituents dissolved from the rock in the water. The degree of dissolution is expressed as *hardness*. High-hardness water has significant amounts of dissolved minerals, a typical example being limestone. High concentrations of minerals means low-quality water. A significant relationship exists between mineral quality and the depth of groundwater; the mineral content of groundwater usually increases with depth, except along the ocean coast. Groundwater quality in many sedimentary basins (where the older and deeper sediments were deposited by oceans) can change very abruptly in mineral content. Poor-quality water can be drawn upward after production begins ("upconing"), even if a production well does not penetrate the formation.

FIGURE 3-11 Top of riverbank filtration well in Louisville, Ky.

The wells can also induce salt-water intrusion into freshwater aquifers by drawing the salt-water front horizontally toward the wells.

The presence or potential presence of synthetic and naturally occurring organic compounds, plus refined minerals and heavy metals, must be considered when evaluating the development potential of a groundwater resource because of the difficulty in removing these constituents. Microbiological organisms, especially in membrane treatment applications, are increasingly a concern. In some cases, construction, maintenance, and operation of facilities to remove these constituents is more costly than finding a new water source.

Bacteria in wells tend to have an impact on the life of the wells, their operation, and the integrity of the materials used to construct the wells as a result of degradation from biofouling. The typical agents for microbiological fouling include iron bacteria and sulfur-reducing and slime-producing bacteria, although many others exist. Iron bacteria, like *Gallionella*, are common in aerobic environments where iron and oxygen are present in the groundwater, and where ferrous materials exist in the aquifer formation (such as steel or cast-iron wells). These bacteria attach themselves to the steel and create differentially charged points on the surface, which in turn create cathodic corrosion problems. The iron bacteria then metabolize the iron that is more soluble in the process. Iron bacteria tend to create rust-colored colonies on the pipe surfaces.

Sulfur-reducing bacteria are often responsible for the release of hydrogen sulfide when raw water is aerated. These bacteria are common where sulfur naturally exists in the aquifer formation, and will tend to form black colonies on pipe surfaces. While anaerobic, they will exist in environments where aerobic conditions exist, which can lead to symbiotic relationships with aerobic organisms.

Slime-producing bacteria are found in surface waters and in soil. The *Pseudomonas* spp. are an example. Members of these genera are often used to protect farm crops from fungal growth and, as a result, are to be expected in groundwater in rural areas. However, these bacteria are highly adaptive; research several years ago indicated that the bacteria would grow in any environment into which they were introduced. The *Pseudomonads* are facultative anaerobes that can persist in oxygen-depleted environments by breaking down complex hydrocarbons for oxygen. In some circumstances, they will even use nitrogen in the absence of oxygen. Given that the *Pseudomonas* spp. are adhering bacteria, they are capable of producing a polysaccharide matrix (biofilm) that acts to protect them from the shearing effect of turbulent flow, to resist disinfectants, and to provide an environment for other species. *Pseudomonas* bacteria can permanently affix themselves to laser-polished 316L stainless steel in a matter of hours, so attaching to steel or lower grades of stainless steel is easily accomplished.

Microbiological accumulations/biofilms pose two significant concerns. First, the accumulations on the metallic surfaces create anodes and, in conjunction with reactions caused by dissimilar metals, can lead to a steady cathodic deterioration of the well casing and column pipe (with or without iron bacteria). Because the *Pseudomonads* are acid formers, ferrous materials are particularly vulnerable to this sort of deterioration, especially in the presence of iron bacteria (which is indicated by iron staining).

Deterioration of water quantity and quality can have grave consequences for those dependent on the water resource, due to higher treatment costs or forced abandonment of the well field. However, changes other than those measured in the field can also be of great concern. Perceived changes may be the result of improved analytical methods (which detect levels of compounds not previously measurable or new chemicals); regulatory priority shifts toward or away from protection of natural resources, including aquifer classifications; new toxicological data or reinterpretation of existing toxicology; or integration of facts regarding groundwater quality into conservative measures for the safe use of all water resources.

Monitoring water quality should occur initially to establish a baseline, which, once established, can permit the utility to reduce the frequency of groundwater sampling (except in areas of suspected contamination), assuming the water quality does not change appreciably. Fortunately, groundwater quality in many locations does not change because the movement of groundwater is very slow, especially compared with surface water quality. Where contamination risks are high, sentinel monitor wells should be installed. These wells, located at various depths, will provide data for the initial groundwater assessment. Sentinel wells also serve as an early-warning system to detect changes in water quality and water elevations before they affect the water supply wells.

Surface Water Supplies

Surface water sources include any river, stream, lake, reservoir, ocean, or other body of water on the ground surface. Surface water bodies have been used for many years as water sources because the water is easy to extract. However, surface water sources are limited in location, although artificial reservoirs can be constructed to hold water (see Figure 3-12). Permitting of new surface water sources, especially where dams (see Figure 3-13) are involved, is a challenge as a result of potential adverse environmental impacts.

FIGURE 3-12 Example of a reservoir behind an earthen dam in North Carolina

FIGURE 3-13 Example of a typical dam

Surface water bodies are recharged by rainfall and runoff from rainfall that runs off the land, reaching streams, rivers, or lakes. Springs are areas where the groundwater table intersects a surface water body, thereby contributing water to the surface water body. Surface water flow is controlled by the topography; ultimately all water bodies feed the oceans. Estuaries are where the surface water bodies mix with salt-water bodies along the coast. Chapter 2 briefly discussed the rules for source water protection plans. The concept of these plans is to create local decision making on land use and other development to limit risks of runoff to water bodies being used for water supply purposes.

Operations

The operation of surface water systems differs from groundwater systems in two respects: (a) the water temperature may vary considerably, and (b) the concentration and number of contaminants may fluctuate, especially during periods of turbidity after spring rains. River systems tend to have snowmelt in the spring (cold water), but warmer rainfall water in the summer. Changes in lake temperatures may cause shifts in stratification, suspending particles that adversely affect treatment. As a result, the intake structures for surface water systems are complex, i.e., they generally have pumps, but often have multiple levels from which water can be drawn, thereby maximizing water quality. Intake structures may be simple pipes in a stream (Figure 3-14 shows a bar screen on the side of stream), a sluice from a dam, or complex intake structures like those seen in the Ohio River or the Great Lakes that permit operations personnel to open intakes at different levels. These structures may be concrete, steel, or some mixture. They are designed to remain in place for very long periods of time.

Water quality considerations

Unlike with groundwater systems, few people believe surface waters are pristine or drinkable in their current condition. While many of the same issues apply, surface water systems are far more susceptible to contamination than groundwater supplies. Synthetic and naturally occurring organic compounds, plus refined minerals and heavy metals, are washed into surface water systems as a result of drainage from contaminated areas. Unlike with groundwater, surface water quality can change rapidly during runoff events. Therefore, current and past land use practices are important when evaluating the development potential of a surface water resource. Surface water bodies are also common wastewater outlets, so bacteria, viruses, and protozoans are a major concern. Viruses and protozoans are resistant to chlorine disinfectants. As with groundwater, monitoring of water quality should occur on a continuing basis. Recreational access to lakes and reservoirs increases the potential for contaminants.

Surface Versus Groundwater Supplies

The following summarizes the benefits and problems with both raw water resources (Bloetscher, 2009). It should be noted that availability is the deciding factor for most utility systems. Benefits of groundwater supplies are as follows:

- Less exposure to contamination (assuming it is not a surficial aquifer)
- Water quality is stable
- Water temperature is stable

FIGURE 3-14 Example of intake bar screen on a stream, Idaho Springs, Colo.

- Water quality changes are slow to occur
- Less treatment is typically required

Disadvantages of groundwater include:

- Difficult to clean up once contaminated
- No early warning of contamination
- Competing users
- Safe yields often are uncertain
- Aquifers are not available everywhere

Surface water supplies have the following benefits:

- Water levels are obvious
- Supplies are generally large and often can be managed from year to year to optimize stored water
- Dilution of contaminants reduces potential impacts of contamination

Problems with surface supplies include:

- Contamination is easy and can occur very quickly
- Additional treatment (filtration) is required
- Evaporation losses are high
- Watershed protection is difficult—recreational users want to share the resource

Land Use Effects on Raw Water Supplies

Agricultural use of land can affect groundwater quality due to runoff containing pesticides, herbicides, and animal feedlot or other animal wastes. Residential land uses with septic tanks may pollute groundwaters and surface waters with household chemicals, microbiological contaminants, salts, hydrocarbons, solvents, endocrine disruptors, and nitrates. At any potential raw water supply site, the pertinent historical land use practices should be reviewed and potential development impacts considered. Unfortunately, private property rights laws in many states may frustrate utility efforts to protect water sources in the future without significant costs for land acquisition.

As land uses change, raw water quality may change because the recharge supply quality changes (with surface water supplies, runoff may cause rivers to flow faster, increasing sediments and nutrients). As a result, understanding the growth, development, and zoning strategy of the area associated with the raw water supply is needed. In some instances computer models can be developed to mimic actual conditions. Examples of activities that may affect raw water supplies include delivery times of the water through piping installed to reduce flooding and/or supply quantity and development replacing irrigated agriculture with paving. Offstream reservoirs are a common solution to deal with potential water quality events. Urbanization reduces recharge to groundwater but increases flows to surface reservoirs and streams (with potential added nutrients and contaminants).

Analysis of Raw Water

Samples taken from monitoring programs should be analyzed for suspected contaminants, including minerals. The mineral quality of water will limit the range of possible water uses and indicate potential treatment options. For example, water containing high concentrations of sodium or boron will be unsuitable for irrigation. Although the biological quality of deeper groundwater is usually better than that of surficial sources, testing for fecal bacteria and other microbiological indicators should be performed regularly.

Water quality sampling is an integral part of proper operations of any utility system. Water samples must be obtained and analyzed periodically by a licensed state- or USEPA-certified laboratory in accordance with applicable permits, federal and state laws, rules, and regulations, and by the utility for quality control and monitoring purposes. The sampling frequency should be determined by the stability of the parameters in the water sources, based on how often, if ever, the parameters change concentrations and the degree of change that occurs. In wells, changes in water quality are an indicator of problems that, if large enough, may disrupt the water treatment process. In surface supplies, raw water quality will change with rainfall, season, and upstream land use practices. Utilities should monitor raw water quality, as certain changes in water quality can cause adverse environmental impacts. ASTM Standard D-4195-88 requires water analyses to be performed on raw water for the parameters in Table 3-1.

Knowledge of probable sources of contaminant chemicals used in the area and selection of key indicator constituents should be used in the design of the sampling and analysis program to reduce cost without loss of study credibility. Guidance for selecting chemicals to be tested may be obtained from state and federal regulatory officials responsible for facility permits. "Priority pollutants," as defined in the Safe Drinking Water Act (SDWA), often can be used to determine the likely presence or absence of chemicals that are a con-

TABLE 3-1 Summary of routine parameters to be analyzed in raw water quality tests

Calcium (Ca)
Magnesium (Mg)
Sodium (Na)
Strontium (Sr)
Barium (Ba)
Aluminum (Al) (total and dissolved)
Manganese (Mn) (total and dissolved)
Iron (Fe) (total, dissolved, and ferrous)
Potassium (K)
Bicarbonate (HCO_3^-)
Sulfate (SO_4^{-2})
Chloride (Cl^-)
Carbonate (CO_3^{-2})
Nitrate (NO_3^-)
Fluoride (F^-)
Phosphate (PO_4^{-3}) (total)
Silica as silica dioxide (SiO_2) (total and dissolved)
Total dissolved solids (TDS)
Total organic carbon (TOC)
Hydrogen sulfide (H_2S)
Free chlorine (Cl_2)
Oxygen (O_2)
Carbon dioxide (CO_2)
pH
Temperature
Turbidity (nephelometric method)
Silt density index (SDI)

cern to raw water development. Unfortunately, analyses to detect all of these chemicals can be expensive, so analytical methods used should be directed toward detection of suspected compounds. It is important to keep in mind that the quality of raw water drives treatment.

WATER TREATMENT SYSTEMS

Treatment of water supplies can vary widely depending on the raw water supply quality. All systems require disinfection, although high-quality groundwater systems may only require disinfection. Small groundwater systems that treat only with disinfection are typical in many parts of North America, but actually account for a small proportion of the people served. Where further treatment is needed, there are two basic types of water treatment plans in common usage—lime softening for groundwater, and coagulation systems for surface waters. Membrane/reverse osmosis systems are in vogue in some locales because of their ability to achieve high water quality at competitive prices from otherwise unusable sources. However, while many options are available, this chapter will focus only on the more common processes and how they are applied.

Disinfection

Disinfection is the process to remove or destroy disease-causing pathogenic organisms, like viruses, bacteria, fungi, and protozoans. The removal of these organisms was one of the initial water treatment goals. There are several ways to do this. Chlorination is the most popular method to remove pathogens in the United States and Canada. Ozone is used in Europe and certain South American countries, and ultraviolet light is also used.

The benefit of chlorine over the others is that there is a residual. Regulations in the United States require that the chlorine residual be in existence in the distribution system at all times. Without a disinfectant, the distribution system would quickly grow biofilms like the ones found in wells. It is impossible to extend disinfection to the distribution system if ultraviolet light or ozone are used to treat the water. Once the finished water is disinfected, it can be pumped directly into the distribution system or stored on-site for later distribution.

Chlorine comes in different forms. The one typically used is chlorine gas. Chlorine gas is a yellowish-green gas and is actually a liquid when stored under pressure. It usually comes in cylinders (see Figure 3-15). The concept is to take the compressed liquid chlorine, allow it to become a gas, mix it with water, and inject it into the water prior to the chlorine contact chamber. Small chemical pumps like those shown in Figure 3-16 can be used for a variety of chemicals, including chlorine solutions. A residual will remain once the reaction has occurred if the proper dosage of chlorine is added.

Chlorine gas (Cl_2) is the most potent form of disinfectant; however, it very quickly changes form to hypochlorous acid (HOCl). Chlorine gas is the most widely used disinfectant in the United States at the present time, although inroads have been made with liquid forms due to the potential dangers associated with the gas. The activity of chlorine is dependent on temperature, pH, and contact time (CT).

The proposed Chemical and Water Security Act, if approved, will require USEPA to establish a risk-based performance standard for utilities that serve populations over 3,300 people. This is in conjunction with antiterrorism goals and provides a way to track chemicals throughout the industry. Currently water and sewer utilities are exempt from this requirement. The law would require utility managers to review the safety of the treatment process and procedures, such as changes in liquid and gas chlorine.

Figure 3-17 shows an on-site chlorine generator. Many utilities have converted to on-site generators due to security risks and the requirement for risk management plans that notify the public about a potential hazard. Other options include sodium hypochlorite (see Figure 3-18) and calcium hypochlorite. Sodium hypochlorite is used as a liquid system and many utilities have converted to this method. Sodium hypochlorite is basically bleach, although in a stronger form than Clorox®. Calcium hypochlorite is a white powder that is used for swimming pools but can also be used for water treatment.

One of the issues associated with chlorine disinfection is that there are two regulated carcinogen by-products as a result of the combination of chlorine with organics: trihalomethanes and haloacetic acids. As a result, systems that have significant amounts of organics tend to require additional treatment to remove them and avoid the formation of carcinogens. For years, most of the focus has been on trihalomethanes. To reduce trihalomethane formation potential, other disinfection agents can be applied. Chlorine dioxide is a strong disinfectant, but produces a by-product (chlorite) that is regulated. Ammonia is used to increase the breakpoint by creating additional chloramines (which provide less disinfection capability, but a longer residual time). Ozone and ultraviolet light (see Figure 3-19)

FIGURE 3-15 Typical 1-ton chlorine cylinders

FIGURE 3-16 Small chemical pumps that can be used for a variety of chemicals, including chlorine solutions

FIGURE 3-17 On-site chlorine generator

FIGURE 3-18 Sodium hypochlorite system

are the other options, although chlorine is also added to water systems when ozone and ultraviolet light are employed to obtain a residual.

Ozone is generated at the site and requires a significant amount of electricity. Ozone is highly reactive, but very short lived. It is capable of oxidizing the organic compounds and removing them so that subsequent disinfection will not create significant concerns with tri-

FIGURE 3-19 Ultraviolet light disinfection

halomethane formation. Ozone creates no residual, but does create oxygen in the water that may encourage biofilm growth; therefore, chlorine in some form is always required.

Ultraviolet light (UV) acts completely differently. Ultraviolet light is electromagnetic radiation that destroys portions of the pathogens, thereby preventing them from being able to reproduce. Depending on the intensity of the light and the amount of contact time, it removes very high percentages of the pathogens. UV dose is related to water clarity. Natural organic matter (NOM) absorbs the ultraviolet wavelengths in the germicidal region, while particulates will scatter the light, negating the dosage. Pathogen inactivation dosages have been well researched. The important ones to develop 99.99% (4-log) removal are as follows:

- *E. coli*—5.6 mJ/cm^2
- *Cryptosporidium*—5.7 mJ/cm^2
- *Giardia*—1.7 mJ/cm^2
- *Poliovirus*—21.6 to 30 mJ/cm^2

Fouling will occur on almost any source, but the fouling rates are slower with hardness less than 100 mg/L.

One of the advantages of UV is that most viruses and bacteria are inactivated. Viruses are more resistant to chlorination than bacteria and, as a result, they can escape the chlorination process. Ultraviolet light has a relatively short contact time, which is beneficial in trying to construct UV systems. Ultraviolet light systems use relatively little space and do not create any disinfection by-products.

The disadvantages are that inadequate UV doses will not be effective. It is possible to foul the tubes or the lightbulbs for the ultraviolet light, and this reduces their effectiveness significantly. Highly turbid waters also reduce UV effectiveness. At present, UV technology is expensive, although the costs are decreasing.

Lime Softening

Hard water does not produce soap bubbles easily and leaves a scum layer. Hard water is typically caused by large amounts of calcium bicarbonate [$Ca(HCO_3)_2$] and magnesium bicarbonate [$Mg(HCO_3)_2$] in the water, created by the dissolution of limestone. Since many productive aquifers are limestone, it should not be a surprise that hard water is a groundwater issue for the most part (however, note that hardness is defined as the sum of all polyvalent ions in the water). With groundwater, calcium and magnesium are normally the largest constituents.

The concept behind lime softening is to precipitate compounds causing hardness by creating a chemical reaction with lime and the hardness particles. Lime softening processes do a relatively good job at removing both the bacteria in raw groundwater and the hardness, the latter manifesting as settled materials (calcium carbonate and magnesium hydroxide). The typical process is shown in Figure 3-20. The facilities consist of lime reactors, followed by filters to remove the remaining particles. A lime silo (Figure 3-21) contains dry powdered lime that is typically blown into the silo by air. The slaker is at the bottom of the silo (see Figure 3-22). The slaker measures out the appropriate quantities of lime. There are two different forms—hydrated lime and quicklime. Hydrated lime is created by mixing quicklime with water. This is a very fast, very hot reaction. The lime slurry must be pumped immediately to the lime reactor or the hydrated lime will stick to the pipes.

Aeration devices are used to remove hydrogen sulfide and other gases from the raw water. Lime reactor units (cone-shaped, open reactor vessels) are used to mix slaked lime with raw water to create a chemical reaction to remove hardness (Figure 3-23). Lime is mixed with raw water where it reacts with the calcium carbonate, making calcium bicarbonate, which makes large particles that settle quickly, thereby removing calcium and magnesium hardness, along with iron and other metals in a clarification zone. The lime reaction occurs as the water moves to the top of the reactor units, reacting with and removing calcium and magnesium hardness which is precipitated as lime sludge. The outside part of the unit is a clarifier, which allows the calcium/lime material to precipitate. These units are serviced periodically to remove excess calcium buildup (Figure 3-24). The softened water then flows out through weirs on the top of the reactor/clarifier units to filters for removal of suspended materials.

Soda ash may be used in conjunction with lime to improve the efficiency of magnesium removal, but it should be used with caution since the concentration of sodium may become an issue. In most cases the amount of magnesium is one quarter to one sixth the amount of calcium, so both lime and soda ash are normally used to reduce both constituents. Lime is not as effective at removing the magnesium portion; it changes from one magnesium bicarbonate form to magnesium carbonate, which does not settle. Soda ash resolves this problem. The addition of soda ash will cause the calcium to react with the carbonate ion and the magnesium to react with the hydroxides.

One of the issues that occurs with lime softening is that it is normally used to raise the pH. In looking at the carbonate cycle, when the pH is greater than 10, carbonate forms (CO_3^{-2}) will dominate. This is beneficial; however, as the magnesium hydroxide begins to

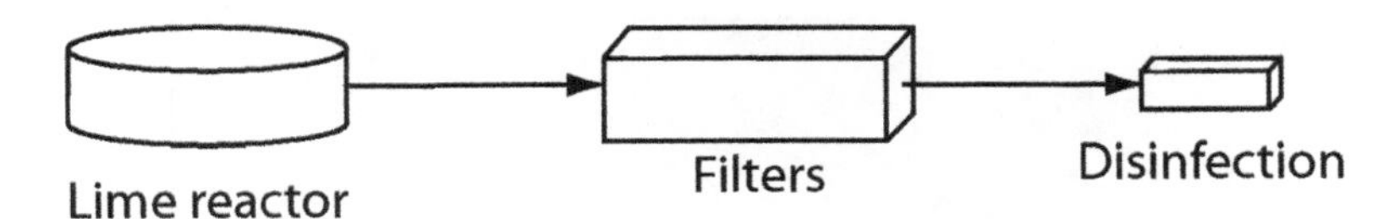

FIGURE 3-20 Lime softening process diagram

FIGURE 3-21 Typical lime silo

settle, the pH starts to slide back to below 10, which will slow the lime softening reaction. If the operators know the milli-equivalence of calcium, magnesium carbonate, and sulfates that exist in the raw water, they can easily determine how much lime is required for treatment. A problem with the lime softening/soda ash process is that it produces a lime sludge by-product that may prove difficult to dispose of after treatment because it dewaters (dries) poorly.

One of the benefits of lime reaction is that it occurs very quickly, so the basins are small. The water immediately moves out into the clarifier portion and is then moved in to the launders. It takes about a minute for the lime sludge to settle to the point that clear water appears to be coming through the surface. As a result, the clarifier portion of this system is usually less than 30 minutes as opposed to 3 to 4 hours for a coagulation system. Launders are the weirs used for removing the clean water (see Figures 3-23 and 3-24). Instead of V-notch weirs, the launders simply have holes in the sides of them to let water in. Periodic testing is simple and fast. Metal launders are potentially subject to severe corrosion. The selection of appropriate metal for the condition of the system is an important issue, especially when making repairs.

FIGURE 3-22 Slaker at the base of the silo

FIGURE 3-23 Photograph of lime softening reactor

FIGURE 3-24 Photograph of empty lime softening reactor

It should be noted that it is not possible to get below 30 mg/L of calcium or 10 mg/L of magnesium as $CaCO_3$. This is because these are the saturation points of calcium and magnesium in water. If one tries to get below the saturation point, then the water will start to pick up other compounds. Excess lime softening to try to reduce hardness further results in a carryover of the active lime into the filters. As a result, the filters will become cemented together very much like an aquifer in a carbonate-rich environment. Many limestone aquifers are exactly this.

Another option is the use of spiractors, which are upflow catalyst systems (see Figure 3-25). The lime sludge plates onto the sand catalyst (just like what could happen with excess lime softening). Both spiractors and lime reactors require filtration.

Coagulation/Flocculation

Turbidity looks bad and interferes with the disinfection process. Turbidity is normally associated with surface waters and, fortunately, is easy to remove. Alum is added to cause the suspended material to coagulate into floc, which then settles in the clarifiers. Figure 3-26 illustrates the coagulation process. Similarities to lime softening are obvious, except that coagulation is typically applied to surface water, not groundwater. Surface waters are usually behind dams. Because rivers are moving water bodies, surface water supplies suffer from suspended solids in the water. There are five types of particles:

1. **Settleable**—solids that will settle with time, but may need chemicals to advance the settleability.
2. **Suspended**—those that are held in the water by the natural action of the water.
3. **Nonsettleable**—small suspended particles that are too small to settle easily. They need chemicals to settle.
4. **Colloidal**—clay size (too small to settle).
5. **Dissolved**—can only be removed with membrane. Salts are an example.

FIGURE 3-25 Spiractors

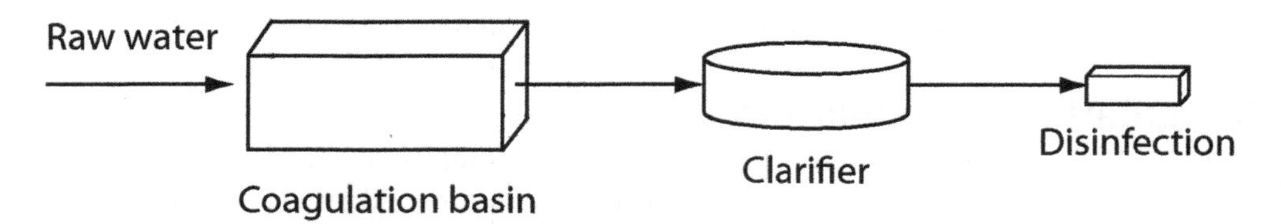

FIGURE 3-26 The coagulation process

None of these particles will settle easily. All but dissolved solids can have their settleability enhanced with coagulation (dissolved solids can only be removed with membrane processes). The problem with fine particles is their very slow settling velocity. The suspended solids only settle when large.

Coagulation and flocculation are designed to have the suspended solids mix with a chemical to create larger particles that settle faster. Flocculation is the slow mixing process. Coagulation is the "sticking together" part of the process that makes the bigger particles. Surface water systems are usually treated with alum addition (also ferric chloride, ferrous sulfate, and polymers), slow mixed for coagulation and flocculation (see Figure 3-27) to coagulate the floc into the larger particles, which then settle in the clarification and sedimentation basins (Figure 3-28). In addition, the variability of surface waters requires much more extensive monitoring of the raw water and ongoing modifications to the treatment process to provide adequate treatment at all times.

FIGURE 3-27 Flocculation panel under water
Courtesy of Bill Lauer.

FIGURE 3-28 Clarification basin
Source: AWWA.

An important issue arises with mixing. Some mixing requires a fast pulse to completely mix chemicals in the water. Lime softening and disinfection are examples. Other chemicals need a quick impulse, but then slow mixing for coagulation. Alum is an example. The goal of alum usage is to create large particles that will stick together and remove turbidity. Alum is very effective in doing this. It does have a chemical reaction that occurs where there is some hardness. However, with surface water systems, hardness is usually a minimal problem, and as a result, is not likely to create any significant issues for the water treatment system. Thus flocculant mixing is slow (for alum) and rapid mix is used for lime and disinfection (as examples).

Flocculation basins are typically larger than those of lime softening systems because the agitation action must be much slower to prevent the floc from breaking apart and keep the initial suspended particle condition persisting. The flocculation basins may be separate from the clarifiers.

The most commonly used flocculation compounds are alum, ferric chloride, ferrous sulfate, and polymers. These typically come in powdered form and must be mixed with water to be used in the treatment process. Polymers would be the one exception to this practice. Polymers tend to be very specific to water supplies and are used in very small quantities.

Filtration

Surface waters are required to be filtered. Most lime softening processes also result in a turbid product, which requires filtration. Filters follow the coagulation and lime softening processes. They are designed to remove all particles that were not removed during the sedimentation/clarification process; to remove bacteria, cysts, and other organisms that are too small to be removed or to react with the chemicals; and to reduce turbidity that can interfere with the disinfection process. Filtration enhances the ability of disinfection to kill the remaining pathogens. Unlike with groundwater sources, the variability of surface waters requires much more extensive monitoring of the raw water and ongoing modifications to the treatment process to provide adequate treatment at all times. Gravity filtration (Figure 3-29) and pressure filters (Figure 3-30) are options .

Filter structures contain a filter media consisting of graded gravel, silica sand, garnet, and anthracite coal, or some combination to remove sand, silt, and some colloidal particles. The filters must be rehabilitated and new media installed every three to five years. Filters will remove particles that are 20 times smaller than the filter media particles. This is because the area and size of the holes between the filter media are much smaller than the media themselves. However, filters only work where there is some uniformity in the filter; otherwise it will plug because the small particles will get caught at the smaller holes. Too much flow will blind the filter by putting too many solids on it. Too much flow will also increase the amount of backwashing needed.

Head loss increases as the filter collects impurities. So as the filter is doing what it is supposed to do, the amount of water that appears to be above the filter will increase with time. This is the built-up head or resistance to flow caused by the material collecting on the top of the filter. To remove the material, filters are normally backwashed every day for about 10 to 15 minutes at between four and eight times the designed filter rate. There are many equations for filtration, most of which are oriented to the size of the diameter of the holes between grains, which is extraordinarily difficult to figure out.

FIGURE 3-29 Gravity filtration

FIGURE 3-30 Pressure filters

The design parameters for a filter are based on usually high uniformity. The more uniform the grain size, the more consistently and longer the filter will run and the fewer problems with head loss. The lower the flow rate (2 to 4 gpm [7 to 12 L/min]), the more likely it is the filter will operate properly. Filters do not need to be very deep; they are typically under 30 in. (0.8 m). The underdrain system removes the water beneath the filter.

Membrane Systems

Dissolved solids usually cannot be removed by chemical processes. As a result, physical straining is required. Membranes are not chemically driven—they are physical separation processes. The process flow schematics for reverse osmosis treatment include pretreatment, membrane treatment, and posttreatment. They are very, very fine filters that normally require water to be under pressure to move through the membrane filter. As a result, they require very high quality water in order to function properly. One of the issues that most operators are surprised to find is that the higher the quality of the water treatment process, the higher the quality of the water it requires and the more treatment it requires in order to get it to that point. As a result, there are significant design considerations that must be accounted for in membrane processes with regard to raw water quality. Iron, organics, and biological fouling are emerging issues. The pH of water must be kept below 6.5 to prevent carbonate fouling of the membranes.

Figure 3-31 shows a typical process flow diagram for membrane processes. Cartridge filtration is essential for removal of suspended particulates larger than 0.45 micrometers (μm) from the raw water. Once the raw water is chemically conditioned and suspended solids are removed by the cartridge filters (see Figure 3-32), it is delivered to the feed pumps as feedwater. A dedicated feed pump supplying each skid increases the feedwater pressure prior to applying it to the membranes themselves. Figure 3-33 shows a typical membrane skid. The membranes are housed in the horizontal "tubes" shown in Figure 3-33. Figure 3-34 shows a cutaway of a spiral-wound membrane. Figure 3-35 details the membrane units and how they work. One side faces the clean water (permeate), while the other side retains the minerals (concentrate).

Figure 3-36 shows a used membrane. The raw water goes in one end of the membrane element. It can only enter the windings through the raw water spacer. Pressure will force some of this water through the membrane where it spirals down to the center of the element to a central pipe. This pipe is permeate, i.e., clean water. The permeate is removed from the element. The water that cannot get through the membrane flows through the

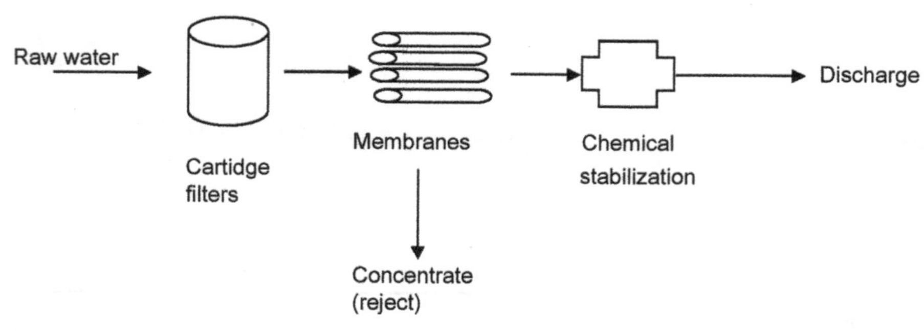

FIGURE 3-31 Membrane treatment process

FIGURE 3-32 Cartridge prefilter installation

FIGURE 3-33 Membrane treatment system

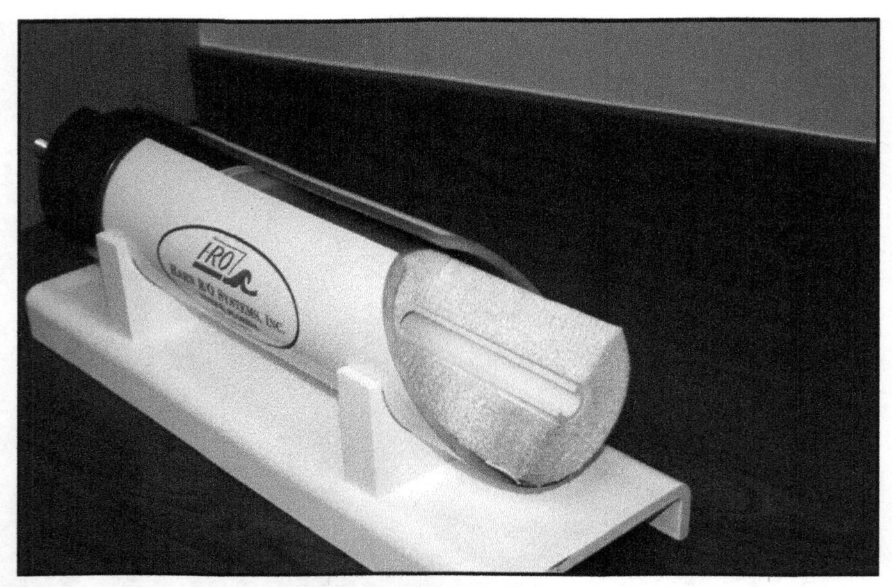

FIGURE 3-34 Cutaway of a typical membrane element

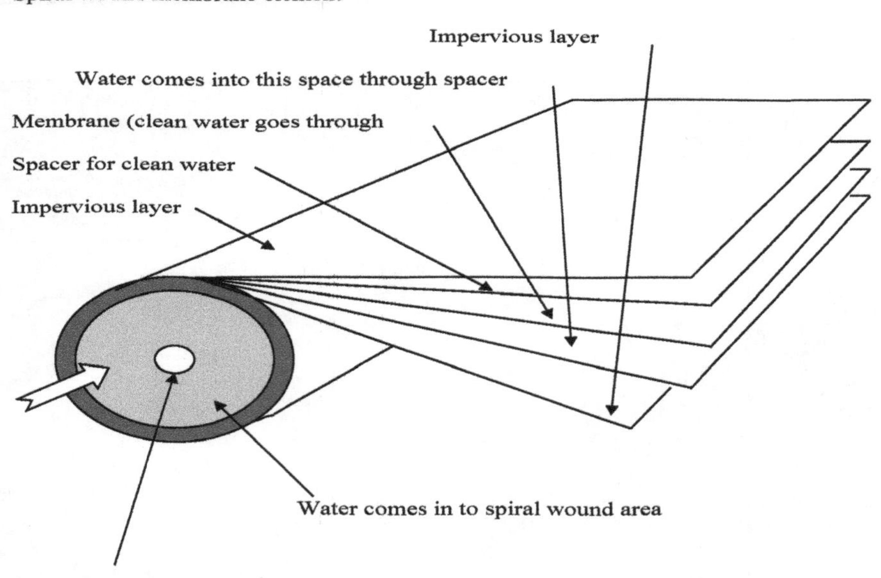

FIGURE 3-35 How the membrane process works

Source: http://images.google.com/imgres?imgurl=membranes.nist.gov/Bioremediation/fig_pages/figures/fig3.gif&imgrefurl=http://membranes.nist.gov/Bioremediation/fig_pages f3.html&h=520&w=703&prev=images%3Fq%3Dspiral%2Bwound%2Bmembranes%26svnum%3D10%26hl%3Den%26lr%3D%26ie%3DUTF-8%26oe%3DUTF-8

FIGURE 3-36 Example of used membrane element

spacer to the next element. Ultimately water that cannot get through any membrane will be termed *concentrate* (concentrated raw water).

The smaller the size of the particle, the higher the pressure needed. As a result, seawater removal starts around 800 pounds per square inch (psi) (5,515 kilopascals [kPa]) and goes upward to about 1,200 psi (8,273 kPa). Meanwhile, freshwaters, from which only organics are being removed, will permit removal at something like 100 psi (689 kPa). Table 3-2 outlines the membrane options that are available. The cost of power is also a concern as higher pressure means higher power demands.

As shown in Figure 3-32, the typical skid accommodates two or three stages of membranes. Each stage recycles the concentrated water from the previous stage. The more stages, the higher the recovery rate. The process recovery rate for nanofiltration (membrane softening—two or three stages) is 85 to 92 percent, while brackish reverse osmosis (salt-water removal) systems can produce only a 50 to 75 percent recovery, as compared to minimal losses in more traditional treatment regimes. The reduced recovery in the salt-water processes increases the quantity of raw water required to produce the same amount of permeate from one process skid, while producing a larger waste stream of concentrate, meaning that the concentrate disposal requirements are larger, and more raw water is needed to serve the same number of people.

Membranes create a concentrate that must be disposed of. This is the biggest barrier to membrane usage outside of coastal areas because there are very few places that will accept the concentrate. The concentrate is acutely toxic to aquatic organisms due to ion imbalances and/or salt content. As a result, it cannot be discharged to a surface water body. A wastewater treatment plant is an acceptable alternative but it assumes that the chlorides remain low and that there is sufficient treatment plant capacity.

TABLE 3-2 Summary of membrane processes

Type of Membrane	Pore Size	Particles Removed	Pressure	Permeate
Microfiltration	1–10 μm	0.5–1 log virus removal turbidity, cysts	5–30 psi	95–98%
Ultrafiltration	1–100 nm 100 K MW	4-log virus removal Macromolecules	7–60 psi	80–95%
Nanofiltration	100 MW	Softening	80–120 psi	70–90%
Reverse osmosis	10 MW	Salt, ions	200–1,200 psi	50–85%

A membrane cleaning/flushing system is required, which consists of cleaning and flushing solution tanks, 0.45-μm cartridge filters, and cleaning pumps. The cleaning pumps are constructed to handle high- and low-pH cleaning chemicals. The cleaning system must be designed to accommodate future needs when the system is expanded. If a membrane system is shut down, then a shutdown flush is required so that any raw water in the membrane elements is replaced with permeate water.

Ion exchange

Removal of metals is not effectively accomplished with coagulants, so other means can be used. Where membranes may be deemed overkill, ion exchange may work. Ion exchange easily removes arsenic, calcium, magnesium, and a host of other metals. Special ion exchange resins can be used to remove organics and fluoride. Ion exchange resins are not effective at straining water, so dissolved compounds are not as easily removed.

The ion exchange process is most often used to remove arsenic, nitrate, hardness, and organics. It is typically used with a resin, which can be anionic or cationic, or both (but in separate chambers). The idea is to move the anions and cations onto the resin and exchange it, typically with sodium.

Cation exchange is designed to remove positive ions, primarily calcium and magnesium, but also iron and organics (e.g., metal ions), and replace them in solution with sodium and hydrogen. Strong acid cation exchange systems are normally used, but regeneration is more frequent than with other ion exchange applications. Cation exchange systems are the most commonly used ion exchange systems (see Figure 3-37).

Anion exchange is designed to remove negatively charged ions like nitrate and sulfate. Weak base resins developed first, but now mostly strong base resins with ammonium are used. Less frequent regeneration is required but the pH range in which they work is limited to 6 to 8. A problem with ion exchange can be the release of sodium or chlorides into the water, so high-chloride-content water cannot use ion exchange. Ion exchange can be used effectively for softening as well as for removal of organics.

Ions are mobile in water. Ion exchange requires a medium, which is the resin, with a periodic salt (NaCl) flush. The surface of ion exchange beads is not the point where ion exchange occurs—99 percent of ion exchange happens inside the ion exchange bead (Wirth, 2010). It releases a non-hardness-causing ion attached to it in favor of the ion to be attached. The previous example is a softening reaction. When the resin collects enough of the unwanted ions, it will need to be flushed with salt, therefore inside the CaX the calcium is attached to the ion exchange resin. When it is flushed with salt, the product will be calcium chloride that will be removed, and the sodium will take the place of the calcium. The same thing happens for magnesium. This way, hardness is removed from the equation.

Depending on the organics to be removed, either anionic or cationic exchanges might be used. Figure 3-38 shows the resin for a cationic ion exchange unit. The resin is a bead, with tiny filaments that are wrapped into the beads. The filaments are a polystyrene chain

FIGURE 3-37 Ion exchange vessels

FIGURE 3-38 Typical Ion exchange resin
Source: US Filter/Permutit.

and divinylbenzene cross-link, with an inconsistent electrical charge throughout. Hence, the resin is charged to remove desired constituents. There is a downside to the process, however, as 1 mg/L of calcium removed will increase sodium in the water by 0.15 mg/L. So if sodium is a concern, ion exchange may pose a problem. The ions exchange and adhere to the beads, releasing the sodium. The beads are relatively small and can plug in a manner similar to a filter.

Weak base resins will work for some humics issues. A number of groundwater sources that have high organics are beginning to use ion exchange as a solution to reduce color and organics. This reduces the potential for trihalomethanes as well. This application is used to remove trihalomethane precursors in Florida, where water table aquifers contain considerable humic material. It is effective at removing humics and color. Strong bases will lose capacity quickly, shortening runs and increasing backwash. What this demonstrates is the need to test ion exchange applications prior to construction to ensure operating parameters can be predicted.

The treatment reaction occurs on the resin beads. Figure 3-39 shows a stainless-steel ion exchange system. Putting salt into a regular steel bed or a concrete bed could create corrosion problems, so stainless steel is used to minimize corrosion. Figure 3-40 shows an ion exchange plant used to remove mine tailing waste leachate. The operation on an ion exchange system involves:

1. Flow through
2. Backwash
3. Regenerate resin
4. Rinse slowly
5. Rinse quickly

Carbon Usage

All natural waters contain some amount of carbon-based constituents as a result of decaying vegetation and soils. This is called *natural organic matter*, or NOM. NOM is mostly a surface water or surficial aquifer problem, but all water has some degree of NOM. In addition there are thousands of synthetic organic compounds that make their way into water through spills and discharges. These include pesticides, oils, industrial solvents, and so forth. Many are carcinogenic, thus are regulated. Others create taste-and-odor problems, and all support bacterial growth. They also contribute to the formation of trihalomethanes resulting from disinfection. Trihalomethanes are regulated contaminants as noted previously.

The major source of taste problems in surface water is algae. Different algae may impart different tastes, but as the algae moves toward blue-green cyanobacteria, the problem worsens. Chlorine increases the taste problem, and warm, stagnant surface water sources are particularly susceptible to this difficulty. Groundwater sources, unless under the influence of surface waters, will not have algae issues, but may have synthetic organic compound contamination or leakage from underground storage tanks (USTs). Air-stripping is one method of removing these constituents, and adsorption may be of benefit. Remnant, decayed vegetative matter (under swamps) may create significant organics issues. Chlorination leads to trihalomethane formation as well.

FIGURE 3-39 Stainless-steel ion exchange vessel

FIGURE 3-40 Ion exchange building that treats metals from abandoned mining area

It should be noted that chlorine, permanganate, aeration, and ozone are used to remove organics, but when there are significant organics that cannot be removed or cannot be removed without creating other problems (like disinfection by-products), carbon adsorption is used.

Carbon adsorption is designed to remove carbon-based constituents from the water based on the concept of adhesion. Carbon adsorption is the collection of contaminants, usually carbon-based, onto the surface of activated carbon. Carbon has an affinity to stick to carbon sources, so adhesion means the organics are sticking to the carbon media. Adhesion forces are complex, and often require a series of chemical reactions as well.

Activated carbon has a huge surface area. The pores and surface area are created through high heat. There are deep crevices and cracks in the media that are microscopic in scale. Since the particles are very small, the surface area for many particles can be very large. Using activated carbon takes two forms—one as a feed, powdered activated carbon (PAC), and the other as granular activated carbon (GAC) filter media.

PAC is used for short-term situations, such as spring runoff events, unusual surface water conditions (algae), or spills. It comes in 50-lb (22.7-kg) bags and is fed into the raw water, thus is removed in the treatment process. A dry feed system will have a steep hopper (60 degrees). The PAC is fed via slurry or dry carbon feeds into the influent water prior to coagulation. This is because coagulants may compete with the humics for available sites on the PAC particles, and thus the PAC will lose its effectiveness. PAC slurry tanks are steel or plastic, with epoxy liners with a vibrating mechanism to feed the appropriate PAC quantity. PAC should never be permitted into the water distribution system. As a short-term option, PAC can be used for extensive taste or odor issues. However, as a long-term treatment solution, it is expensive. The bags must be stored in a dry area. Since there is potential for fire, sprinklers and other fire suppression should be considered.

The density of PAC is light, 20 to 45 lb/ft^3 (0.4 to 0.7 kg/L) or 15 to 30 percent the density of sand, with a surface area of 21,000 ft^2/lb (500 to 600 m^2/g) of media. The particles are 0.004 in. (0.1 mm) in size. The dry feed rate must be very accurate and the PAC feed must meet the demand. Because the density of the PAC is low, there is a tendency for the small particles to float on top of the water, which is not helpful to treatment. Instead the water should be rushed across the PAC or a mixer should be used to fully disperse it.

In comparison, GAC operation is very much like a filter, including the need for backwashing. Backwash should expand the GAC bed by 50 percent, but note that the media has a lower specific gravity than sand. Therefore, backwash rates vary according to media size, and the manufacturer or plant engineers should be consulted on backwash rates. Carbon loss is a concern in backwashing, requiring periodic replacement of GAC media. The media depth should be tracked after every cleaning, and replacement media added to ensure the bed is approximately the same depth at all times. GAC comes in 60-lb (27-kg) bags. GAC has a density of 26 to 30 lb/ft^3 (0.4 to 0.7 kg/L) and a surface area between 25,000 and 60,000 ft^2/lb (650 and 1,150 m^2/g). Grains are 0.04 to 0.06 in. (1.2 to 1.6 mm). The media provides consistent removal for a long period of time; depending on the organic load to be removed, GAC may last from months to years.

GAC can be reconstituted with high heat, but the cost is high. Regeneration requires that the GAC be removed and reheated to the original GAC processing temperature so that organics stuck to the GAC are "burned" off. About 5 percent of the GAC is lost in this process. If there is a source of power on-site that can reach the required temperatures, then regeneration can be done on-site, however, this is only cost-effective for large treatment plants.

Because GAC and PAC are organic, and have a lot of powder associated with them, safety and fire precautions are needed. Both GAC and PAC should be stored on pallets in clean, dry places that allow air to circulate beneath the pallets. The bags should be stacked in single or double rows with access rows around every stack to allow inspection and handling. Never store over 6 ft (2 m) high. PAC and GAC will burn like charcoal without producing any flame, so these are hard fires to detect if they start. The storage area should be fireproofed and have self-closing fire doors to isolate it from the rest of the plant. If a fire occurs, it should never be sprayed with water jets; instead, a fine mist should be applied. The spray will tend to cause spattering that may spread the fire.

Aeration

Water contains a variety of dissolved gases, such as carbon dioxide and hydrogen sulfide, that may interfere with the treatment or maintenance performed at a water plant. Aeration is a gas transfer process used in water treatment to remove gaseous compounds from water. This removal is called *stripping*. Aeration systems cascade water through air. Aeration can be used to remove certain taste-and-odor-causing compounds, carbon dioxide, hydrogen sulfide, iron and manganese, radon gas, and volatile compounds. Aeration is not efficient in removing inorganic constituents, but is very effective at stripping certain gaseous constituents and small amounts of other compounds. The air interacts with the constituents in the water to release the constituents.

Types of aerators and the concentration of constituents they strip include the following:

Cascade—limited concentrations (see Figure 3-41)

Cone—limited concentrations

Tray—limited concentrations

Draft—medium concentrations

Spray—medium concentrations

Towers—large concentrations, and tough constituents—pack atomized water to increase exchange rate

The packed tower stripper combines many of the features of prior strippers in an effort to make the water particles as small as possible to permit stripping the maximum amount of contaminant. The influent is at the top of the unit. Spray nozzles are oriented downward, to create fine bubbles of water (see Figure 3-42). The unit is filled with a plastic media. When the water drops hit the media, they "bounce" and make even smaller drops. Water is forced up through the bottom with blowers. The interaction of the air and the fine bubbles is highly effective. This type of stripper permits the highest removal percentage. Figure 3-43 shows an example installation.

Aeration of water with significant amounts of hydrogen sulfide can create odor problems and a highly corrosive environment for plant and equipment. The introduction of air into water containing microbiological populations can also increase biological activity, potentially

FIGURE 3-41 Cascade aerator

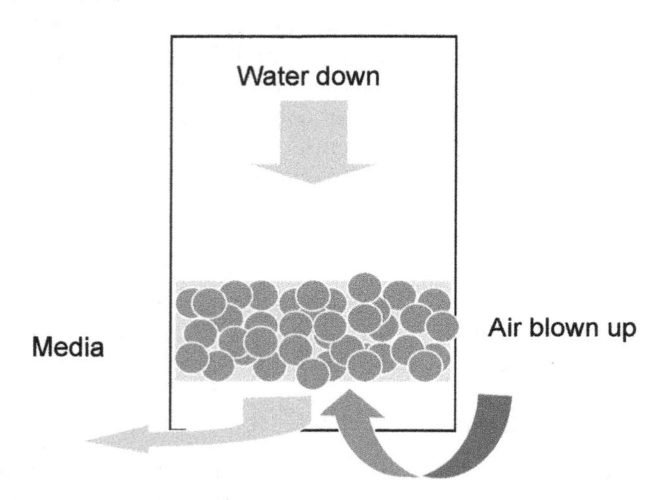

FIGURE 3-42 Basic operation of a packed tower

affecting downstream water quality. Aeration should never precede membrane processes due to the potential for increased microbiological activity and air binding in the membranes.

Posttreatment Processes

Posttreatment processes can involve a variety of measures, including aeration and chemical addition. Corrosion can affect both the structural capacity of the pipe and the quality of the water. While external corrosion can lead to leaks, internal corrosion can result from metabolic (microbial) activity, chemical dissolution, or physical abrasion by excessive fluid velocities. In addition, corrosivity of the water may leach metals such as lead and copper from metallic pipelines, lowering water quality. The federal lead and copper rules require

FIGURE 3-43 Packed tower application

that water be chemically modified to reduce the potential for corrosivity. This is typically accomplished with the addition of polyphosphates and pH adjustments. Chemical addition to limit corrosion is especially important with membrane treatment that will tend toward undersaturation, which requires postchemical addition of caustic and other chemicals.

Iron and magnesium removal can be done with potassium permanganate, which is a purple compound that will stain everything. The dosage of potassium permanganate that is added depends on the amount of iron or manganese that is in the water. One milligram of potassium permanganate will remove 1.06 mg/L of iron or 0.52 mg/L of manganese. It should be noted that potassium permanganate is often in a granular form and is dissolved in water and then applied as a liquid compound that needs to be properly cared for. Consult the material safety data sheet (MSDS) provided by the manufacturer.

Fluoride is usually added to water supplies up to a concentration of 1 mg/L, as this level has been shown to reduce dental caries. Normally fluoride comes as a powder, except fluorosilicic acid (H_2SiF_6), which is a liquid. In all cases, fluoride compounds are added as a liquid (see Figure 3-44). The chemicals are mixed with water and pumped into the water being treated. The pumps feeding the chemicals are flow paced, meaning that a precise amount of chemical is added based on the amount of water being treated (hence the need for meters). Fluoride, like chlorine, is a highly corrosive substance and, as a result, must be properly contained.

High-service pumps are used to keep the water in the distribution system at a desirable pressure (see Figure 3-45). High-service pump capacity is typically twice (or more) the maximum plant capacity. Storage at the plant site is typically a minimum of one half of the average daily flows from the plant.

FIGURE 3-44 Example installation for posttreatment chemicals—fluoride and polyphosphate

FIGURE 3-45 High-service pumps

WATER DISTRIBUTION

The water distribution system delivers water to the residents, and provides for fire-fighting water. The system consists of a large number of pipes, some pumps, and limited storage reservoirs. Table 3-3 is an inventory of one community's water distribution system. Where possible, this survey should be updated continually and should include the age and material of pipe. Unfortunately few systems have this detail, which makes long-term capital replacement planning more difficult.

TABLE 3-3 Example summary of water system components

Water System	Inventory	Units
2-inch and under water main	987,800	LF
4-inch water main	198,600	LF
6-inch water main	650,800	LF
8-inch water main	427,800	LF
10-inch water main	6,900	LF
12-inch water main	254,400	LF
14-inch water main	600	LF
16-inch water main	84,700	LF
18-inch water main	5,700	LF
20-inch water main	0	LF
24-inch water main	73,200	LF
30-inch water main	5,200	LF
36-inch water main	100	LF
48-inch water main	100	LF
Elevated water tank	2	
Water treatment plant (WTP)	38	mgd
Land	40	acres
Membrane WTP	18	mgd
Wells—Biscayne Aquifer	28	
Wells—Floridan Aquifer	4	
24-inch raw water line	8,200	LF
12-inch raw water line	4,200	LF
Floridan Aquifer well raw water line	3,750	LF
⅝- or ¾-inch services	33,100	
1-inch services	1,530	
1 ½-inch services	970	
2-inch services	585	
3-inch services	160	
4-inch services	58	
6-inch services	36	
8-inch services	3	
Backflow devices	200	
Fire hydrants	2,150	

Note: LF = linear feet

Pipe Materials

The pipe materials used in water distribution systems vary from galvanized iron to asbestos concrete to PVC and ductile iron, depending on the age of the system. AWWA Research Foundation (ARF) funded a study on water main replacement strategies in the late 1990s. They analyzed 28,543 breaks and determined what the breaks, and hence likelihood of failure or shortened life, were related to (Cromwell et al., 1997). The results included the following:

- Age of pipe
- Pipe material
- Pipe diameter
- Length of pipe
- Traffic on the surface
- Soil type
- Acidity of the soil

The length of time the pipe will last also involves another variable: the construction methods used. Wood might be the oldest pipe material (see Figures 3-46 and 3-47). That said, old cast-iron pipe seems to have a life that may exceed 100 years in normal soils, but the lead joints break down in high-traffic areas after about 40 years. In high-traffic areas, it could be assumed that pipes over 50 years old may be deteriorated due to vibrations from heavy traffic disrupting the joints and bedding.

Pipe age is especially critical if for portions of the year the pipes are partially submerged. In salty soils, the pipe will deteriorate faster, although the damage may be limited to small sections that need to be replaced (300 ft [100 m] or so at a time in contact with salty water) every 25 to 30 years, in some cases. An investigation of the condition of pipelines in submerged conditions should be undertaken periodically (including comments recorded during repairs) to evaluate the state of deterioration of older pipelines and the priority for replacement. The situation with pipes submerged in salt water is especially acute.

Newer ductile iron pipe seems to have long life, although the product has only been in wide use for about 40 years. Ductile iron pipe was the replacement pipe for cast iron (see Figure 3-48). The joints are rubber so unless there are extreme concentrations of chloramines in the distribution system, the pipe should last 60 years. There are two exceptions. First, if the pipe is unlined, then it should be replaced now as it is creating biofouling conditions in the water distribution system. Unlined pipe should be assumed to have a 20-year life because of the potential for biofouling (and it really should not be installed under any circumstances). Second, in salty soils the pipe will deteriorate faster, although the damage may be limited to small sections that need to be replaced, as with cast-iron pipe. Cromwell et al. (1997) found that higher grades of ductile iron pipe were less than half as likely to fail as asbestos–concrete pipe in their system.

Asbestos concrete was a new material in the 1960s and early 1970s. Asbestos–concrete (A–C) pipe is an inert material that has a long life if not disturbed, but tends to fail via shear breaks when disturbed because it is brittle. The potential for release of asbestos fibers is overstated from a water quality perspective, but may be a significant exposure issue for field

FIGURE 3-46 Old wooden water line—Leadville, Colo.

FIGURE 3-47 Existing wood water main—in service

FIGURE 3-48 Ductile iron (smaller, black pipe on far left) and PVC pipe

crews cutting A–C pipe. The pipe contains asbestos fibers, so it requires special handling to protect workers during sawing and repairing operations. Appropriate masks must be used.

Because of the asbestos, a plan for dealing with asbestos pipe repairs and disposal should be developed if a significant amount of it exists in the system. A plan to evaluate its eventual replacement should also be pursued. Many utilities are evaluating the replacement of A–C pipe over the next 30 years. USEPA may accelerate this but, to date, utilities have not seen a lot of problems with asbestos fibers in water (its impact is more airborne).

Galvanized pipe will last 10 to 30 years depending on soil conditions. It is typically used for service lines and small-diameter water lines. From a potable water utility perspective, it has not shown itself to be reliable in most communities, because it simply does not hold up over time. There are significant quantities of galvanized pipe in the Southeast. It deteriorates readily in all soil types and is prone to corrosion due to discontinuities in the surface of the pipe and dissimilar-metals issues. Salty soils will accelerate deterioration. Poor backfilling practices will also reduce its life. Galvanized iron pipe is routinely connected to other metallic materials (e.g., copper, lead, and iron), creating ongoing galvanic action that accelerates deterioration. Utilities should evaluate all galvanized pipe to determine if it should be incorporated into a replacement program. Sandy or clayey soils are not conducive to indicating small leaks. As a result, these galvanized lines could be a source of significant leakage in the system if these soils are present.

PVC and other plastics have a shorter history of usage, although two things are already known. AWWA Standard C900 indicates PVC pipe that appears to hold up for at least 60 years if properly installed. Most utilities east of the Rocky Mountains view it as being the equivalent of ductile iron, minus the potential for corrosion. There is no reason it should deteriorate in the soils. It will fail if backfilling is improper (backfill containing rocks, failure to properly tamp the pipe, and so forth). It will also fail if it was left in the sun too long—it gets brittle and will shatter (but this usually happens quickly). No PVC pipe that is colored should be installed if the surface shows any bleaching.

Lower grades of PVC, such as 2241 PVC, appear to have a much shorter life than C900 PVC as a result of significant sensitivity to improper backfilling. It is thin-walled and the egg shape caused by improper backfilling creates pressure points that will fail. Several eastern North Carolina utilities only obtained 20 years of service from the thin-walled pipe before having to replace it. Longitudinal breaks are common. The material is cheaper than C900, but not very much. Most cities avoid 2241 pipe. Note that there is also insulated piping for water distribution systems (see Figure 3-49).

If the oldest water lines exceed 50 or 60 years, they are likely beyond their useful life (unfortunately, piping over 100 years old is common in urban areas). The conditions of the soil and the installation methods are critical variables, so depending on these circumstances, water line life may be more or less. Ductile iron and PVC have been demonstrated to last in excess of 60 years given all the right conditions. Public officials should be wary of those making claims of extended life for their pipe or fittings; they are potentially passing the repair costs on to future ratepayers. Failures of these pipelines, especially large ones, could cause road and property damage as well as service disruptions, so a proactive approach is needed.

High-density polyethylene (HDPE), polyethylene (PE), polybutylene (PBE), and similar pipes are usually used for service lines. They are flexible, but construction practices may overestimate their flexibility, potentially causing failures. Otherwise these pipes appear to have a life of 60 years or more (although current experience with this pipe is insufficient to test this life expectancy). In service lines, it is the copper fittings (or galvanized iron, if still used) that are more likely to fail.

Individual service lines attached to the cast-iron lines may be constructed with galvanized fittings. Such service lines are subject to severe corrosion and may also be a source of water leaks and of lead leaching into the water supplies. They should be replaced at the same time as the rest of the pipes. The same is true for the few remaining lead goosenecks used on service lines before 1960. Standard materials are PVC C900 and ductile iron pipe for water distribution systems, polyethylene tubing for service lines, and brass/bronze fittings. All are appropriate materials. Galvanized iron, lead fittings, and low-grade PVC are not.

FIGURE 3-49 Insulated pipe

Water mains are normally installed in a grid pattern in the streets (see Figure 3-50). This idealized version of a series of city blocks shows the developed block as solid rectangles and water mains and valves installed in a grid pattern. The water lines are shown as the solid lines in the street. Valves are typically located on the various branches. If there are *n* branches, there should be $n - 1$ valves. So in an intersection like those shown in the figure with four connections, there should be at least three valves.

Rarely are easements used to access water lines, although alleys are not uncommon. If little has been done by the utility to exercise its easement rights (a common problem where alleys contain water lines), it may be that many of the pipelines and services are not accessible. This kind of situation requires attention so that a future leak event that cannot be accessed does not subject the utility to undue criticism. Strong winds may topple trees that have roots wrapped around pipelines, a potential outage problem that has no short-term solution.

Larger-diameter water mains are also constructed of reinforced concrete, steel, or fiberglass, depending on their intended use, type of installation, and soil conditions. AWWA publishes installation and materials standards that should be followed for each type of pipe and fitting to ensure long, low-maintenance life. Prestressed concrete pipe (C303) and steel are used for large transmission piping, but not for local water mains.

Trench technology

Trench technology is a means to construct or replace water and sewer lines that are not required to be laid on grade. For example, replacing a large-diameter water main in a downtown area would be disruptive to commercial operations. This might be an opportunity to use trenchless technology to avoid digging up the roadway. Similarly, trenchless technology works for pipelines under sensitive areas, like wetlands or under water bodies. Technology has improved over the past 20 years, allowing trenchless technology companies to directionally drill in from point to point, but not necessarily in a straight line. The piping can be a variety of materials, but normally the piping is HDPE pipe, which is usually pulled back through the drill hole. Service lines are commonly directionally drilled, either by drill or compression hammer.

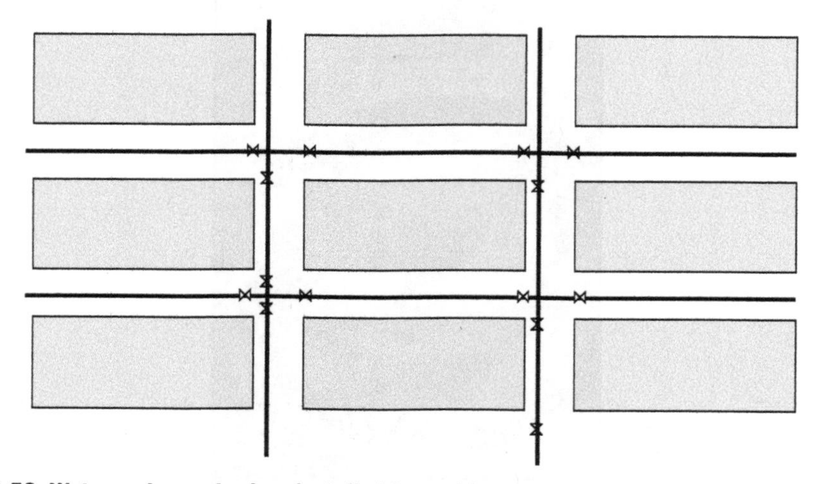

FIGURE 3-50 Water mains and valves installed in a grid pattern

Depth of Pipe

The critical concerns in burying pipe are the prevention of freezing, minimization of temperature fluctuations (hot or cold), and protection of the pipe. As a result, the depth at which water mains are buried varies greatly throughout the United States. Water mains can be buried quite shallowly in warm southern states (minimum 30 in. [0.6 m] in most cases) because the only concern is physical damage. Mains are buried deeper where there is moderate ground frost, e.g., up to 8 ft (2.5 m) in northern states where temperature changes and frost may expand and contract the pipe.

Valves

Valves are installed at intervals in water main piping so that segments of the distribution system can be shut off for maintenance or repair. Valves should be located close enough so that only a few homes or businesses will be without water while a leak or break is being repaired. Valves should be installed in the grid system shown in Figure 3-50, in which mains are in all the streets and run in every direction of the grid. Valves should be placed where water mains intersect. The best practice is to install valves on all but one of the branches.

Each valve should be installed with a valve box that extends to the ground surface and has a cap that can be removed so that a valve key can be used to operate the valve. Valves should, if possible, be located where the box is easily located and where damage by snowplows and other equipment is least likely.

A valve exercise program (opening and closing them at least once per year) should be in place to ensure that valve boxes are exposed and have not been filled accidentally with dirt or damaged by paving or snow removal activities, to ensure that the valves are open and work properly, and to loosen up the valves so that they will operate more easily. Systems with a large number of valves often purchase power valve-turning equipment to speed the job of exercising the valves. When valves are operated, the number of turns should be counted to make sure they are fully operated in both directions. (Note: the size of the valve is found by the number of turns, minus three, then divided by six, e.g., 21 turns would indicate a 6-in. [152-mm] valve.) Valves that do not operate properly should be dug up and repaired as soon as possible. There are both right- and left-hand valves (indicating the way they turn on). Valves should be standardized on any utility system; otherwise people may be turning valves on when they think they are being turned off, and vice versa, leading to a potentially dangerous field situation.

Leaking or damaged valves should be replaced when discovered to minimize outages. Valve exercising programs may be the most ignored program for water systems—very few systems actually exercise valves because few local officials recognize the importance. Until a major break occurs, none of the necessary valves may work properly and large sections of the system may be shut down to make repairs. A typical valve is shown in Figure 3-51.

There are a variety of types of valves on the water (and wastewater) system, each with different purposes, although each in some way can restrict or completely shut off the flow of water (Pensic, 2010), including those listed below.

- **Gate valves** are in the street, designed to operate infrequently, but available to shut down areas of the distribution system for repairs of breaks or construction activity. Otherwise they remain open for many years (hence the need to operate them periodically). Gate valves

FIGURE 3-51 A distribution system valve

have existed for over 100 years and are the most common large valves in any utility system. They are reliable but are large and, for large pipelines, their dimensions may become too great for burial in the street. Gate valves only obstruct flow when closed or partially closed. The valves close as a disk is lowered vertically from the top of the valve housing.

- **Butterfly valves** are used in plants, on large pipelines, and where gate valves are impractical. The valve mechanism pivots across the flow stream, which limits the ability to pig pipelines to remove tubercles and scale.
- Sanitary sewer systems may favor **plug valves** over gate valves, although functionally they accomplish the same thing. The plug rotates horizontally to close the opening in the plug body.
- **Check valves** operate constantly to ensure flow goes one direction. They are usually found in pump stations and wells. They are rarely buried except in vaults.
- **Corporation stops** and **meter valves** are small brass or bronze valves used on service lines (generally under 2 in. [51 mm]). Corporation stops are used when tapping the main for a new service. Otherwise they are rarely touched. Meter valves permit water to be shut off to a customer for maintenance or nonpayment of the bill. In many cases these valves are operated infrequently but the concept is that a quarter turn will seal the pipe.

AWWA has standards for valves (including AWWA Standard C500) that should be used on water systems. Coatings, materials, and applications are included in the standard.

Fire hydrants are of two general types. A wet-barrel hydrant is full of water at all times and can only be used in parts of the country where there is no danger of freezing. Dry-barrel hydrants have the valve located at the bottom of the barrel and are operated by a long shaft extending down from the operating nut on the cap. Dry-barrel hydrants also have a small valve connected to a weep hole at the bottom that allows water to drain from the barrel when the hydrant valve is shut off (see Figure 3-52 and AWWA Standards C502 and C503).

FIGURE 3-52 Typical fire hydrant installation

Hydrant locations should be selected carefully. They should be readily visible and located near a paved surface where they will be accessible by fire-fighting equipment. They should also be placed where they are protected from damage by vehicles and are least liable to be covered by plowed snow.

Utility officials should always insist that police enforce parking restrictions adjacent to fire hydrants so they will not be blocked if needed. Police should also be reminded to watch for vandalism and unauthorized use of hydrants and to report incidents to the water system manager. Frequent painting with bright paint protects hydrants from rusting and makes them easy for the fire department to find. Well-maintained hydrants also project a positive public image of the water system. Fire departments will often color code hydrants based on the available flow. No hydrant should ever be permitted to draw negative pressure. Thus all fire trucks should be fitted with valves that cut off if the pressure drops below 20 psi (138 kP) (the minimum accepted health standard) to protect the integrity of the utility system and the health of the public.

Service Lines

The small-diameter pipe used to carry water from the water main connection to an individual building is referred to as a *water service pipe* (Figure 3-53). A water service pipe may range from ¾ in. (20 mm) in diameter for a small home to 6 in. (150 mm) for an apartment building. Large buildings and industries often have service and fire sprinkler pipes that are even larger. AWWA Standard C800 was designed for application to service lines (AWWA Standards C901 and C903 apply to plastic pipe).

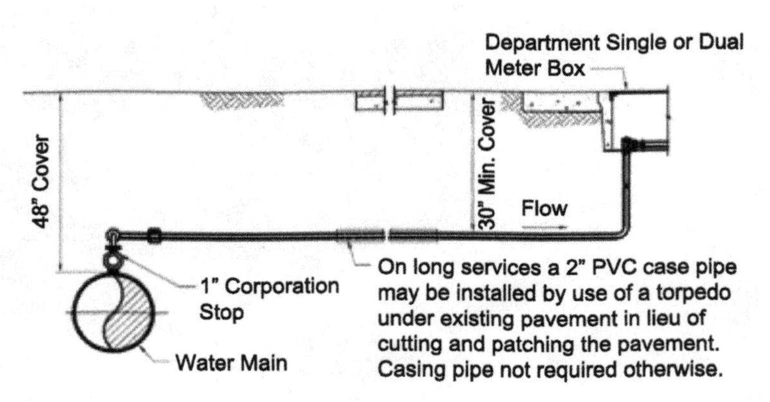

FIGURE 3-53 Water service pipe

Each water service pipe usually has a buried valve called a *curb stop* inserted in the line at a point at the edge of the public street or alley right-of-way or an easement. Where curbs and sidewalks exist, water system policies generally standardize the curb stop location at a set distance between the curb and sidewalk or at the lot line. The buried valve is fitted with an adjustable service box (curb box) that extends to the surface and has a removable cap so a valve key may be inserted to operate the valve. The curb stop is primarily used to shut off the service if the building being served is vacant or repairs are needed. It is also a way of discontinuing service for nonpayment of the water bill. Fittings should be bronze or brass.

Water service pipes are typically made of lead, galvanized iron, copper, or plastic. Lead was the best material available for small pipes when the first water systems were developed. Lead service lines are relatively flexible and resist corrosion but gradually become more likely to leak or break as they get older. Lead is no longer used for service lines, but many may still exist in water systems. New federal regulations, designed to protect the public from the danger of lead in drinking water, require systems to ensure that leaching of lead from water services is minimized. Systems with aggressive water that tends to dissolve lead may have to install additional chemical treatment (typically phosphates) to meet the requirements. Systems that cannot adequately control the leaching of lead may be required to remove existing lead service pipes and replace them with other material.

Galvanized iron pipe, used for water service piping for many years, corrodes very quickly in some types of soil. This material should be replaced because it is a source of leakage in the system. The same issues apply for galvanized service lines, except that the service lines are smaller and tend to deteriorate more quickly than galvanized mains.

Copper pipe came into use in the early 1900s and gradually became the preferred material in many parts of the country. Copper is flexible, fairly easy to install, resistant to corrosion, and lasts almost indefinitely under most water and soil conditions. However, copper can also leach from the service line to water supplies where the potable water is aggressive. Lead solder was also used until the mid-1980s for connecting copper pipe. This lead solder was noted as a source of leaching lead in water systems. As a result, it is no longer used in the United States for copper service line solder.

Plastic pipe has been used for water service lines since shortly after World War II. Polyethylene and polybutylene are the most common materials. Both are lightweight, easy to install, flexible, moderately priced, and resistant to corrosion. In some areas, plastic pipe has been used almost exclusively for years. There are many types of plastic, but only certain types and grades are approved for potable water use. AWWA has standards that define these materials, including:

- C901 Polyethylene (PE) Pressure Pipe and Tubing, ½ In. (13 mm) Through 3 In. (76 mm), for Water Service, and
- C903 Polyethylene–Aluminum–Polyethylene & Cross-linked Polyethylene Composite Pressure Pipes, ½ In. (12 mm) Through 2 In. (50 mm), for Water Service.

Plastic pipe must be tested for durability and be free from constituents that might cause tastes or odors or release toxic chemicals. Only pipe that has the seal of an accredited testing agency printed on the exterior (typically NSF) should be used for potable water purposes.

Cross-connections

The purpose of a cross-connection control/backflow prevention program is to isolate the potable water supply system from the possibility of backflow, as backpressure and/or backsiphonage may occur because of the existence of cross-connections between potable and nonpublic systems. Ordinances and other codes prohibit cross-connections of private systems to the utility's potable water system except when and where approved and when appropriate backflow prevention devices are installed, tested, overhauled, and maintained to ensure the health and well-being of the community and the potable water system. In areas with a large number of fire, agricultural, or commercial systems, cross-connection should be a priority. Figure 3-54 shows a standard backflow device on a fire line to a building.

FIGURE 3-54 Backflow devices

WATER STORAGE FACILITIES

Stored water can be used to maintain pressure for a period of time if a well or pressure pump should fail or lose power. The primary reason for providing storage of treated water is to have a reserve supply readily available during fire fighting, emergencies such as repairs to treatment facilities or pumps, loss of water supplies due to pipe breaks or contamination, or for periods of heavy water use. Water demands vary throughout the day, while treatment plants should run constantly. When customers are using water at a low rate, excess water can be stored. When use is high, stored water is used to meet the demand without having to alter the operation of the treatment plant. Figure 3-55 shows the typical peaks in the morning and evening, and low flows at night of a plant's tank and reservoir.

Figure 3-56 shows a typical storage calculation graph. The cumulative flows are shown vertically, by hour (horizontal axis). The peaks are matched to determine the amount of flow required. Usually this amount is multiplied by 1.2 to get the minimum storage required. The quantity of water storage that should be provided on a system is usually based on the amount of water required to meet domestic and fire flow needs, which is normally well above the storage required in Figure 3-56. Other instances for storing more water include the need for storage to last one or two days for systems that depend on a single, long, transmission main for source water and systems that have periodic episodes of temporarily poor-quality source water, in which enough storage is needed to allow them to avoid taking water until the source water quality improves.

Types of Storage Facilities

Water storage facilities fall into the categories of elevated tanks, standpipes, hydropneumatic systems, and ground reservoirs. Elevated tanks are the most familiar because they are visible in prominent locations in most communities. Elevated tanks are normally constructed of steel, with the tank portion supported on legs or a pedestal (see Figure 3-57). Tanks are generally located on the highest ground that is available and acceptable to the residents. The public is not generally bothered by an existing tank located in a residential neighborhood but usually will not want a new tank erected near their homes.

An elevated tank normally fills and empties in response to demands on the water system (referred to by operators as *riding the demand*), and the elevation of the water in the tank determines the water pressure on the system. A 1-ft elevation change is equal to 0.43 psi, so a 100-ft tank will pressurize the system to 43 psi (1 psi = 2.31 ft). A signal indicating the water level in the elevated tank is commonly used to vary the operation of the pressure pumps supplying the system. When the water level is near the top of the tank, the supply of water is reduced or stopped before the tank overflows. When the water level falls to a predetermined point in the tank, flow to the tank is increased. Occasionally, a water system must operate at a pressure that would cause an overflow of the elevated tank. In this case, water is admitted to the tank by an automatic valve that shuts off flow before the tank overflows.

In water systems, a standpipe typically refers to an aboveground tank that is the same size from the ground to the top. Standpipes are primarily used where they can be located on a high point of land so that all or most of the stored water will furnish usable pressure to the water system (see Figure 3-58).

Hydropneumatic systems have been developed primarily to serve small systems where an elevated tank is not practical. A large pressure tank is buried or located above ground

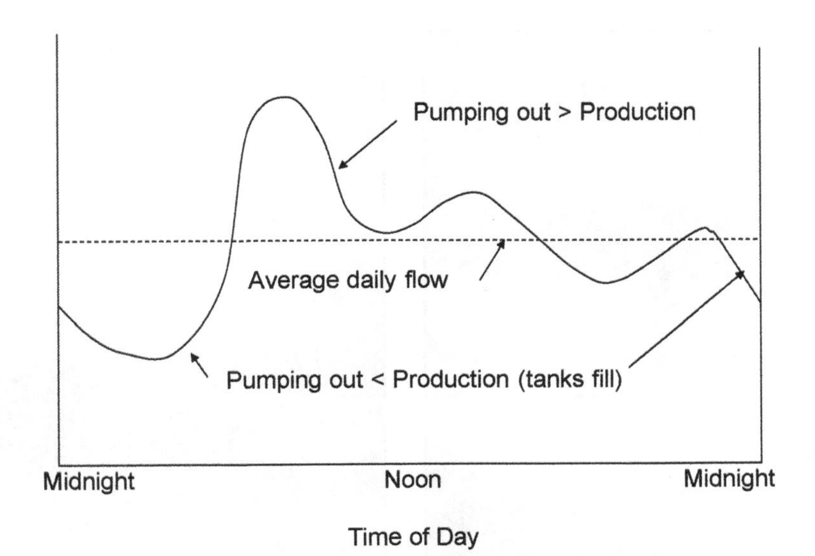

FIGURE 3-55 Typical filling and emptying cycles of a tank and reservoir

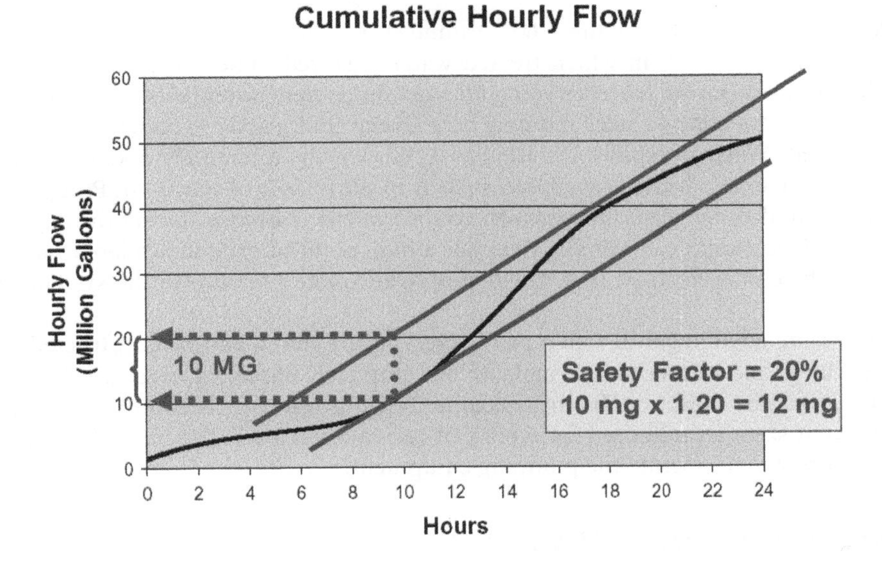

FIGURE 3-56 Typical storage calculation graph

and kept partly filled with water and partly with compressed air (see Figure 3-59). The balance of compressed air against the water maintains the desired pressure in the system and forces water out of the tank when needed. An air compressor is required to maintain the proper air-to-water ratio.

FIGURE 3-57 An elevated tank

FIGURE 3-58 Standpipe

A water reservoir (see Figure 3-60), ground storage tank (see Figure 3-61), or wet well is normally a large tank in which treated water is stored under no pressure. The water must be pumped out of the reservoir and into the system when needed. Reservoirs are constructed of concrete or steel and may be aboveground, partially underground, or completely buried. Water is usually admitted to a reservoir by a remotely operated valve during times when excess water is available, such as in the middle of the night. Pumps are then operated to add water from the reservoir to the system as needed during the day or in an emergency. Occasionally, a water system has a high point of ground available where a reservoir can be constructed so that it will supply adequate pressure to the system without repumping.

The prime advantages of this type of reservoir are that it can be constructed to store relatively large quantities of water and can be completely buried, whereas an aboveground structure could be objectionable to residents. When a reservoir is completely buried, the land above it is sometimes used for a park or recreational area. The main disadvantage is the cost of power to operate the pumping equipment.

WASTEWATER COLLECTION

The collection system consists of the gravity pipes, manholes, service lines, and cleanouts. The manholes and cleanouts are required for access to and removal of material that may build up in the piping system. Collection system piping throughout North America prior to 1980 was constructed predominantly of vitrified clay. Since that time, various grades of PVC have been used. Ductile iron is rarely used due to the potential for corrosion from hydrogen sulfide gas.

FIGURE 3-59 Hydropneumatic tank

FIGURE 3-60 Water storage reservoir in West Palm Beach, Fla.

FIGURE 3-61 Ground storage tank

Vitrified clay pipe has been used for well over 100 years. The pipe is resistant to deterioration from virtually all chemicals that could be in the water, and from soil conditions. It has a long service life when installed correctly and left undisturbed. However, vitrified clay pipe is brittle, so settling from incorrect pipe bedding, surface vibrations, or freezing can cause the pipe to crack. There are also limitations on pipe size. Temperature differences between the warm wastewater and cooler soils can cause the exterior pipe surface to be damp. The dampness encourages tree roots to migrate to the pipe, where they may wrap around the pipe. Where cracks occur, roots will enter the pipe. Over the long term, the pipe will become broken and damaged from the roots, vibrations, and freezing. Where the water table is above the pipe level, significant infiltration can occur that reduces the capacity of the wastewater treatment plant.

A second concern with older vitrified clay pipe is the short joints used. These are as small as 2 ft (0.6 m) prior to 1920, and 4 ft (1.2 m) prior to 1960. Field joints were made prior to 1920, and even later. The joints were sealed with cement and cloth "diapers" wrapped around the joint. However, concrete is not waterproof and will crack with time. The combination results in piping with many joints, each of which has the potential to leak. Even today, the joints are short compared to PVC and ductile iron (20 ft [6 m] and 18 ft [5.5 m], respectively), although the joints and material have improved substantially. Vitrified clay remains the choice of material for use in industrial areas where pipe protection is required. Vitrified clay pipe can be lined with many products, thereby extending the life of the pipe.

Concrete pipe is used for large-diameter sewer lines. Concrete has the benefit of durability and structural strength. Concrete piping as large as 96 in. (2.5 m) is not uncommon. Concrete is the only material made in these diameters (usually the pipe is prestressed when this large). Concrete piping suffers one significant problem, which is its vulnerability to hydrogen sulfide. As sewage remains in the piping system, if air is not entrained, the sewage will become septic. Septic sewage is black, and smells heavily of hydrogen sulfide (rotten eggs). Hydrogen sulfide, when attached to pipe surfaces, will react with water to form sulfuric acid, which is deleterious to the pipe surface. The area just above the water line and the crown of the pipe is most vulnerable. Without proper inspection, maintenance, and care, the pipe will fail at the top. Lining the pipe is beneficial, but the lining must be monitored to ensure it remains intact. High levels of hydrogen sulfide can reduce pipe life by 20 or more years if left unchecked (Yamanaka et al., 2002).

There are three ways to reduce hydrogen sulfide levels in a collection system: chemical additives, biological treatment, and aeration. If the sewage is not allowed to become septic, which is a difficult thing to do in large systems or where significant pumping is involved, hydrogen sulfide is not produced. Chemical additives can be put into lift stations and personnel access openings (manholes) to react with the hydrogen sulfide. This can be an expensive process.

Ductile iron piping used in sewer systems is similar to that for water distribution piping. One area of difference is that wastewater piping is usually lined with polyethylene instead of cement to prevent hydrogen sulfide deterioration. Ductile iron, like concrete, is susceptible to hydrogen sulfide corrosion.

PVC (Figure 3-62) is the choice for most other applications. The color is always green. A variety of PVC grades have been used for sewer lines, including C900 and SDR 35. More utilities appear to be favoring the heavier grades of C900 pipe as a result of concerns during construction (bedding, in particular, and backfill methods). PVC pipe is excellent for resi-

FIGURE 3-62 PVC pipe for sewer lines is typically green as opposed to blue for water lines

dential services, but the longevity of the pipe rests on having proper bedding and backfill methods. Improper backfill will crush, crack, or deform the pipe. C900 is more resistant to damage, and the increase in cost is minimal. All PVC sewer lines should be green.

Personnel Access Openings

Personnel access openings (manholes) are access sites for workers, for changes in direction of the pipe, or every 400 ft (125 m) to permit the pipe to be cleaned if grease or blockages impact flows. Personnel access openings are traditionally precast concrete or brick. Brick was the method of choice until the 1960s. Brick personnel access openings suffer from the same problems as vitrified clay sewer lines, i.e., the grout is not waterproof so can leak significant amounts of groundwater. The access opening cover may not seal perfectly, becoming another source of infiltration. Precast concrete access openings limit the number of joints. Many utilities will require the exterior of the access openings to have coal tar or epoxy covering the exterior, which helps to keep water out (Figure 3-63). Elastomeric seals are placed between successive access opening rings. Elastomeric seals are also used to minimize inflow from the pipe entering the access opening (Figure 3-64). All access openings have a flow zone and a bench (an area on which to stand, see Figure 3-65).

Personnel access opening covers are designed to be traffic-bearing (Figure 3-66). They weigh 90 to 200 lb (40 to 90 kg). Care should be taken to ensure they are properly fitted so

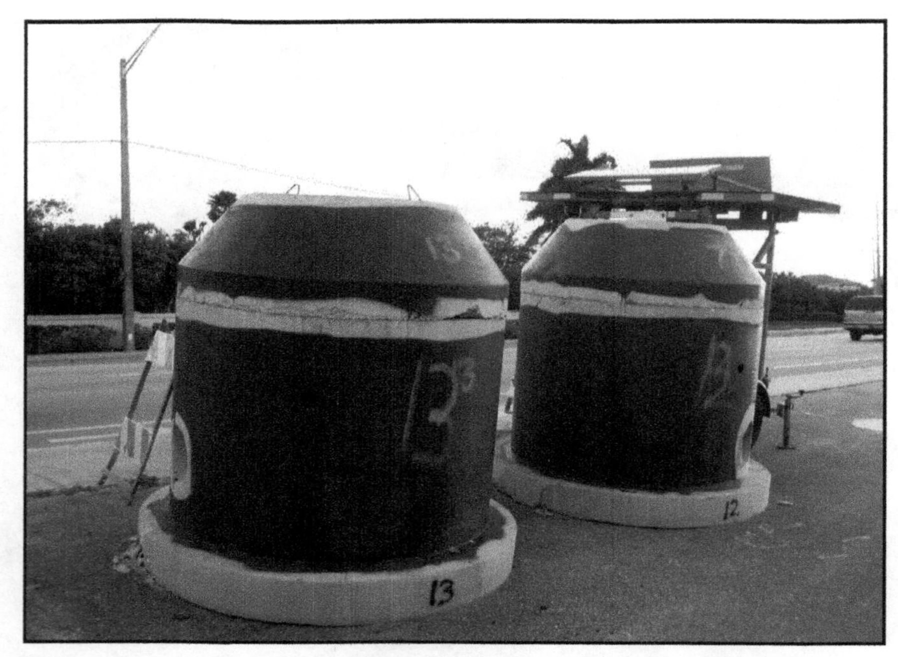

FIGURE 3-63 Exterior of personnel access opening with coating to prevent infiltration

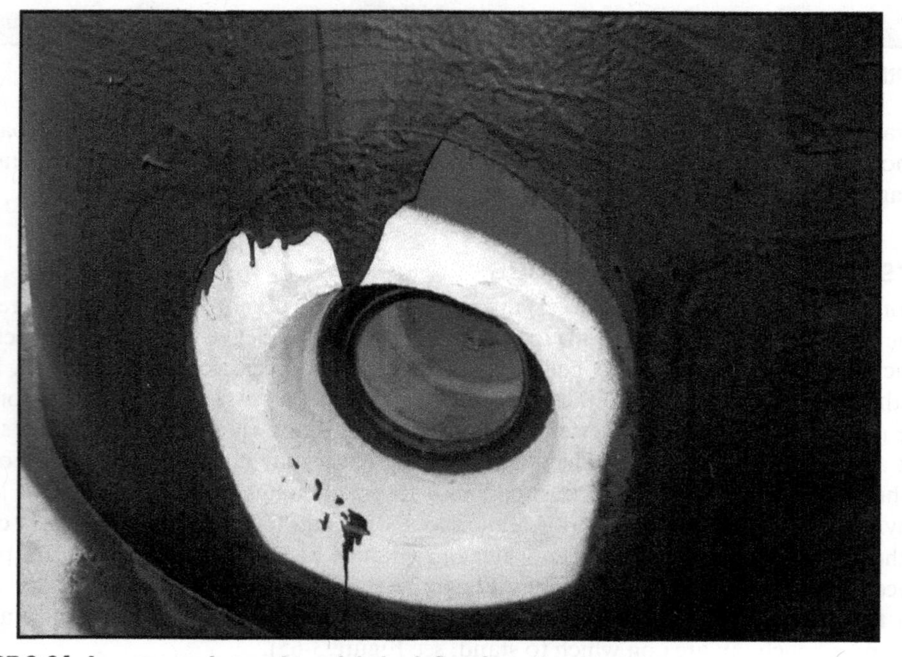

FIGURE 3-64 Access opening seal to minimize inflow from the pipe joint

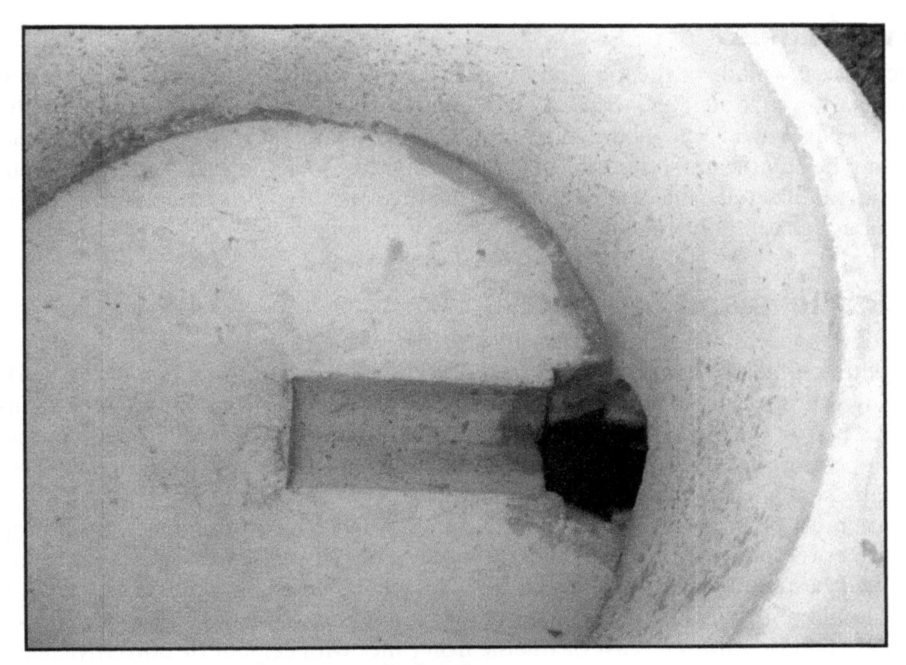

FIGURE 3-65 Access opening flow zone and a bench

FIGURE 3-66 Access opening covers

they do not come loose with passing traffic. Operations personnel should also observe areas where residents may remove covers during heavy rains to reduce yard or road flooding as this can overwhelm the treatment plant, pipeline, and lift stations and create sewer overflows.

Service Lines

Service lines are made of the same piping as is used for the collection system. Services lines are normally 4 in. or 6 in. Schedule 40 PVC and vitrified pipe are the most commonly used materials, although Orangeburg pipe may also be found on older service lines. Service lines may be significant sources of infiltration due to roots and breakage. Unfortunately, the majority of service line length is on private property. Requiring homeowners to repair these pipes is often politically difficult.

Lift Stations and Telemetry

Most sewer collection systems are gravity systems, whereby the wastewater flows downhill to a treatment facility or collection point. The gravity lines can do much of the work, but for many utilities it is not possible to have wastewater flow all the way to the treatment facility. Thus, at the low points, a lift or pump station is installed. The lift station is a large access opening with pumps in it. The waste leaving a lift station is under pressure, although usually less than for water systems.

Care must be taken when considering pumps and pump station materials. Pump stations are corrosive environments and the water may contain solid materials. Usually a pump will be required to pass a 2- or 3-in. (5- or 7.5-cm) ball without clogging. Force mains should not be less than 4 in. (10 cm) and should use the same piping materials as for water distribution pipe, i.e., cast or ductile iron (lines with cement or polyethylene), PVC, or asbestos concrete. Valves and fittings are also the same as for water distributions systems. A major issue to be concerned with is that tools used on the sewer system should not be used for water system repairs.

For busy lift stations, controls are helpful. The controls can alert operators to problems immediately, saving the time required to check stations frequently. Figure 3-67 is a typical lift station control box. Figure 3-68 shows the inside of a lift station wet well. Lift stations are major causes of odor complaints by the public. Figure 3-69 shows an example of an odor control system at a lift station. Another means to control odors is to use in-line booster stations, which are pumps within the pipeline (Figure 3-70), or to inject odor control chemical into the sewer system upstream of the lift station.

Ongoing maintenance and rehabilitation are important, as is telemetry for tracking data. It is especially of concern if the utility purchases bulk wastewater service. Electricians are needed to maintain lift stations to ensure they do not fail or overflow. Pump damage from cavitation or low fluid levels is a potential problem that control systems can rectify. Monitoring flows will indicate if there is a need to review reducing infiltration, and aid in identifying locations and preventing excess inflow.

WASTEWATER TREATMENT

Unlike water treatment, most wastewater treatment systems are neither chemical nor physical. Instead, they are primarily a biological phenomenon, and therefore have a different set of challenges and capacity for disruption. The level of treatment is defined by the method of disposal of the treated wastewater. If the wastewater effluent is to come into contact with humans, then the treatment is higher. The minimum required treatment scheme in the United States and Canada is secondary treatment, but it is by no means the only option. The following paragraphs describe the process requirements at various levels

FIGURE 3-67 A typical lift station control box

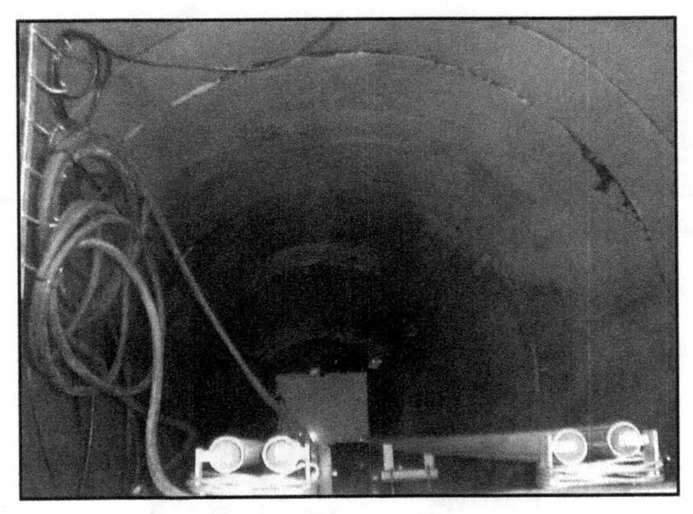

FIGURE 3-68 The inside of a lift station wet well

of treatment. USEPA has made nutrient removal a priority in freshwaters and estuary systems, which will significantly increase the amount of treatment required for those systems. Reuse of the wastewater for irrigation also prompts more extensive treatment.

Primary Treatment

Primary treatment is defined as the use of the treatment system to accomplish the removal of a portion of the suspended solids and organic matter prior to discharge into the receiving water (Figure 3-71). No biological processes are assumed to occur. Typically the treatment consists of settling basins (or primary clarifiers) and macro-scale screening (i.e., bar racks or screens—see Figure 3-72). These processes remove only the largest constituents (and those most likely to clog pumps and pipes). Thus, the effluent

FIGURE 3-69 An odor control system at a lift station

FIGURE 3-70 In-line booster pump
Source: http://www1.eere.energy.gov/industry/bestpractices/pdfs/boosterpump.pdf

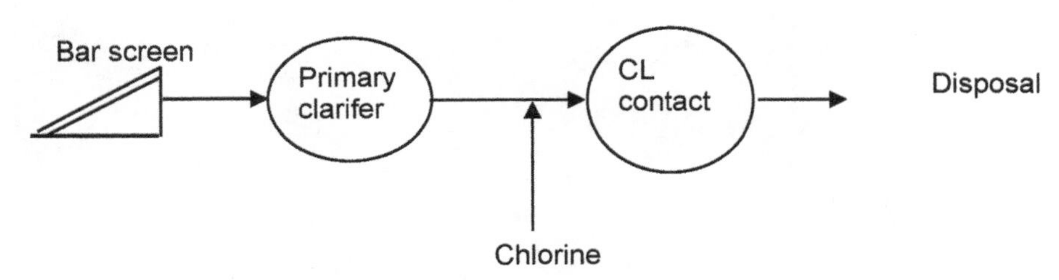

FIGURE 3-71 Primary wastewater treatment

FIGURE 3-72 Bar screen at head of treatment plant

will have a high concentration of carbonaceous biochemical oxygen demand (CBOD) and organics. Primary clarifiers (see Figure 3-73) are designed to remove 50 to 70 percent of the suspended solids and 25 to 40 percent of the CBOD.

Since primary treatment by itself is not permitted in the United States, such processes typically precede biological processes, and can be used for flow equalization in secondary treatment facilities. Primary clarifiers have a detention time of only 10 to 30 minutes, hence

FIGURE 3-73 Primary clarifier installation

the low removal rates compared to secondary treatment. Most primary facilities have been phased out and replaced with secondary treatment systems in the United States and Canada.

Secondary Treatment

Disposal of the treated wastewater and the potential receiving water's use indicate the extent to which treatment is required. Secondary treatment is the minimum standard in the United States and Canada. Most secondary wastewater is discharged to moving rivers or streams where the natural processes can further treat it. Keep in mind that the stream may be a downstream user's water supply.

The basic treatment units include aeration basins, clarifiers, and disinfection facilities (Figure 3-74). The concept of secondary treatment includes biological activation of bacteria to treat the wastewater in the aeration basin. Organically based wastes come into the treatment facility with trace amounts of minerals, metals, and other contaminants. The aeration basin contains bacteria that have been "trained" to use the organic wastes as food. The amount of "food" for the bacteria is measured as the carbonaceous biological oxygen demand (CBOD), typically measured over a five-day period ($CBOD_5$). The bacteria require oxygen, so air is pumped into the aeration basin. The food:microorganism ratio is important in maintaining healthy bacterial populations. The goal is to keep the bacteria close to starvation so they consume the waste and reproduce efficiently.

In the aeration basin, there will be active areas of bacteria growth, and areas where the organics have been removed from the water. The "cleaned" water is decanted to the clarifiers where the remaining materials are allowed to settle to the bottom and then recycled back to the aeration basin as incoming organics. In the process, the bacteria age and die, descending to the bottom of the aeration basin or clarifier. When sludge is removed from the facility, it is really dead bacteria and minerals—not the organic material that comes into the wastewater treatment plant.

Typically a secondary treatment facility will have a bar screen (see Figure 3-74), and may have primary clarifiers or an equalization basin (see Figure 3-75) ahead of the biological treatment process. Secondary treatment is directed principally toward the removal of biodegradable organics and suspended solids. Biological processes include aerated lagoons for very small systems (see Figure 3-76), extended aeration systems (Figure 3-77), contact stabilization (Figure 3-78), activated sludge (Figure 3-79), sequence batch reactors (SBRs) (Figure 3-80), and fixed film reactors.

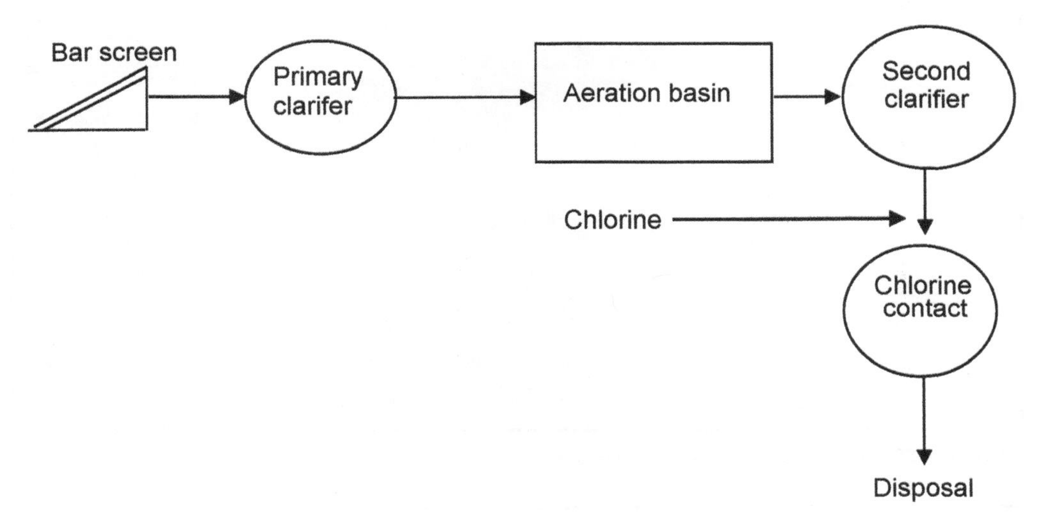

FIGURE 3-74 Secondary wastewater treatment

FIGURE 3-75 Equalization basin

FIGURE 3-76 Aerated lagoon for small system

FIGURE 3-77 Extended aeration system

FIGURE 3-78 Contact stabilization

FIGURE 3-79 Activated sludge system for large system

FIGURE 3-80 Sequence batch reactor

The concept is that during the biological process, air is introduced into the wastewater in order to increase the food:air ratio to a point where the optimum number of bacteria will consume the incoming organics and use up the air. The result is that the incoming organics are converted to new bacterial cells, and the older cells will be removed from the process. Careful operation is needed to make sure that the ratios and sludge age are optimized. Large amounts of inflow from storms, failure to provide enough air or too much aircan disrupt the process. An example is shown in Figure 3-81. It may take up to 30 days to correct wastewater processes that are disrupted, thus operations staffing is very important. All treatment methods employ secondary clarifiers after the biological process. Clarifiers in the wastewater process work precisely the same way they do for water systems, except that the solids are mostly bacteria that settle to the bottom of the clarifier and are removed. Figure 3-82 shows a large wastewater clarifier. It takes longer for the sludge to settle than for lime softening, so the clarifiers are generally large. Figures 3-83 and 3-84 show another clarifier—both full and empty.

During the secondary clarification process, the dead/dying bacteria, along with nonconsumed organics and metals, will settle to the bottom while the cleaner water overflows the weirs (see Figure 3-85) and flows to the next process. Disinfection via chlorination is typically employed (see Figure 3-86). The sludge is scraped to the center for removal or recirculation. The removed sludge is disposed of via the sludge rules (discussed later in this chapter).

Secondary plants are designed to achieve an effluent prior to discharge containing no more than 30 mg/L $CBOD_5$ and 30 mg/L total suspended solids (TSS), or 85 percent removal of these pollutants from the wastewater influent, whichever is more stringent. The requirement is 20 mg/L $CBOD_5$ for injection wells. Appropriate disinfection and pH control of the effluents is normally required. Coastal freshwaters will have more stringent effluent limits and will require more treatment. Figures 3-87, 3-88, and 3-89 show open and covered sludge digesters.

FIGURE 3-81 Treatment plant with operational issues

FIGURE 3-82 Large clarifier

FIGURE 3-83 Secondary clarifier

FIGURE 3-84 Empty clarifier

FIGURE 3-85 Overflow weirs on clarifier

FIGURE 3-86 Chlorine contact chamber

FIGURE 3-87 Open digester at a small wastewater plant

FIGURE 3-88 Covered digester tank

FIGURE 3-89 Open aerated digester tank

Reclaimed Water (Advanced Secondary) Treatment

Reclaimed water is a means whereby water supplies are reused for other purposes. Reclaimed water may come into contact with people, so the water is required to have additional treatment. The treatment program for reclaimed water is also called *advanced secondary treatment* (Figure 3-90). It requires the use of all secondary processes, plus filtration and high-level disinfection, and residual over 1.0 mg/L after a given period of time (typically 15 to 30 minutes). Typically the filtration step uses gravity sand/anthracite filters. Traveling bridge filters (Figure 3-91) and cloth disk filters (Figure 3-92) are common. The high-level disinfection is accomplished with high-service pumps located afterward (see Figures 3-93 and 3-94).

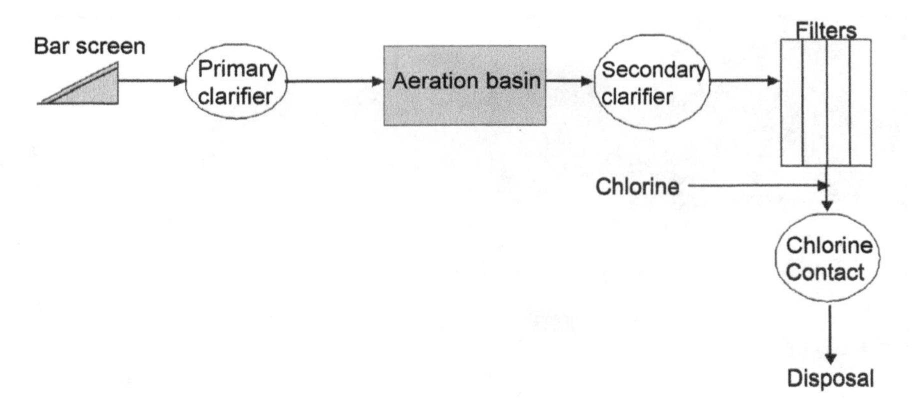

FIGURE 3-90 Advanced secondary wastewater treatment

FIGURE 3-91 Traveling bridge filter

Reclaimed water is used for many purposes. There are over 1,000 sites nationally that apply reclaimed wastewater for irrigation, often for golf courses or crops. Reclaimed water has also been used to irrigate parks and fountains and as cooling water for power plants and air conditioning systems. Advanced secondary treatment is often confused with tertiary treatment. Advanced wastewater treatment assumes nutrient removal, which does not occur with filtration. The nutrients are generally preferred by the reuse recipient (common for golf courses).

FIGURE 3-92 Cloth disk filters

FIGURE 3-93 Pumps after chlorine contact chamber

Advanced Wastewater Treatment

In near-coastal estuaries and slow-moving rivers, the nutrient load in wastewater can overwhelm the natural processes, leading to algae outbreaks, water quality issues, eutrophication, and loss of active biota. The reason that slow-moving waters have this problem is because they do not move fast enough to entrain oxygen, so their dissolved oxygen levels are low. Hence many utilities on slow-moving rivers, near the Gulf of Mexico, and that discharge to estuaries are required to remove the nitrogen and phosphorus from their effluent. This requires significantly more processing, usually termed *advanced wastewater treatment* (AWT).

Advanced wastewater treatment includes all secondary treatment processes, filtration, disinfection, plus removal of nitrogen and phosphorus (see Figure 3-95). AWT wastewater limitations indicate that the wastewater discharges may not contain more, on an annual average basis, than the following concentrations:

- Carbonaceous biochemical oxygen demand ($CBOD_5$): 5 mg/L
- Total suspended solids: 5 mg/L
- Total nitrogen (as N): 3 mg/L
- Total phosphorus (as P): 1 mg/L

FIGURE 3-94 High-service reuse pumps

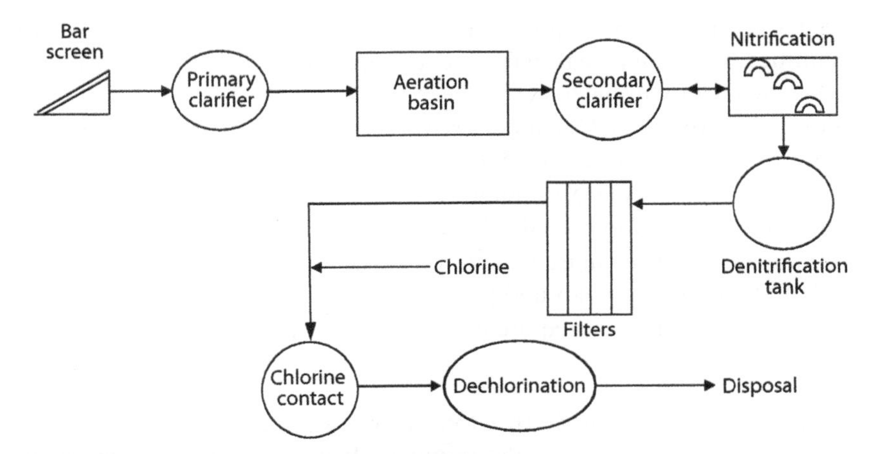

FIGURE 3-95 Advanced wastewater treatment (AWT)

TSS removal can be accomplished via the filtration step in advanced secondary treatment. Nitrification/denitrification is commonly accomplished with methanol and bioreactors or rotating biological disks. Phosphorus is commonly removed by the addition of alum coagulants (which increases sludge volume significantly). This level of treatment is commonly applied to nutrient-sensitive surface waters, groundwater recharge programs, and indirect potable reuse projects. Dechlorination is often required where chlorine is used for disinfection.

Full Treatment and Disinfection Requirements

Where water is to be recycled back into the drinking water, much more treatment is required. Only one place consciously does this today—Orange County, Calif.—although other agencies are looking into the concept. As currently practiced, the idea is to return wastewater to the groundwater, where utilities downstream of the discharge point retrieve it. This concept is employed in groundwater recharge programs in California.

Full wastewater treatment systems include all the treatment steps contemplated under AWT, plus reverse osmosis and/or activated carbon for removal of the remaining organics to reduce total organic carbon (TOC) and total organic halogens (TOX) and to accomplish some pathogen removal (see Figure 3-96). Dechlorination is often required. The process results in what is akin to distilled water. The requirements are as follows:

- The primary drinking water standards are applied as maximum single-sample permit limits, except for asbestos.
- The primary drinking water standards for bacteriological parameters are applied via the disinfection standard.
- The primary drinking water standard for sodium is applied as a maximum annual average permit limitation.
- Except for pH, the parameters listed as secondary drinking water standards are applied as maximum annual average permit limits.
- All pH observations must fall within the pH range established in the secondary drinking water standards.
- Additional reductions are required of pollutants that otherwise would be discharged in quantities that would reasonably be anticipated to pose a risk to public health because of acute or chronic toxicity.
- Total organic carbon (TOC) cannot exceed 3.0 mg/L as the monthly average limitation; no single sample can exceed 5.0 mg/L.
- Total organic halogens (TOX) cannot exceed 0.2 mg/L as the monthly average limitation; no single sample can exceed 0.3 mg/L.
- The treatment processes must include processes that serve as multiple barriers for control of organic compounds and pathogens.
- Treatment and disinfection requirements are additive to other effluent or reclaimed water limitations.

This is clearly a very expensive treatment to pursue and therefore would only be an option in areas having very limited water resources. Power costs are also very high.

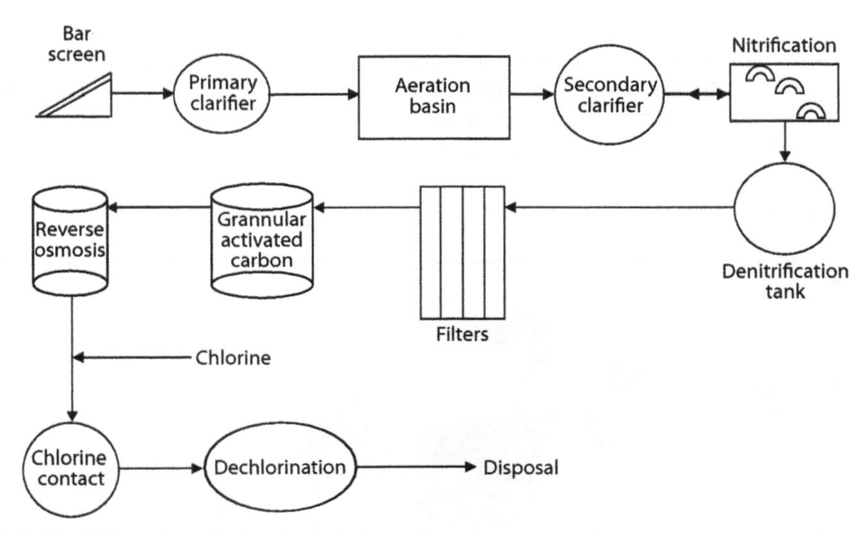

FIGURE 3-96 Full wastewater treatment

SLUDGE DISPOSAL

Both chemical and physical processes create waste called *residuals*. These residuals must go somewhere. The more of them there are, the harder it is to get rid of them, but there are lots of options and the goal should always be to reduce the weight as much as possible by removing water economically.

For water plants, sludge is the by-product of a chemical process, i.e., treatment with calcium carbonate, alum, and so forth. It is hard to dewater. For wastewater plants, sludge is bacteria removed from the aeration basin via the clarifier. Sludge is not easy to pump despite the fact that it is mostly water. Pumping solids is harder than pumping liquids, so a lot of power is used for residuals management. In general, when the solids are less than 1 percent, (10,000 mg/L) pumping can occur, but special pumps are needed (see Figure 3-97). With higher-percentage solids, special pumps are required due to the friction caused by the sludge moving through the downstream piping. Sludge should be pumped short distances.

Many water systems attempt to dewater sludge in settling basins (see Figure 3-98) that can recapture water back to the head of the plant, and permit sludge to be scrapped up and hauled to the landfill or other disposal site. A gravity thickener is a basin that allows the sludge to settle (see Figure 3-99). This process may require the use of polymers to thicken the sludge. These options are common to water systems.

Gravity belt thickeners (see Figure 3-100) allow water to flow through a cloth filter to thicken the sludge. Thickeners normally can produce 2 to 4 percent solids, after which the sludge is often pressed. The best type of thickener depends on the characteristics of the sludge and the ultimate method of disposal. A belt press can be used to squeeze water out of the sludge (see Figures 3-101 and 3-102). A belt press produces a cake that is 20 percent solids and that can be hauled off in a truck (see Figure 3-103). This is a common process for wastewater because wastewater sludge is easier to press than alum sludge, which tends to hold water.

Drying beds work in dry climates with lots of evapotranspiration (see Figure 3-104). Very cold climates can also be effective since they can take advantage of freeze–thaw cycles. Injection wells for concentrate from membranes is the most exotic waste disposal option. Sludge should not go down any injection well, although the concentrate residuals from a membrane plant are appropriate materials for an injection well. This is the normal solution for concentrate that is regulated as an industrial waste product. Regulatory requirements may limit disposal options.

FIGURE 3-97 Special pumps needed for liquids with solids less than 1 percent

FIGURE 3-98 Settling basin for dewatering sludge

FIGURE 3-99 A gravity thickener basin

FIGURE 3-100 Gravity belt thickeners

FIGURE 3-101 Belt press used to squeeze water out of the sludge

FIGURE 3-102 Close-up of a belt press

FIGURE 3-103 A belt press produces a cake that can be hauled off in a truck

FIGURE 3-104 Drying beds

REFERENCES

Bloetscher, F. 2004. Memorandum to Jeff Hughes, Institute of Government, University of North Carolina at Chapel Hill.

———. 2009. *Water Basics for Decision Makers: What Local Officials Need to Know about Water and Wastewater Systems.* Denver, Colo.: American Water Works Association.

Cromwell, J., D. Own, J. Dyksen, and E. Means. 1997. *Managerial Assessment of Water Quality and System Reliability*. Denver, Colo.: AWWARF.

Pensic, J. 2010. Vital Valves Keep Water Systems Flowing. *Valve Magazine*, 22(2):25–28.

Wirth, M. 2010. It's Tough Being a Resin Bead. *Water Conditioning and Purification*, 52(2):46–49.

Yamanaka, T., I. Aso, S. Togashi, T. Minoru, K. Shoji, T. Watanabe, N. Watanabe, K. Maki, and H. Suzuki. 2002. Corrosion by Bacteria of Concrete in Sewerage Systems and Inhibitory Effects of Formates on Their Growth. *Water Research*, 36:2636–2642.

CHAPTER 4

Planning for Operations and Maintenance of Assets

Chapter 3 outlined the utility operations and assets. Significant investments are made in these assets. They are expected to have long lives and to function properly. Regulatory requirements place penalties on utilities that do not function as expected. The public is even less forgiving, making utility officials risk-averse (e.g., a two-hour outage in a utility system can create quite a local controversy, negating the fact that the operation was successful over 99 percent of the year). Systems must be maintained to ensure they operate constantly as expected.

Management of water and sewer infrastructure involves multiple parties, including operations, management, finance, and the engineering/construction group. Planners are usually incorporated as a part of the engineering constituency. The building blocks of an infrastructure management system involve construction, as related to a capital improvement program, and ongoing operation maintenance and rehabilitation, which are usually a part of the field operations system. An inventory of parts would be on hand to facilitate that operation, and some form of financial system in place to track expenditures. Closely related is asset management, which is the financial understanding of the constructed infrastructure. Together these parts should be integrated into an interactive organization.

The annual budget is designed to plan for expenditures, but things happen and expenditures may be more or less than anticipated. At some utilities, the budget, work orders, and capital construction are viewed as separate items, but in reality, they should be linked. With better data management systems in the long term, the impacts of one will be readily transferred to other areas (e.g., once a capital project is completed, the maintenance management system will immediately pick up the need for ongoing maintenance, the value will be added to the asset management system, and ongoing depreciation will occur on an annual basis).

Needs assessments come out of the asset management system. This involves examination of the age and potential deterioration of infrastructure and combines that with the work orders and other input from the maintenance system that would indicate infrastructure that is nearing the end of its life or has excessive amounts of repair occurrences. Tracking from maintenance programs and work orders will help the utility identify where priorities may need to change for infrastructure improvements.

The proper maintenance of infrastructure systems is important to maintain customer service. Providing adequate customer service is perhaps the most important issue affecting water and wastewater utility managers. Therefore, infrastructure management systems are designed to improve customer service, identify potential infrastructure needs to be addressed as a part of a capital program, track costs, identify where excessive maintenance may occur, and demonstrate compliance with regulations. The focus of this chapter is to outline risks of operations, evaluating condition of assets, and maintaining the utility assets inventory.

RISK

When infrastructure systems fail, there are significant risks (Molak, 1997; Bloetscher, 2001). These include risks through property damage, public health, regulatory responses, moratoriums, and wasted dollars. Therefore, the benefits of having an infrastructure management system embedded within the utility are to lower the risks, improve system reliability, decrease operation maintenance expenditures, and improve safety. Implicit in this concept is the idea of optimizing the materials involved in the system to minimize long-term operating costs. It is sometimes difficult for officials to understand that spending money up front to improve the quality of a product likely saves money in the long term.

There are no zero-risk options in the world, a concept with which water and sewer utility system operations and management personnel are well acquainted. The old axiom "if something can go wrong, it will" is frequently supplemented by water and sewer personnel to include "the most likely time for the failure is during rush hour on Friday evening, a holiday, or after midnight." While obviously not always true, the reality is that these are often the events that are remembered by the public. The others are resolved within the normal course of business.

As a result, utility personnel are aware of the potential risks of providing services to the public and often include redundancy, additional personnel, and extra parts and sensors to maintain equipment, and they are prepared for the inevitable failures that will occur. Customers and elected officials expect that the utility system will function 100 percent of the time. The two-hour outage noted earlier means the system operated 99.95 percent of the time, which most enterprises would deem an outstanding success rate. The public, however, might consider this a failure.

Risk is a difficult issue to address as a part of any set of criteria on which a capital evaluation in the public sector is considered. However, risk of failure and vulnerability of the asset are crucial to the decision-making process. To minimize the potential for failure, utility managers must assess the potential risks and the likelihood they might occur. Risk assessments generally include two parts: the scientific investigation and the management portions. Scientific assessment methods include measuring effects of exposure of the activity to the ecosystem or humans, determining the level at which the impacts are negligible, and creating methods to replicate and measure the impacts. Management of the risk includes taking the steps necessary to limit exposure. For local officials these steps include the proper training of employees, having appropriate equipment on hand, redundancy, maintaining appropriate records of operations, and providing those facilities and tools needed to minimize risks to the community.

Elected and appointed officials need to recognize that it is the lack of spectacular failures of water and wastewater systems that should be recognized and appreciated by the public, not the infrequent failures that no system can avoid. Very few public health incidents occur in the United States and Canada in any given year, and few serious accidents occur despite their significant potential.

Risk is defined as the probability of some adverse impact. In the water and wastewater field, risks are related to outages, public health and safety, and injury to employees working around large equipment or within road rights-of-way. Risk assessments and/or vulnerability assessments involve identifying where the most likely negative impacts can occur and taking steps to ensure that a failure cannot affect the public. To this end, risk assessments involve four steps:

1. Hazard identification
2. Response relationships
3. Exposure assessments
4. Risk characterization

Risk management is associated with minimizing some hazard or occurrence that would damage the utility or would cause it to be unable to fulfill its mission of providing reliable water or sewer service and minimizing the potential impacts to public health. *Vulnerability* is something for which there is a potential for an unintended consequence. The vulnerable assets need to be identified and, while some assets can be hardened to lessen their vulnerability, many assets can only be prioritized as to which are the most vulnerable and most likely to fail, versus those that are less likely to be problematic.

Water supplies for many utilities are important and vulnerable assets and, as a result, measures are taken to duplicate systems (redundancy) to ensure water supplies are available. For example, if there is only one pipeline from the wells and it breaks, the utility will not be able to get water from the wells to the water plant; therefore, they may not be able to provide service until the pipe is fixed. If the utility runs out of water in storage, then the result will be failure to deliver safe potable water to customers.

Hazard Identification

To understand the potential hazard, the hazard needs to be assessed. For example, there may be potential vulnerabilities that have a very low probability of happening but have fairly dramatic consequences. The risk of failure of small water lines on individual streets is perceived to be much higher than that of major transmission lines, but the consequences of those individual small leaks is relatively minor in the scheme of things. Each of the pieces of the utility system should be evaluated from this perspective. Vulnerability assessment is one method for doing this. Vulnerability assessments were required of utilities by the federal government a number of years ago, and added vulnerability assessments are likely to be required in the coming years. The vulnerability assessment process identifies and quantifies the risks of the utility.

The hazard identification step involves identifying hazardous situations. Examples include the following:

- Toxic chemicals (such as chlorine and acids) used in the treatment processes that affect workers or get into the distribution system;
- Hazardous work environments (roadways for the repair of pipelines; confined spaces with trapped gases, such as personnel access openings and lift stations);
- Pathogenic organisms (wastewater treatment plants, sewer lines, personnel access openings, and lift stations); and
- Contamination of the water system (treatment plant failure or, more likely, source contamination).

Response Relationships

Response relationships involve determining the toxic effects that might occur in each of the hazardous situations. In traditional risk assessments, a dose–response function is usually developed to determine the levels at which the public might be affected, but in a utility environment the presence of, for example, toxic chemicals in significant quantities or large vehicles in roadways are obvious hazards (it only takes one vehicle to kill a worker). Most of the Occupational Safety and Health Administration (OSHA) rules are aimed at minimizing the response relationship to hazardous situations.

Exposure Assessments

The risks should be related to the actual vulnerability of infrastructure that the utility has in place. Once the hazards have been identified, they need to be quantified, which refers to the realistic probability that a specific action can occur. This is termed *exposure assessment*. The exposure pathway may consist of a variety of steps through which the end result (failure) can actually occur. Exposure assessments and risk assessments in general were originally developed as a part of the nuclear industry in an effort to ensure that the probability of failure in nuclear power plants was minimized. The same concept lends itself to utilities through the hazard identification and exposure assessment steps.

The exposure scenario may be better defined if examples have already occurred (in the example of minor water mains and the potential impacts on customers.) Unlikely scenarios, or those that have never occurred within the utility, are much more difficult to quantify. Determining the probability of occurrence may need to be performed in conjunction with staff and an expert in dealing with exposure assessments.

Exposure assessments are geared toward measuring the potential impact on the population. Water distribution system contamination would be the appropriate arena for this type of analysis. In a well-known Walkerton, Ont., incident in 2001 (see chapter 16), and a Milwaukee, Wis., incident in 1993, the reason the problems were detected was that numerous residents became sick, while no one in the neighboring cities served by other utilities was affected. In both cases, the source water became contaminated and the treatment plants failed to fully remove the contaminants, which is why source water protection regulations are imposed.

Risk Characterization

Once the response and exposure assessments are completed, the risk can be characterized. The most significant risks utility managers face include service outages and worker safety. The result of the analysis should be hazard mitigation steps. The following are examples that most utility managers will recognize as necessary hazard mitigation procedures:

- Chlorine gas is an obvious risk to operators and, since cylinders are worked on routinely, the risk of incident may be significant. Yet only one death has been attributed to chlorine gas exposure at utilities in the United States since its use began. Utilities recognized its risk and developed plans and procedures to limit risk to workers.

- Redundant high-service pumps are routinely installed because of the likelihood of a pump failing, and the public health risk of not being able to supply water for service, for example, during a fire.
- Likewise, mechanical equipment in a treatment plant may fail periodically, which may have a potentially significant health risk. To protect the public health, the utility typically will have sized basins and treatment equipment to permit full capacity with certain components out of service.
- Water lines and sewer lines will fail routinely. Typically, the risks are low, but because of the frequency of failures, the utility will have crews, equipment, and parts ready to make these routine repairs.
- The risk of contamination of groundwater is low (albeit difficult to clean up) but the risk of surface water contamination may be significant. Treatment plants are designed to treat source waters, but the design engineers have to assume certain average and peak concentrations of contaminants. The likelihood of the maximums being exceeded should be low, but because the potential impact to the public may be very high (Milwaukee and Walkerton, as examples), source water protection is nonetheless recommended and pursued by utilities despite its very high cost.

Determining the characteristic of the risk indicates something about what the potential impact might be. The risk characterization and consequence determination provides management with a synopsis of the available data and the consequences of action or inaction. The result of the risk characterization and consequence determination step is that the utility can begin to address responses should an incident occur or steps to take to prevent the consequences from occurring. For example, if this were a health impact, the assessment would need to determine what the probability of health impacts would be on the customers, the rate at which they would occur, and the probability of occurrence. The level of detail of the risk assessment depends on the likelihood of failure, hence, a more complex system that has significant consequences would require more time and understanding as to the response.

The vulnerability assessment tools required of the utilities by the federal government were intended to provide the utility with a response plan should various events occur. Vulnerability assessments can be used both as a baseline estimate of the existing risks and to determine potential steps that can be taken to reduce those risks.

Typical problems in trying to evaluate risks are lack of operator information, undocumented procedures, lack of staff training, and inadequate documentation of existing procedures. The risks are ultimately health safety issues, which are the most important for the utility to deal with, and include performance failure, construction or maintenance failures, artificially created or natural disasters, employee problems (typically safety issues related to employees performing a given response, such as working in the street or dealing with chemicals), and liabilities to the utility as a result of failures such as cave-ins under roads that cause accidents.

The list of plans, procedures, protocols, work-space rules, and so forth is extensive. Some will seem unnecessary, but utility managers are notoriously risk-averse, as are most regulatory personnel. The situation has served the utility industry well as, according to the USEPA, fewer than 30 people each year in the United States are affected by waterborne contamination. This is the best rate in the world.

MAINTENANCE

The key to appropriate infrastructure management within the utility is that appropriate capital expenditures are made and that those investments are properly maintained. When there are limited dollars available from customers, spending large sums of money on capital infrastructure that may be of limited value in the future reflects poorly on utility management.

Long-term costs are reflected in two ways. First is the lifetime of the asset. Second is the annual cost associated with the asset. Figure 4-1 shows two curves. The lower curve shows a lower initial cost, but the life expectancy of the asset is significantly less than the upper curve. The upper curve reflects a higher cost, but the asset lasts nearly twice as long. From a risk management perspective, the second option reduces risks of failure.

There is an incentive for using cost for capital to improve quality to decrease long-term asset requirements. A present-worth analysis of the system shows that the annual cost of the higher-quality items has a lower life cycle cost than cheaper materials that need to be repaired or replaced more often.

Figure 4-2 shows a typical cost curve for depreciation. The top curve is actual value, while the straight line is the commonly used concept of straight-line depreciation, where asset value decreases the same amount each year. This figure shows that while most finance experts depreciate the infrastructure using straight-line depreciation, the reality is that the in-service condition generally stays significantly above that, as illustrated by the curve on top. Moreover, as a result of the deterioration occurring more slowly than anticipated with straight-line depreciation, the ability to maintain infrastructure condition if improvements or rehabilitation are made at the appropriate time is more likely (see Figure 4-3).

Rehabilitation never returns the system to the initial value of the asset; only replacement would accomplish that. With pipelines, this may not be as significant a concern, given that the technology does not change significantly. In the treatment plant, however, upgrades usually involve replacing older, outdated mechanical equipment and controls with new equipment, which allows the systems to be operated more efficiently. In general, upgrades are required to deal with changes in technology, address concerns with age, comply with regulatory requirements to improve treatment quality or reliability, and improve the operation and maintenance of the system, usually by lowering costs and making things simpler and less likely to need those repairs.

Ultimately, the utility will need to gauge the condition of its infrastructure, something typically known only by utility staff. Nationwide, however, there are indications that significant infrastructure problems exist, particularly in older developed areas in the Northeast and Midwest. Newer systems in the Southeast and Rocky Mountain states are not as critical as older systems, but their time is coming. The infrastructure condition is especially relevant when one considers that there are over 20 million personnel access openings (manholes), 250 million mi (400 million km) of sanitary sewer pipe, and over 3 billion mi (4.8 billion km) of water pipe in the United States. All this infrastructure begins to deteriorate once a capital project is completed.

Table 4-1 shows a scenario where the capital improvement project is completed and immediately input into the asset management system, which then tracks its depreciated cost with time and is input into the maintenance management system. Both of these lead to repair and replacement cost numbers that, once reaching a certain point, lead back to a new capital project. Implicit in this cycle is the need to perform the following functions: asset management, planning, data management, operations system management, public communications, and finance.

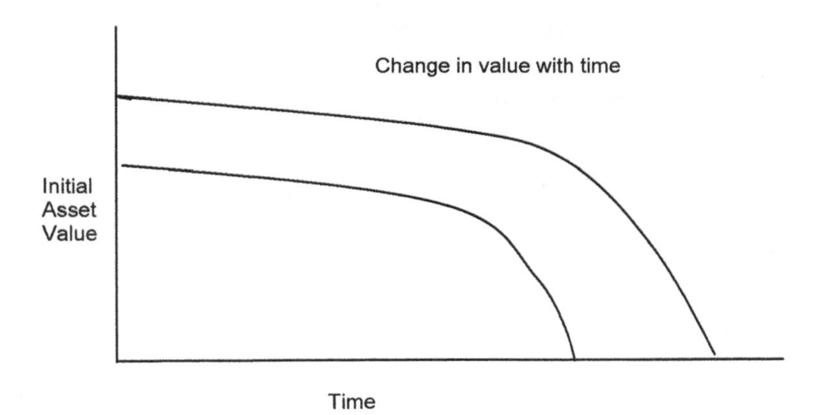

FIGURE 4-1 Initial cost versus life expectancy

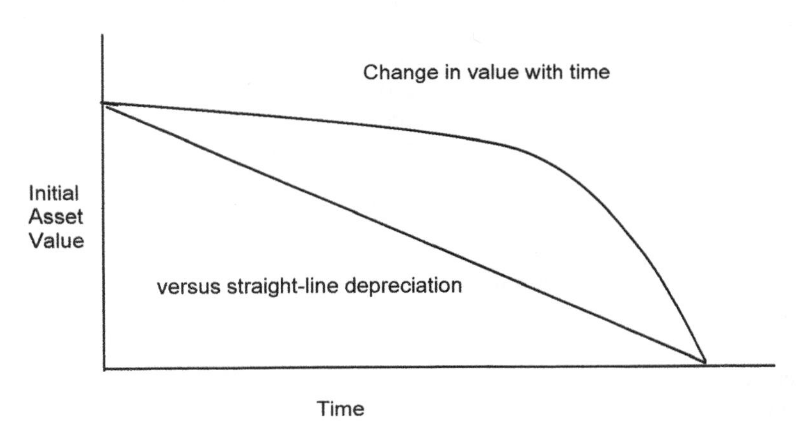

FIGURE 4-2 A typical cost curve for depreciation

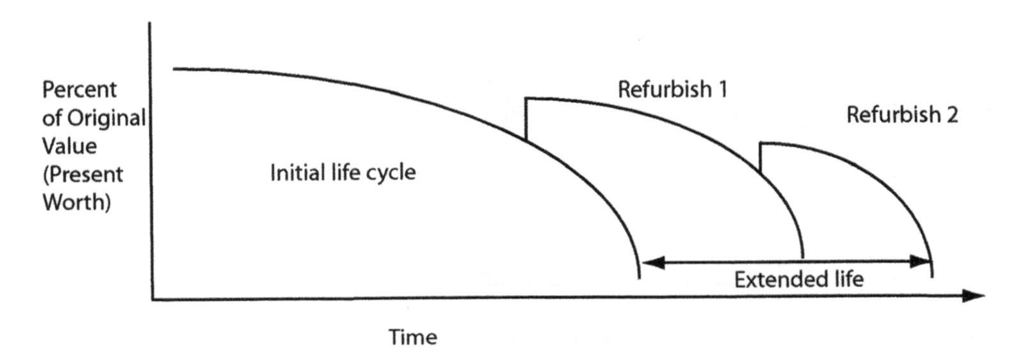

Note: Refurbishment may occur later in life before full devaluation takes place.
Refurbishment never reaches initial condition.
Extension of life of asset decreases with each refurbishment.

FIGURE 4-3 Extension of life of assets where refurbishment can occur

TABLE 4-1 Extension of life of assets where refurbishment can occur

Asset	Pump Station 1	Year	Pump Station 1	Pump Station 1 Upgrade	Total Asset Value
Asset value	$1,524,383				
Year	1992				
Life expectancy	20				
Annual depreciation	$76,219.15				
Asset value year		1993	$1,448,164		$1,448,164
		1994	$1,371,945		$1,371,945
		1995	$1,295,726		$1,295,726
		1996	$1,219,506		$1,219,506
		1997	$1,143,287		$1,143,287
		1998	$1,067,068		$1,067,068
		1999	$990,849		$990,849
		2000	$914,630		$914,630
		2001	$838,411		$838,411
		2002	$762,192		$762,192
		2003	$685,972		$685,972
		2004	$609,753		$609,753
		2005	$533,534		$533,534
Asset	Pump Station 1 upgrade				
Asset value	$373,655				
Year	2006				
Life expectancy	10				
Annual depreciation	$37,365.50				
		2006	$457,315	$373,655	$830,970
		2007	$381,096	$336,290	$717,385
		2008	$304,877	$298,924	$603,801
		2009	$228,657	$261,559	$490,216
		2010	$152,438	$224,193	$376,631
Note here original depreciation is zero...		2011	$76,219	$186,828	$263,047
		2012	$0	$149,462	$149,462
		2013		$112,097	$112,097
		2014		$74,731	$74,731
		2015		$37,366	$37,366
		2016		$—	$—

The operations and maintenance portion that feeds to the maintenance management system is significant in that the condition of the system normally is only determined once pipe has been dug up and the conditions documented. For example, a piece of pipe might not be dug up for 20 years after installation, and that one instance may provide the only real indication of the quality of the pipeline. While a break may be a part of ongoing maintenance, by looking inside the pipe, one can tell if there are tubercles or other indications of long-term deterioration. (As an aside, a lot of the old cast-iron pipe installed during the Works Progress Administration [WPA] days in the 1930s, when dug up, is found to be in excellent condition. Meanwhile, asbestos concrete and especially galvanized pipe installed 20 to 30 years later is in significantly poorer condition.) As a result, the maintenance management system and the asset management system need to be tracking that data, which they can only obtain through field investigations and reports from work order systems.

Maintenance systems are divided into four areas: corrective, routine, preventive, and predictive. Corrective maintenance occurs in response to an event, such as a water main break, sewer blockage, or motor failure in a pump station or treatment plant. Most corrective maintenance issues are not immediately predictable, instead occurring on older or stressed parts of the system, or because of acts of nature or human activities (accidents). Budgets should include amounts for dealing with ongoing routine maintenance of infrastructure. Maintenance crews to address these issues are needed on all utility systems, and appropriate repair parts must be immediately available (on-site or locally from dealers or other utilities). Figures 4-4 through 4-12 show various damages to utility system components caused by corrosion, much of it as biofouling-induced corrosion (in wells), exposure to difficult environments (wastewater plant or aerators), dissimilar metals, failure to remove heat tint on welded stainless steel, materials left in the pipe, rough spots in the pipe, and handling damage. All these can contribute to creating cells where biological agents can attach and attack the pipe surface. Most of the examples shown are 316L stainless steel, which many engineers think is the best-quality pipe for use at a treatment plant.

Routine maintenance of the water system is designed to ensure the availability of operable equipment necessary for the treatment and distribution of potable water to meet demands continuously, or for the collection and treatment systems for wastewater. This is achieved through the design and execution of corrective and preventive maintenance programs.

Preventive maintenance normally applies to routine tasks like greasing bearings, replacing packing material, and painting and cleaning pumps. Most of this information is clearly laid out in manufacturers' literature, which is found in the operations and maintenance (O&M) manuals at most plants. The O&M manuals for every piece of equipment a system has should be kept in an accessible place, and the preventive maintenance suggestions of the manufacturer strictly followed. This includes greasing bearings, replacing seals, and other mundane tasks that take a limited amount of time to accomplish but are often neglected.

Following the preventive maintenance suggestions of the manufacturer will minimize breakdowns. However, a system to track the thousands of pieces of equipment that need work done on them requires a sophisticated work-order tracking system that will automatically print work orders for preventive maintenance purposes. A number of manufacturers have created maintenance management information systems (MMIS) to track what kinds of maintenance occur and to create work orders to tell operators that certain maintenance should be performed. This is helpful so that busy schedules do not cause preventive maintenance tasks to be skipped and demonstrates to management the need to ensure that the appropriate amount of maintenance monies are in place to keep up the preventive maintenance. Neglecting these simple preventive maintenance issues due to lack of time or staff often leads to much larger and more costly maintenance problems.

Predictive maintenance recently has been touted by private interests as a benefit for utilities. There is some indication that the use of infrared sensors on pumps and motors

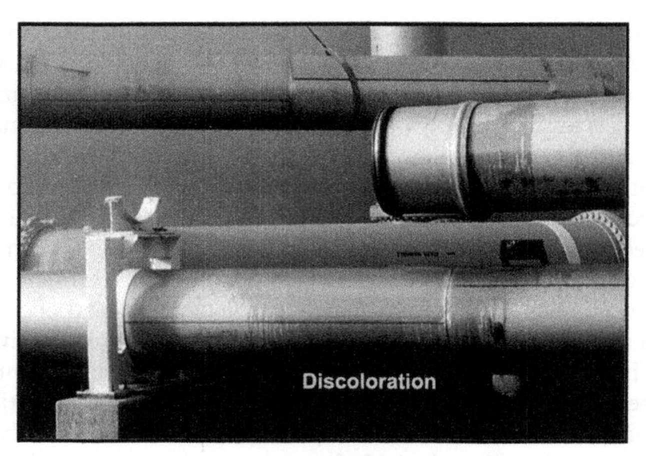

FIGURE 4-4 Areas of potential damage to stainless-steel pipe (Failure to remove heat tint is a common problem [discoloration]. Heat tint changes the metal, which encourages galvanic activity.)

FIGURE 4-5 316L stainless-steel column pipe (Bacterial activity sheared it after 18 months.)

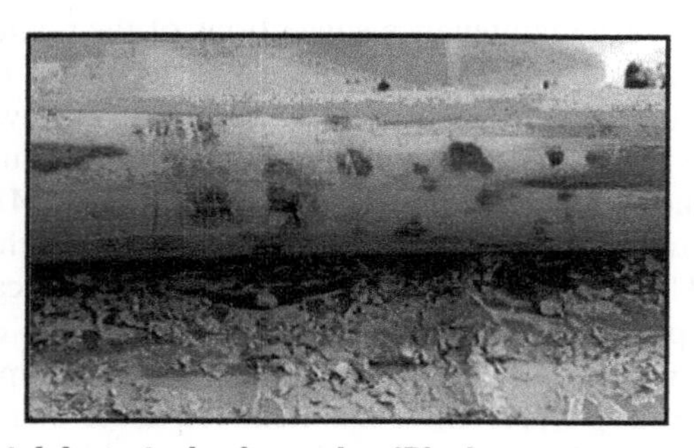

FIGURE 4-6 Damage to stainless-steel column pipe (Black areas are caused by corrosion.)

FIGURE 4-7 Damage due to corrosion of clarifier launder

FIGURE 4-8 Microbial activity on the well column pipe

will indicate if the bearings are wearing, which may foretell failure of the pump or motor in the foreseeable future. However, most other predictive measures involve simple inspections of the system to determine where hydrogen sulfide is causing pipe, personnel access opening, or concrete deterioration in the sewer system; where galvanized or other inappropriate materials may lead to breaks; and where inspections of metal parts at treatment plants should take place—all things utility staff or their engineers can do easily.

Implementing Maintenance Responsibilities

Maintenance responsibilities at the plant include keeping the structures clean, safe, and presentable, and carrying out projects to maintain or improve the treatment system to ensure the quality and/or quantity of water available to the plant. Such maintenance and improvement projects include pump and motor repairs and testing, greasing, and painting equipment and structures for protection against the corrosive treatment plant environment. A part of any

FIGURE 4-9 Corrosion damage due to contact of steel with concrete

maintenance program is to foster safe practices in the execution of the projects. A work-order system should be used to make sure work takes place and to track the nature and frequency of repairs so that replacement of worn-out equipment can be planned.

Pumps are among the most important components of any utility system. As a result, adequate staffing and funding must be provided to ensure that pump repairs occur when it

FIGURE 4-10 Corrosion damage to metal at a wastewater plant

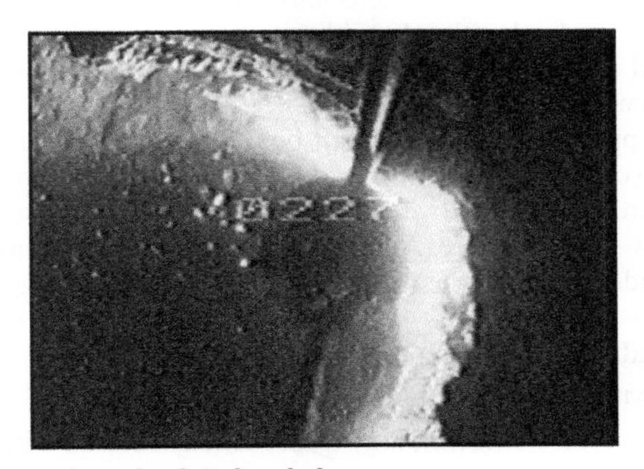

FIGURE 4-11 Well with sand coming into borehole

FIGURE 4-12 Column pipe with bacteria growth

is convenient to do so, not when failure occurs (which is rarely a convenient time). Ignoring pumps because they seem to operate without problems disregards the need for semiannual care to check O-rings, gaskets, packing, amperage, and flow rates. O-rings, gaskets, and packing will wear, causing the pump to leak. Amperage will wander when the pump motor begins to fail or the pump is overloaded. Worn pumps and those operating outside their designed pump curves will be inefficient, increasing operating and power costs.

Wastewater treatment plant maintenance is similar to water treatment plant maintenance. Maintenance crews address all equipment, structures, and grounds maintenance needs of the wastewater treatment plant on a routine basis, and as necessary. The work performed encompasses mechanical repair and installation, plumbing, carpentry, painting, masonry work, machining of parts, electric and electronic installation, and repairs. A comprehensive stock of spare parts and equipment should be maintained to respond to emergencies.

Corrosion is a major maintenance issue that can affect both the structural capacity of the pipe and the quality of the water. While external corrosion can lead to leaks, internal corrosion can result from metabolic (microbial) activity, chemical dissolution, or physical abrasion by excessive fluid velocities. In addition, corrosivity of the water may leach metals such as lead and copper from metallic pipelines, lowering water quality. Hence the water treatment process may be a concern and need monitoring.

Older pipe and pumping equipment can be more subject to failure than new equipment simply because of age. Steel is damaged by chlorides and microbial activity, and both steel and concrete can be damaged by hydrogen sulfide. Older infrastructure requires more monitoring than new infrastructure. As infrastructure ages, decisions to repair or replace capital assets are made based on one or more of the following characteristics:

- More economical to replace than repair
- A critical asset with a history of failure
- Obsolescence as a result of technology (more efficient to run with new technology)
- Repair parts unavailable
- Water quality goals are not met

In established or stable utilities, when one of the five characteristics listed exists, the utility must pursue replacement of the capital asset. These established utilities will be faced with many such replacement projects simply due to age, but the number and magnitude of the projects will increase if the utility has not maintained its systems adequately due to funding or other factors. The utility can control the timing of its own projects to a degree, but must establish a program to ensure that the proper projects are constructed to meet operational timelines. As a result, with time, the amount of money spent for new growth-related infrastructure will decrease, while expenditures for repair and replacement will increase.

Clearly, materials can be an important issue for plant design and maintenance. Galvanized iron, lead goosenecks, and a variety of pipe materials used in the wastewater system in the past are no longer appropriate. Current materials may be inappropriate as well; e.g., hydrogen sulfide damages ductile iron fittings and concrete pipe. Heat and light damage PVC pipe. Stainless steel may not hold up to microbial activity any better than plain steel.

Decisions to replace assets should be based on more than just cost, although cost is an important consideration. An asset may be old but may continue to provide good service with a minimum of downtime. Operator comfort and minimizing disruption are worthy goals to consider along with economics.

Every asset has an installation date and a useful life, whether it be a water plant, a pipeline, or a pump. Mechanical equipment has a shorter life than pipe. Over that life, the asset deteriorates and requires periodic repairs; the older the asset, the more repairs are needed. Figure 4-13 shows a graph of two typical assets' value/condition with time—a pipeline and a pump. Some assets can be refurbished substantially, such as a water tank or the concrete structure of a treatment plant. Such refurbishment will restore some of the expected life, but the asset can never be made new, as noted previously in Figure 4-3. Other assets, like water pipelines, are much harder and more expensive to renew, so replacement is the only worthwhile option. Annual maintenance programs should include the replacement of worn-out equipment and piping.

The most insidious maintenance problem is deferred maintenance. This is maintenance that is delayed pending replacement with another capital investment. It makes the assumption that the capital investment will actually be made. Unfortunately what typically happens is that, due to the need to control rates or to cut budgets, capital items are deferred and maintenance budgets are cut. Utility systems have traditionally fared poorly in replacing aging infrastructure because of rate impacts to customers.

Because most utility systems are underfunded, most have deferred maintenance needs that are unmet. Most have limited understanding of their asset values, life, and condition, and most struggle with maintenance as this is usually the first item cut in the budget because it is very difficult to quantify. There is an assumption that the operations staff will work to prevent those failures, which in many cases does occur, furthering the perception that the infrastructure is in better condition than it actually is. As a result, maintenance or replacement of items is deferred, sometimes indefinitely, which means the likelihood for failure increases significantly. However, at some point, the asset will no longer be able to act as it needs to and failure is likely to occur. The longer the deferred maintenance, the more likely it is that the maintenance issue will be catastrophic. Once monies are allocated for these items, there is a responsibility to make sure the programs and projects are implemented by managers and operators.

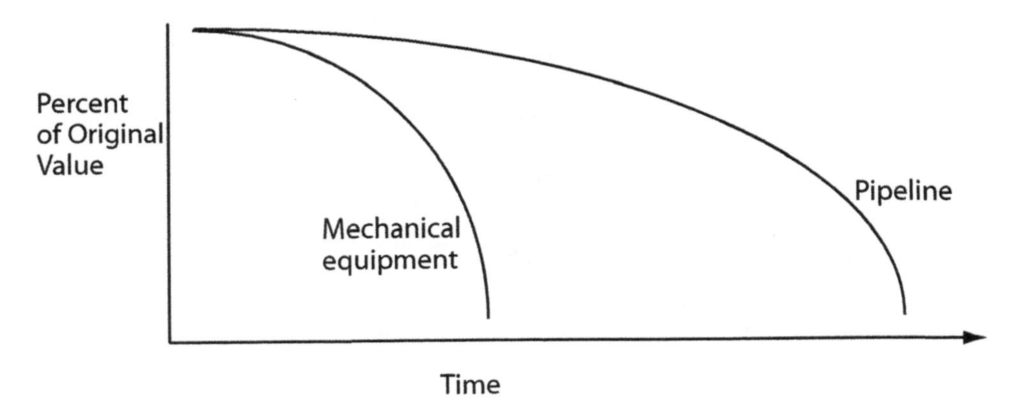

FIGURE 4-13 Typical decline in value of two capital assets

Management of infrastructure integrity is important to maintain the life cycle of infrastructure and ensure reliability. Items that affect infrastructure reliability are the quality of the equipment that is initially installed, the quality of the construction and construction supervision, adequacy of the design, the liability mechanisms built into the project, assessment of ongoing condition, and the potential for public health risks.

There are a number of ongoing issues in regard to infrastructure maintenance. These include the human resource aspects of both operations and engineering (a disproportionate share of both operators and engineers will likely retire over the next 10 to 20 years, taking with them a lot of knowledge of the system); perception by the public; the impact of water conservation, especially as it relates to rates, whereby the loss of revenue resulting from water conservation actually works against those conserving because their rates will increase. Utilization of SCADA and other technology will become a more permanent aspect of most utility systems.

Maintenance Needs of the Water Distribution System

The water system should strive to account for all water and to track water loss. The water loss is usually attributable to system leakage and slow meters, although some theft could be involved from unmetered or unapproved taps. Metering all customers solves two problems. First, it allows the water system to account for the water that is sold, and second, it allows the water system to bill people on the basis of their actual usage. A residual benefit is that metered services encourage customers to use water wisely. Estimates of water used for fire flows, hydrant tests, and water main flushing programs cannot be easily quantified, but should be estimated. Water systems can obtain revenue from street cleaning, construction, and other minor water uses through the use of temporary meters.

Except in areas of very porous soil, water leaks tend to come to the surface. Water system and local government employees should be encouraged to report leaks so that prompt repairs can be made. Leaks can damage pavement, create sinkholes, and cause excessive flows in sanitary or storm sewers. Some leaks may be detected by sewer maintenance crews noticing excessive flow in a sewer line. Suspected leaks also may be located by using a listening device that amplifies the noise made by escaping water. Auditing the water system via leak detection programs is a method useful to water systems in need of water conservation measures. Any water system with an excess of 15 percent unaccounted-for water should employ an ongoing leak detection program. Professional leak survey firms can help water systems start a leak detection program. The period of return of investments of such programs is usually very short.

Some of the problems that are caused by poor system maintenance include customer complaints of poor water quality or lack of pressure, difficulties in repairing water main leaks, and inadequate or unreliable water availability for fighting fires. Despite being buried, water distribution valves, piping, and equipment should not be neglected, since they are needed to properly operate the system. All portions of the distribution and collection systems should be in good working order to minimize response time to emergencies and to restore service as soon as possible.

Valves

Valve maintenance is one of the more overlooked maintenance activities (Owens, 2010). There are two parts to a valve maintenance program: locating all the valves and annually operating them. The goal is to make sure that every valve on the water (or sewer) system operates properly and that they can be found when needed. Most of these valves are rarely required to operate unless there is a break in the main. As a result, if they are not exercised, then they may "freeze" and not operate when needed. If they do not work, or cannot be found, the break cannot be shut off as quickly, which increases the likelihood of property damage and increases the number of people who will be without service while the break is fixed. Unfortunately valve maintenance programs are people intensive—they cannot be tested remotely—and there many valves on every utility system. When budget reductions are required, this program is easy to cut since the benefits are not obvious to the public.

Hydrants

Fire hydrants should be operated and tested at regular intervals, because many small systems can go for years without having to use them for fire fighting. Many systems have a yearly program that is a combination of flushing water mains and testing hydrants. Records should be kept of the static pressure, flow test results, and any repair work performed on each hydrant. In freezing climates, whenever a hydrant is used, an inspection should be made to make sure that the barrel has fully drained.

Hydrants provide the water system with the ability to check flow and pressure, flush the system, and find closed or broken valves as a part of the valve exercise program. The three most common reasons for a hydrant not functioning properly are (1) unnoticed damage, (2) improperly drained or frozen barrel, and (3) closed valves in the distribution system. All of these problems can be readily identified and corrected by a regular program of hydrant inspection and testing. Failure of hydrants to operate properly during a fire could leave water system personnel and managers liable to a suit for damages, in addition to causing needless risk to life and property.

Flushing

Flushing is more appropriate for water systems, although sewer lines need cleaning as well. All water systems should flush the piping at least once per year to remove sediment and stagnant water. When flushing the system, notify residents so that they are not alarmed if the water appears murky. The chlorine residual should be increased to help retard biofilm growth in the pipelines. If significant problems are found, pigging is an alternative. Flushing the system gives operators the opportunity to test fire hydrants' work valves (the blow off points) and detect areas where valves may be shut or otherwise not working.

The buildup of rusty sediment in the bottom of mains is a common problem in water distribution systems. This sediment can be caused by iron or manganese that was in the source water and has precipitated in the pipe, or from the presence of iron bacteria in the system. It can also be from rusting of old cast-iron pipe. If the problem is not too severe, some systems have found that a thorough flushing of the system once or twice a year is sufficient. Dead-end water mains will need to be flushed more often. While the sediment is normally not harmful to health, it may cause a few customer complaints when it is disturbed. If water flow is suddenly increased, such as from use of a fire hydrant, then the sediment will be disturbed and can turn customers' water slightly rust-colored or even dark brown. Customers will be reluctant to drink and use the water, and the water may badly stain laundry items. Where the problem is ongoing and severe, professional advice should be obtained on the best method of correcting the condition. Residents should always be notified if flushing is planned.

Old water mains that become encrusted to the point of seriously restricting flow can be cleaned using a power rodding machine or by pushing a flexible "pig" through the line using water pressure. In most cases, the encrustation will quickly re-form on the interior of a cleaned pipe, so consideration should be given to applying a cement lining on the pipe interior or replacing it.

Lead and copper

Lead and copper contamination received significant attention in the early 1990s when regulations were promulgated by the USEPA that required all water systems to monitor for the presence of lead and copper in drinking water. Excessive quantities of lead and copper are associated with a number of health effects. In particular, children who have had excessive exposure to lead are likely to experience a delay in mental and physical development, impaired mental abilities, and behavioral problems.

While there are many sources of lead and copper in the environment, including paint, soils, dust, eating utensils, and gasoline, the first draw of water systems with corrosive water and lead and/or copper piping tended to contain significant quantities of these metals. Rarely is there an appreciable amount of lead or copper in either groundwater or surface water, so it was determined that aggressive waters dissolve minute quantities of these metals from the pipe during low-use periods. The principal sources of lead contamination are lead pipe, lead–tin solder used for joining copper pipe, and brass plumbing fixtures. Copper contamination can be caused by corrosion of copper pipe. The concentration of lead or copper that may be present in drinking water is affected by a number of factors, such as the contact time, the corrosiveness of the water, and the age of the piping.

Most systems completed their lead and copper studies and implemented actions to control corrosion. However, ongoing monitoring is required for all systems. The testing requires selecting customers who have plumbing systems vulnerable to lead and copper contamination and collecting samples of water that has been in the piping for at least six hours. Many water systems began programs to remove lead jointed pipe, lead goosenecks, and similar materials from the distribution system. Such programs, because of the age of the piping, may be more beneficial than extensive corrosion-control measures. Each utility should evaluate this on an ongoing basis.

Cross-connections

In regard to cross-connections, AWWA has a policy that states that

> in the exercise of the responsibility to supply potable water to their customers, utilities must implement, administer, and maintain ongoing backflow prevention and cross-connection control programs to protect public water systems from the hazards originating on the premises of their customers and from temporary connections that may impair or alter the water in the public water systems. (AWWA, 2005)

Water distribution systems are under pressure, so the only way a cross-connection works is for there to be backpressure from the customer or a drop in pressure in the water distribution system due to system damage, breaks, or repairs. The purpose of a cross-connection control/ backflow prevention policy is to isolate the potable water supply system from the possibility of backflow, as backpressure and/or backsiphonage may occur because of the existence of cross-connections between potable and nonpotable systems. Both are possible.

Backsiphonage can occur when there is reduced pressure or vacuum found in the water system that will allow a foreign substance to flow into the system. Backpressure is the situation where contamination is forced into a potable water system through a connection that has a higher pressure than the water system. There have been many cases recorded of contamination of a public water system as a result of backflow of contamination through a cross-connection, including pathogens, pesticides, herbicides, and nontoxic materials. The conditions under which contamination occurs from a cross-connection are often unusual.

Water age

Water age in the distribution system relates to water demand (Wang et al., 2009). Water age increases as water remains in the system longer, usually where there are large numbers of part-time residents and when water conservation measures are implemented during droughts. While increased conservation may create funding concerns, it also creates a maintenance issue due to greater degrees of stagnation that may increase water quality problems in the distribution system. Wang et al. (2009) showed that when the city of Raleigh, N.C., experienced up to a 50 percent decline in total water sold after water conservation requirements were implemented during a recent drought, it experienced increases in positive coliform hits.

Water Main Upgrade Programs

Water main upgrade programs are designed to replace current pipelines—those that are small, galvanized, or provide insufficient service and/or no fire protection. Typically a water utility replaces these small lines with 6- or 8-in. (15-cm or 20-cm) pipelines made of PVC C900 or ductile iron, to provide fire protection. Leaky pipes and salt-immersed pipes should be a priority for replacement with appropriate pipe materials. Small PVC C900 and ductile iron pipelines generally cost under $50 per foot and can often be installed by in-house utility crews. Other programs should address low-flow or low-pressure problems in the system. Existing developed areas are the first priority. Another priority should be on completing loops that will address pressure or low-flow issues. Both of these programs have the benefit of increasing water sales to the areas where the loops are made or the lines upsized because flows are no longer restricted.

Maintenance Needs of the Sanitary Sewer Collection System

There are three major concerns with the sewer collection system: odors, blockages, and infiltration and inflow. Public concerns are over odors that occur when wastewater cascades into a pipe or wet well, volatizes the hydrogen sulfide and mercaptins, and escapes to the areas surrounding treatment facilities and lift stations. Because a significant component of the odors is typically related to sulfur compounds, solutions for odor control must address the control of hydrogen sulfide off-gassing.

The sulfides formed are composed of dissolved and precipitated sulfide. Depending on the concentration of cations in the wastewater, the dissolved fraction may amount to as much as 80 to 90 percent of the total sulfide; and it is only the dissolved portion that can further create the formation of H_2S. Testing of the wastewater is required to confirm the actual values of dissolved sulfides. The wastewater pH dictates the relative amount of bisulfide (HS^-) ion and H_2S gas. At a pH of 7.0, the amount of HS^- ion versus H_2S will be equal. As the pH increases, the portion will favor the HS^- ion. At a pH greater than 9.5, negligible amounts of H_2S will be present in the wastewater, but a wastewater pH of this level is rare.

Elimination of corrosion resulting from hydrogen sulfide is difficult to accomplish. Sulfide is present in wastewater as a result of anaerobic conditions resulting from force main utilization. Anaerobic conditions occur due to lack of oxygen in force mains that flow full and under pressure. Anaerobic bacteria within biofilm layers on the pipe and tank walls reduce sulfates in the wastewater, forming sulfide. Transport of sewage in sewer pipes leads to the release of hydrogen sulfide through aeration when the pipe discharges to a bigger pipe or wet well. In such cases, the equilibrium of the H_2S in solution with H_2S gas above the liquid interface is altered, creating a release of H_2S (aeration). The points where the force mains discharge to concrete are of special concern. The release of dissolved hydrogen sulfide from the sewage in large quantities will cause it to combine with water on the pipe surface to form sulfuric acid, which dissolves the crown of the pipe.

The second method of release is via metabolism of sulfur compounds by bacteria. In domestic sewage, sulfur compounds originate from animal or vegetable proteins or washing powders. Typical concentrations in sewage are 3 to 6 mg/L of organic sulfur (Estoup and Cabrillac, 1997). The release of hydrogen sulfide originates from the release of soluble sulfides through the decomposition of sulfur compounds resulting from anaerobic bacteria (Estoup and Cabrillac, 1997). Biological generation of sulfate ions and transport of the hydrogen sulfide to the corrosion points can be modeled (Bohm et al., 1998). Bohm et al. reported significant corrosion at the water level in the sewer. Conversion of sulfur to elemental form is possible, but the process creates a solid by-product. Solutions to solve elemental sulfur deposition require additives to the collection system.

Impacts on the collection system are twofold. One impact includes effects on day-to-day operations of the collection system. The other impact is permanent damage to the piping, wet wells, pumps, and appurtenances of the sewer system (see Figure 4-10 for an example). Vitrified clay is not as likely to be attacked by corrosion as steel or concrete. However, the wet wells, large sewers, and many personnel access openings are concrete or contain concrete grout. Damage to these facilities is greatest at the access opening rings, at the water levels, and at the top of the wet wells.

Concrete pipe in sewage systems is corroded in a shorter period than expected based on durability estimates. Concrete pipe deterioration is slightly different than steel pipe because concrete is not impermeable. Hydrogen sulfide contact with concrete alone has no appreciable deterioration effect, but in the presence of water, it reacts to create sulfuric acid. This type of corrosion is normally slow, usually taking three to four years to show its presence. During this period, sulfur-reducing bacteria are proliferating on the pipe (colonizing within days).

In contrast to the initial deterioration, in the presence of sulfur-reducing bacteria, the water and bacteria combine to create sulfuric acid that directly attacks the concrete, creating significant deterioration in short periods of time. Analysis of concrete corrosion indicates gypsum is initially released through the following chemical reaction (Estoup and Cabrillac, 1997):

$$Ca(OH)_2 + H_2SO_4 \rightarrow CaSO_4 + 2\,H_2O$$

Later tricalcium aluminate in the cement is attacked to form ettringite, which creates swelling in the concrete that leads to spalling (Estoup and Cabrillac, 1997). Mori et al. (1992) demonstrated that the corrosion rate of hydrogen-sulfide-exposed concrete pipe was 0.2 in. (5 mm)/year at the sewage water level, while the deterioration at the crown was only 0.06 in. (1.4 mm)/year. Idriss et al. (2001b) showed that hydrogen sulfide penetrated deeper into steel-reinforced pipe with time. At 360 days, the penetration was twice that of 90 days. Idriss et al. (2001a) showed that the failure time for pipe-reinforced concrete pipe exposed to hydrogen sulfide was twice as fast as for pipe that was not exposed to hydrogen sulfide.

Concrete deterioration is accelerated when the pH is below 9 in the concrete; with sulfuric acid being formed, it is virtually impossible to maintain a pH above 6, let alone 9. Bohm et al. (1998) noted that the pH of the corrosion surface is under 3, and that corrosion of 4 in. (10 cm) can occur within 20 years of installation of the pipe. Yamanaka et al. (2002) found that in sewer systems with a hydrogen sulfide concentration of 600 ppm, the pH at the pipe surface was 2, a situation likely replicated at the water line in the pipe. Yamanaka et al. (2002) reported that concrete coupons were reduced by 90 percent within one year of exposure to the sewage environment (600 ppm hydrogen sulfide). This low pH was attributed to the ability of the bacteria to colonize the pipe surface and survive off the hydrogen sulfide in the adjacent water. The sulfuric acid created within the bacterial layer penetrated the concrete pipe, leading to corrosion. High concentrations of hydrogen sulfide have been found 0.75 in. (15 mm) into the pipe surface (Mori et al., 1992).

Iron pipe deterioration is a function of reactions of the hydrogen sulfide and sulfur-reducing bacteria with the iron in the pipe. The deterioration effect of biological elements is important in predicting the life of iron-based pipe materials. Sulfur-reducing bacteria transform sulfates into sulfides, which has a deleterious effect on pipe surfaces. Ma et al. (2000) note that under certain conditions, hydrogen sulfide can act as a barrier to inhibit corrosion. Below 0.04 mmol/m^3 of hydrogen sulfide, iron sulfate crystals may form, which acts as a barrier to corrosion. However, the presence of sulfur-reducing bacteria and water will disrupt the crystals with time, thereby accelerating corrosion significantly under the right environmental conditions.

In acidic mediums, hydrogen sulfide above 0.04 mmol/m^3 will accelerate corrosion of iron. In the presence of organics, certain sulfur-reducing bacteria enhance corrosion through metabolic processing of hydrogen in the organics (Keresztes et al., 2001). Below pH 6.5, FeS compounds are not protective of pipe surfaces (Pourbaix et al., 1993). It has already been noted that the pipe surface pH is generally lower than 6.5, so FeS compounds are not useful for protection in the wastewater environment.

Three mechanisms are purported to explain the high corrosion rates of steel in the presence of sulfur-reducing bacteria: the oxidation of sulfides to elemental sulfur and back; the reduction of thiosulfate to activated sulfur compounds; and acidification resulting from the oxidation of sulfides to sulfates or thiosulfate (Pourbaix et al., 1993). Werner et al. (1998) report that hydrogen sulfide enhances pitting of stainless-steel pipe more than the sulfur-reducing bacteria might suggest. They suggested this might be a result of the gas being trapped in smaller pits in the steel. They also reported that the sulfur-reducing bacteria colony was established within eight days of placement into service.

Tekin et al. (1999) investigated the kinetics of iron corrosion in the presence of hydrogen sulfide. They found that sulfate ions are not generally detected in the corrosion process of iron pipe in accordance with the expectation of the following simple first-order reaction:

$$H_2S + 2\ Fe^{+3} \rightarrow {}^1/_8 S_8 + 2Fe^{+2} + 2\ H^+$$

Instead they found a series of reactions that they believe explain the process.

Tekin et al. (1999) determined reaction constants for each of these reactions, based on pH of the environment. All of the rate constants were above 12, meaning the products are favored and the reaction happens quickly. In each case, the aqueous iron is used as a catalyst to modify sulfur compounds to yield the elemental form, and the iron sought an additional electron, which was provided through iron-oxidizing bacteria.

The majority of force mains are small cast or ductile iron piping. These pipes carry large quantities of aged sewage from lift station wet wells to the gravity system. Since damage is known to exist in the wet wells, the situation cannot be better in the force mains. Unfortunately corrosion is not as noticeable in buried pipe until it fails. PVC is a better option for new piping because it avoids the corrosion problem.

There are five options for odor control within a wastewater collection system:

- Prevent dissolved sulfide formation
- Remove dissolved sulfide after formation
- Prevent the release of dissolved sulfide as hydrogen sulfide gas
- Control sulfide-oxidizing bacteria and/or neutralize formed sulfuric acid
- Treat odorous air

Most wastewater treatment system operations use a combination of these approaches as an effective sulfide odor and corrosion prevention program.

Blockages are a routine maintenance item, but should be investigated further if they occur regularly in one place. It may be the material flushed into the collection system does not break down correctly (e.g., paper towels), or there may be damage to the pipe. Maintenance and repair of the gravity collection system includes cleaning and televised inspection of the gravity lines and personnel access openings to look for leaks and breaks and to respond to complaints about stoppages. Repairs include excavation and repair to access openings, gravity piping, and service connections. Many of the same issues involved with field crews apply to a wastewater collections system, although the federal Capacity, Management, Operation, and Maintenance (CMOM) program adds to these requirements. The CMOM program is a federal requirement intended to ensure that sewer collection systems, pumps, and wet wells are properly maintained in order to eliminate sanitary sewer overflows from plugged pipes or lack of pumping capacity in lift stations. Pipe is inventoried while cleaning and repair work is tracked. Maintenance logs are required for lift stations.

Maintenance and repair of lift stations and force mains includes pump and motor repairs (as in treatment plants), removal of grease in the personnel access openings, cleaning and adjustments to the force main air release valves, and repair of force main breaks. Work crews also install new service connections and repair old ones, which may be a significant source of infiltration. New gravity mains and force main piping are usually installed by construction contractors with large-scale equipment to correct problems.

Infiltration/inflow reduction

Utilities can avoid many headaches with sanitary sewer systems by planning, budgeting, monitoring, maintaining, and inspecting the system as a part of the CMOM program. This is a formal program with requirements to put dye in the water, review flooding areas, inspect personnel access openings, run closed-circuit television, and conduct flow monitoring. The CMOM programs, developed as a part of rulemaking efforts under the Clean Water Act, made each utility responsible for maintenance of its own lift stations and collection systems. Because keeping excess flows down benefits the utility financially, correction of leaks and infiltration should be priority projects.

Initially the Clean Water Act was meant to clean up the nation's rivers and streams through the removal of untreated industrial and domestic wastewaters from the waters. Once treatment plants were constructed, the priority shifted slightly to combined systems that had a propensity to overflow during rain events due to hydraulic limitations of the piping systems (sanitary sewer overflows [SSOs]). A major focus was on separating wastewater and stormwater systems, most of which were located in the industrialized Northeast and Midwest. As these combined systems were eliminated, the focus turned toward infiltration through pipelines.

Collection system piping in North America prior to 1980 was predominantly vitrified clay (VC). VC pipe is resistant to deterioration from virtually all chemicals, has a long service life when installed correctly and left undisturbed. It is brittle, so settling from incorrect pipe bedding, surface vibrations, or freezing can cause it to crack. Temperature differences between the warm wastewater and cooler soils can cause the exterior pipe surface to be damp, which encourages tree roots to migrate and possibly wrap around the pipe. Where cracks occur, roots will enter the pipe. The short joints compound the problem. Over the long term, the pipe will become broken and damaged from these forces. Where the water table is above the pipe level, the pipe will leak constantly, resulting in significant infiltration that can reduce the capacity of the wastewater treatment plant.

This constant flow of water is termed *infiltration*. Infiltration increases the base flow and will indicate low-strength wastewater during routine tests. It rarely leads to peak flows. Lining vitrified clay pipe is possible with many products, thereby extending the life of the pipe. The major focus to remove infiltration has been, and continues to be, lining gravity pipe. The process includes a significant amount of televising to find leaks.

Where there are peaks in wastewater flows that match rainfall, inflow would appear to be a more likely cause than infiltration into pipes that are constantly under the water table. Figure 4-14 shows common infiltration points. It also shows a bigger problem, i.e., inflow points where water from rainstorms runs into the system and often overwhelms the treatment plant. The spikes seen on Figure 4-15 are inflow, not infiltration. Correction of inflow is much less costly than of infiltration and often resolves immediate concerns. Simple methods can be used to detect inflow, and they should be part of ongoing maintenance efforts.

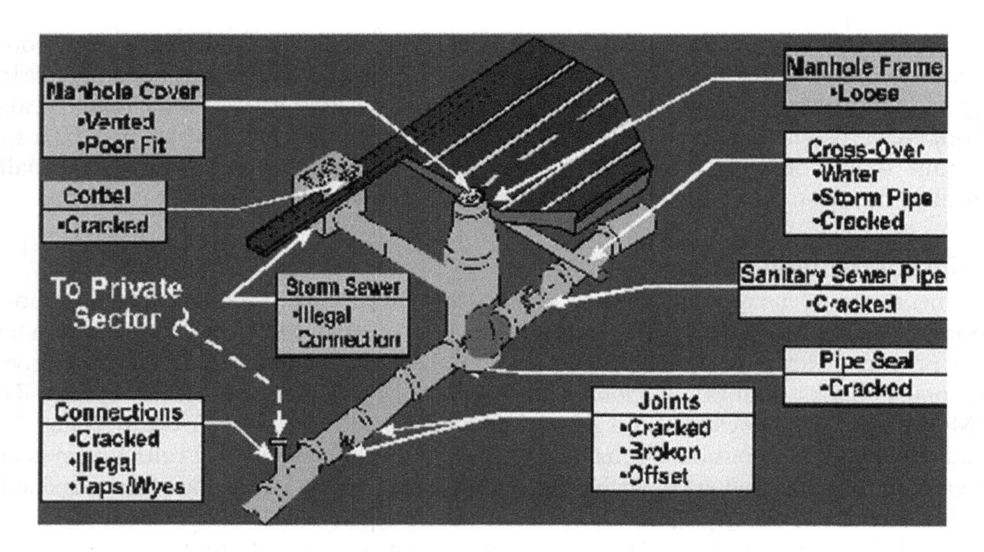

FIGURE 4-14 Potential infiltration and inflow areas
Source: Bloetscher (2009).

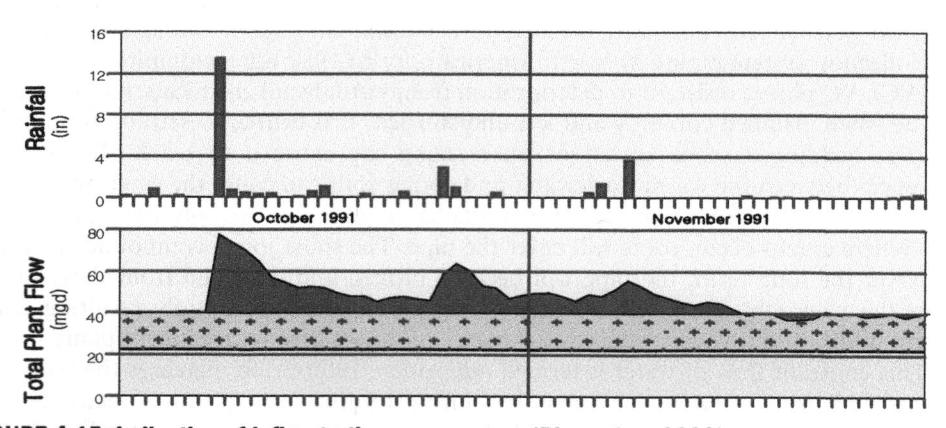

FIGURE 4-15 Indication of inflow to the sewer system (Bloetscher, 2009)
Source: Bloetscher (2009)

The following outlines a basic program for inflow detection evaluation of the utility system:

- Inspection of all sanitary sewer personnel access openings for damage, leakage, or other problems.
- Repair of benches in poor condition or exhibiting substantial leakage.
- Repair of access opening walls in poor condition or exhibiting substantial leakage.

- Repair/sealing of chimneys in all personnel access openings to reduce infiltration from the street during flooding events (Figure 4-16).
- Installation of dishes in all personnel access openings to prevent infiltration (Figure 4-17).
- Installation of LDL plugs where personnel access openings in the public right-of-way or other portion of the utility's system are damaged (Figures 4-18 and 4-19).
- Smoke testing of sanitary sewer system (Figures 4-20 and 4-21).
- Low-flow inspection event.
- Documentation of all problems in a report to utility that identifies problem, location, and recommended repair.
- Personnel access opening inspection and dish replacement. This is for access openings where the repairs have previously been made and only the inspection and dish replacement occur.
- Identification of sewer system leaks, including those on private property (via location of smoke on private property).
- Submittal of ArcGIS database of access opening and repairs with integrated photography and GPS locations.

The costs for this type of program will be on the order of $500 per personnel access opening plus repairs to at least 15 percent of service laterals (2,300 at a cost of up to $500 each). The program could be funded with state revolving funds (SRF) or other sources of borrowing, but ultimately should be part of on ongoing maintenance program. Repairs to pipes and laterals are estimated to add another $200 each based on experience elsewhere. These repairs should be followed up every five years. The benefit of this program is that it would keep excess water out of the sewer system, especially salt water from coastal areas. The drawback is the ongoing maintenance costs, however, this program has a low initial cost and high rate of return compared to other options. If inundation of roadways occurs as a permanent issue, then those affected areas would likely need to be abandoned, since there is no fail-safe way to prevent water from seeping into the sewers under these conditions.

Ongoing testing of the influent and monitoring of the lift stations by the utility provide measures to determine whether or not inappropriate amounts of infiltration are going to the wastewater plant. This testing takes a variety of forms. The first is a review of personnel access opening run times, followed by analyses of the influent wastewater quality. Low-strength wastewater is an indication of infiltration and inflow problems.

Remote sensing in personnel access openings can provide utility staff with useful information about gases in the openings, water depth to identify blockages, potential sanitary sewer overflows, and even entry. Sanitary sewer overflows are the biggest risk; USEPA estimates that 40,000 SSOs each year discharge between 3 and 10 bil gal (11.4 and 37.8 GL) of raw sewage into the service area (Quist et al., 2009). SSOs are a violation of the Clean Water Act and therefore a fineable offense. Most utilities see the value in having supervisory control and data acquisition (SCADA) to detect the potential for this problem so the incident, cleanup, and fines can be avoided.

Most personnel access opening telemetry is placed under the opening lid or on the wall of the opening to send a signal. The signals feed into the system's SCADA system via low-band radio signal or cell phone. Because personnel access openings have no power to them, batteries are used to power the system, and chips are used to collect, store, and

FIGURE 4-16 Repair/sealing of chimneys in all personnel access openings to reduce infiltration from the street during flooding events
Courtesy of USSI.

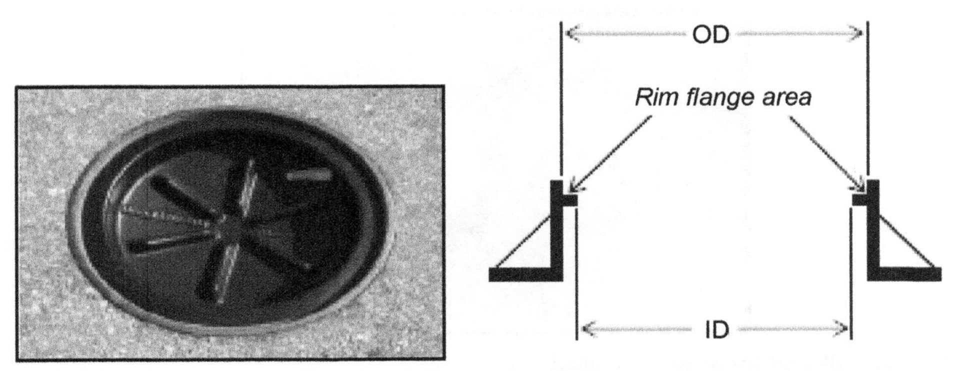

FIGURE 4-17 Installation of dishes in all ersonnel access openings to prevent infiltration
Courtesy of USSI, LLC.

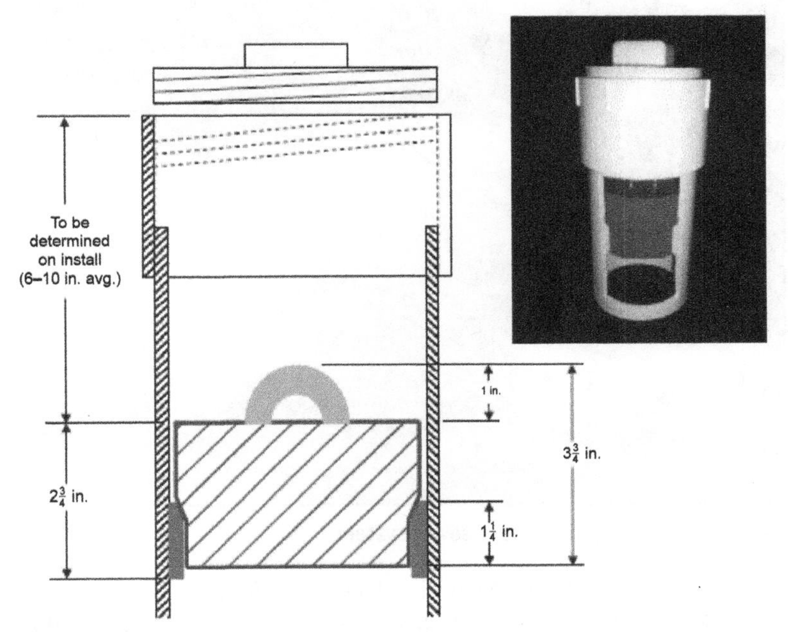

FIGURE 4-18 LDL plug design
Courtesy of USSI, LLC.

transmit the information (Quist et al., 2009). Sonic level indicators are preferred to sense water depth. Receiving towers need to be located to collect the data into the larger SCADA realm. Long-term trending can indicate problem areas in the system; usually SSOs and blockages occur in the same areas repeatedly, so if the problems persist, additional efforts can be undertaken to resolve them.

Confined space entry on sewer systems is especially hazardous. Current OSHA regulations require that utility personnel be trained on confined space entry procedures, that a minimum of two people be present for any entry, that an entry permit is acquired, and that

FIGURE 4-19 LDL plug installed in cleanout
Courtesy of USSI, LLC.

FIGURE 4-20 Smoke testing of sanitary sewer system

the appropriate gantry and lift tools are in place to retrieve the persons entering the confined space. Risk hazards include sewer gases such as hydrogen sulfide and methane, slippery surfaces, crumbling concrete, and damaged stairs and steel fixtures. Each year a number of people die in confined spaces. Some of the deaths are of people trying to rescue fallen co-workers.

Maintenance of SCADA systems

Using much the same technology as automatic meter reading, SCADA can also be used in the distribution and collection systems. Water telemetry is mostly focused on pumps and pressure, while sanitary sewer telemetry is focused on pump station operations, as noted in chapter 3. SCADA systems have inherent vulnerability. While SCADA is helpful in monitoring the utility system, it is an electronic system that requires maintenance and may periodically provide spurious data. A new

FIGURE 4-21 Smoke exiting sewer system during smoke testing of sanitary sewer system

threat, hacking from remote sites and losses of computer operations, may incapacitate the SCADA system and leave the utility blind to field operations. Methods to protect SCADA include limiting internet access to the SCADA system and limiting suppliers and vendors from having a direct link to the SCADA system. SCADA systems should be checked regularly for Internet connection, firewalls, and connectivity issues. Staff should have access to alarms, on-call dial-up systems, and encryption services to protect the SCADA system (Brilliant and Mains, 2010).

Work Order Systems

In a perfect world, the utility will have complete, accurate system maps; detailed maps to locate all piping (indicating pipe size, installation date, and material); valves and fire hydrants, personnel access openings, cleanouts, services, and details on all repairs made to them and the conditions surrounding those repairs; pipe condition reports, data reports on all new pipes installed, and conditions found during excavation; a spreadsheet file of all personnel access openings, fire hydrants, and valves with a number assigned to each and records for each that include installation date, type, repairs, and results of hydrant flow tests and exercise or flush dates; and list of all equipment in the treatment plants. Rarely does a utility have all of this data or the ability to obtain it.

Good record keeping should be started as soon as practical to build a base of knowledge that will improve the efficiency of repairs. Staff must be assigned to perform the record keeping, data entry, and mapping. Water distribution and sewer collection system maps and other information can be maintained on a geographic information system (GIS) or other computer system. Information that should be obtained and recorded before the pipe is covered in the trench are measurements locating the water main tap, the type and size of pipe, burial depth, measurements to the curb stop or meter pit, cleanout locations, location of the pipe at various points, and the location where the pipe enters the building. Photographs are always useful, especially digital photos that can be put into report files and stored on CDs.

The sewer collection system map should include the location and size of gravity collector mains, force mains, lift stations, personnel access openings, and valves. More detailed records should track information such as measurements from each valve to such aboveground features as trees, curbs, and extended lot lines in the event the valves are buried during street paving activities or disturbed by snowplows. Lift station records should include the make and model of pumps, operating conditions, and pumping rates. Pipeline information should be as detailed as possible with indications of size, type of material, depth, periodic location measurements, and date of installation, where possible.

A computerized work order tracking system should be used to record pipe system repairs and the condition of both the interior and exterior of pipe and appurtenances. Whenever a piece of pipe is removed, information should be recorded about the pipe, location, and conditions of the soil, and a coupon should be photographed and compiled for later use when assessing pipeline conditions, potential for failures, tubercles, and wear. Some utilities may wish to keep a small piece of pipe tagged with the date and location.

Service line information should be recorded as installations or repairs are made. A metallic water service pipe with an inadequate record can usually be located using an electronic pipe locator, but the process takes longer than using good recorded information. A pipe locator will not usually work on plastic pipe, so finding a pipe with no record can be difficult unless metallic tape was also installed (the normal practice).

Crew productivity is difficult to measure without tracking the work they perform. A work order tracking system is required to accomplish this goal. While many utilities have not implemented work order tracking systems, new regulations are pushing utilities to do so. While a program similar to CMOM does not currently exist in the federal rules for drinking water treatment and distribution systems, it is only a matter of time given the potential health impact of contaminated water systems.

Treatment Plant Laboratory

All utilities need access to a laboratory. The laboratory is needed to monitor a treatment plant's water quality for those parameters required by the federal, state, and local regulatory agencies. Additionally, the laboratory should monitor plant processes to ensure optimal treatment efficiency. Many municipal laboratories can be state-certified for some parameters. For contaminants that the laboratory is not equipped to analyze, such as metals, the samples are sent to a contract laboratory.

Pretreatment Programs

All utilities are required to meet stringent limitations on the quality of treated wastewater that is disposed of. The intent of an industrial pretreatment program is to help prevent business enterprises from discharging substances into the wastewater treatment system that cannot be treated, will cause the treatment organisms to die, will cause the effluent limitations to be exceeded, or will otherwise be detrimental to the plant process or operators. Substances regulated include oil, gas, metals, cleaning fluids, paint, and industrial process water. Pretreatment is mandated under the Clean Water Act and may be a condition of the utility's National Pollutant Discharge Elimination System (NPDES) permit if perceived by the regulatory agency to be an issue. System size, type of customers, and effect on receiving waters will determine this requirement. The impact of pretreatment programs can affect the ability of the community to attract businesses, so careful consideration of effluent limits and treatment efficiency is required.

Staffing

Management of operations requires an understanding of maintenance and operating needs of the system, and having personnel, equipment, and funding in place to allow operations personnel to do their jobs. Productivity is an important aspect of operations maintenance. It should be noted that increases in productivity may concern employees in regard to job losses, but better productivity should translate to the ability to accomplish more work. The economic growth of the United States in the late 1990s was in many ways directly related to an improvement in productivity among American workers. The utilization of technology helped spur this growth because it made it easier to collect data needed to react to a given situation.

Operations management requires that there be a clear set of goals and objectives or a vision of the utility and where it is going. Having accomplished a vision, effective operations require that an appropriate organizational structure be in place and that appropriate lines of communication to facilitate productivity within and between those work groups be maintained. As a part of operations maintenance, targets for productivity of workers are required. In other words, if the meter reader is expected to read 200 meters per day, someone needs to track to see if he/she actually reads that many. At the same time, managers and supervisors need to determine if the goals are too easy. For example, at one utility solid waste workers used to take eight hours to accomplish their jobs. Once they were converted to a system where they could go home when they were done with the work, workers were suddenly done in three or four hours. Clearly the managers and supervisors were not gauging whether or not the employees were productive. The result is that much more work could get done and, in a utility where staffing is typically limited, the ability to do more work with existing staff is an important consideration.

A work plan system needs to be developed by management and the supervisors for implementation by the line workers. A lack of understanding of what tasks need to be accomplished and what planning needs to be done for those tasks leads to inefficiency, confusion, and competition for the same tools and equipment. These are supervisory issues (see chapters 5 and 6).

Operational audits may be a means to help utilities that are dealing with issues associated with productivity and employees. The operational audit will initiate an investigation of what problems may exist within the utility and why work is not accomplished, and will identify the planning and management needs. On-site evaluation of the services and supervisors is required with the individual findings developed into a series of recommendations to improve operational efficiency. Some of these may target individuals, supervisors, or others within the organization who do not have the proper skills for their jobs, but more commonly they identify where training is required to make workers more productive and how to coordinate work efforts and plan daily activities in a more consistent manner.

Use of work orders is a major step in improving utility operations efficiency. More recent developments include the incorporation of computerized maintenance management systems designed to optimize when certain maintenance activities should take place to ensure the long-term life of infrastructure systems. Maintenance management is linked to the fiscal side through asset management practices, whereby the asset manager looks at the additions, subtractions, and depreciation of the physical assets with time.

Operations personnel have a variety of jobs to do on an ongoing basis, including the following categories and descriptions:

- **Day-to-day**, such as testing water, monitoring chlorine residuals, and preparing daily log sheets and information for monthly reports.
- **Public health and safety**, including activities such as chlorine residual tests in the distribution system, ensuring disinfection requirements are met, preparation of reports on water quality, water quality testing (in-house and contract), and maintenance of facilities, which involves general cleanup, maintaining generators and backup systems, ensuring the availability of extra chemicals, and maintaining emergency procedures.
- **Periodic jobs**. These include accepting delivery of chemicals, parts, or supplies; changing chlorine cylinders; pump repairs; ordering supplies; monthly operating report preparation; periodic water quality analyses; pump tests; well or intake cleaning; communicating with supervisors; and budget items.
- **Plant safety,** such as proper signage on the site, maintenance of personal safety equipment, self-contained breathing apparatuses (SCBAs), masks, alarm systems, and generators.
- **Maintenance**, such as mowing grass and sweeping, painting the plant and appurtenances, and making repairs to broken items, including pumps.

Field crews spend time on the following activities:

- Travel to job sites
- Excavation and pipe repairs
- Preventive maintenance of pumps and motors
- Emergency repairs
- Customer response

The crews normally meet at a central site where they receive work assignments, materials, and supplies to accomplish the day's work. They then leave for the job site, which may be some distance away. Having the appropriate supplies and materials is important to maximize field crew productivity.

Field crew productivity is dependent on the following:

- Optimal crew size. Crews that are too large or too small create inefficiency. Two-person crews are often all that is needed for many jobs. One-person crews are not recommended for safety reasons.
- Having materials and parts in inventory and not delayed in arriving at the job site.
- Proper scheduling of work and workloads.
- Flexible organizational structures among crews.
- Supervision to ensure crews arrive at the job site in a timely manner and that downtime is minimized.

It should be noted that, although several workers may appear to be idle at a given point in time during an excavation, the repair may need four or five people when fully unearthed (one or two in the trench, one operating equipment, one retrieving materials on the surface, and one watching traffic). Assigning workers elsewhere may reduce efficiency and lengthen the time needed for the repair. In a high-traffic area, this could pose a serious problem.

Due to the variety of tasks possible at a job site, a variety of job classes should be expected to be present. Most utilities will have some form of a supervisor or foreman for a crew. Within the crew there may be laborers, drivers, equipment operators, and perhaps electricians. Not all crews have all these people, and crews may be created on a daily basis to do certain types of jobs. In looking at skill sets, it requires more than the day-labor ditchdigger to lay pipe, concrete, or sewers or to do other general construction or repair work. The appropriate people need to be in place. Good concrete finishers will save a significant amount of time and effort for the utility by ensuring that concrete looks appropriate. Likewise, pipe layers who have experience in laying pipe will be both efficient at laying the pipe and will do it correctly, limiting needed follow-up and repairs required due to construction deficiencies.

Sewer line work requires a different crew than for a water main. Water mains are not required to be on grade, whereas sewer mains must be on grade and require a crew that is used to dealing with lasers and grade lines. Likewise, roadway crews require personnel familiar with grade lines and surface paving, but they do not need excavation experience. The appropriate construction equipment operator is required for each. An equipment operator may be able to operate several different types of equipment, but is unlikely able to operate all that is required for a utility. Training and ongoing reviews are likely needed for pieces of equipment such as backhoes and excavators to ensure that the proper skills remain in place. For less frequently operated pieces of equipment, refresher classes may also be beneficial.

The responsibility for quality on a project falls on the supervisor and management. The responsibility for the quality of the specific tasks falls on those doing the work. Thus the utility needs to instill within its workforce and its consultants the need for quality work. To ensure quality, the utility needs to set standards for planning, design, and construction. Inspection is required to verify the quality of the project.

There are a variety of standards that are useful for setting quality parameters. The American Water Works Association (AWWA), the American National Standards Institute (ANSI), and the American Society for Testing and Materials (ASTM) all have standards for dealing with different types of equipment. The benefits of maintaining quality and meeting industry standards are less work, lower cost, and better schedule performance, which lead to better owner satisfaction. The added costs of adhering to quality standards is offset by savings by avoiding added maintenance and operating costs, added repairs, shortened asset life, unreliability, and other aspects of dealing with work that does not meet the requirements. This may mean a little more cost up front, but life cycle savings. Poor work quality costs the owner significantly over the life of the process and has a much higher life cycle cost than the work up front. Monitoring the project ensures that specific results are obtained to comply with quality standards and identify ways to eliminate causes of unsatisfactory results. This will include inspections, inspection reports, and planning and quality control logs to indicate where corrective actions must be taken or have been taken.

CONSTRUCTION EQUIPMENT

The main reason that construction equipment is important is that worker productivity is related to having the appropriate equipment on-site. It is important to understand what the limitations of the equipment are. Most projects progress from some form of underground construction to aboveground construction (e.g., buildings) to some sort of surface treatment, such as paving and ultimate completion of the project. Different equipment is needed at each stage of the project, and each piece of equipment has a different purpose. The intent of this section is to aid the utility in understanding the uses and limitations of a variety of pieces of construction equipment.

One piece of equipment required in all piping systems is a tamp. The tamp is shown at the bottom of an excavation in Figure 4-22. The tamp is used to compact dirt. It is run by hand and usually is used to compact dirt above and below pipe, all the way to the surface. Many contractors attempt to avoid tamping, which means settlement will occur in the future. Stone is often used for bedding purposes. Stone does not compact like sand or soil. The intent is to lay the pipe in the bedding so that it has the appropriate strength from below and is not crushed. Likewise tamping above the pipe ensures that the pipe is not point-loaded (i.e., that the load from above is not carried on a small portion of the pipe), which may lead to future failures.

More significant equipment used on the construction site include the following:

- Backhoe
- Excavator
- Front end loader
- Trencher

FIGURE 4-22 Hand compactor, or tamp, to compact soil in a pipe in a trench

- Bulldozer
- Road grader
- Asphalt machine
- Vibrator/roller
- Cranes
- Trucks

Backhoe

The backhoe is the most common piece of equipment found on the construction site, especially for a utility. It is also one of the most versatile. It is a small piece of equipment, usually on a series of rubber tires so it can be moved across roadways. The uses of the backhoe include making trenches for piping utilizing the rear boom, loading materials using the front bucket, leveling, and materials placement. A good backhoe operator can dig the trench to lay 1,000 ft (305 m) of pipe in a day without a significant amount of trouble; however, a backhoe is limited to fairly small pipe due to the reach and carrying capacity of the boom. It's important to note that, as the boom is stretched out, the amount of weight that it can hold decreases (see Figure 4-23). The front end bucket can be used to cart up to a couple of cubic yards of material or parts needed for installation. Therefore a backhoe can provide both the trench and the laying of the material for bedding, and it can cover the excavation at the end, making it a highly useful piece of equipment for small pipelines. The benefits are its versatility, limited amount of pavement damage, ease of operation and maneuverability, and the ability to ride on roads. The limitations have to do with the size of the equipment and the depth of excavation; rarely can a backhoe dig down more than 10 ft to 14 ft (3.5 m to 4 m), depending on how far the boom is out.

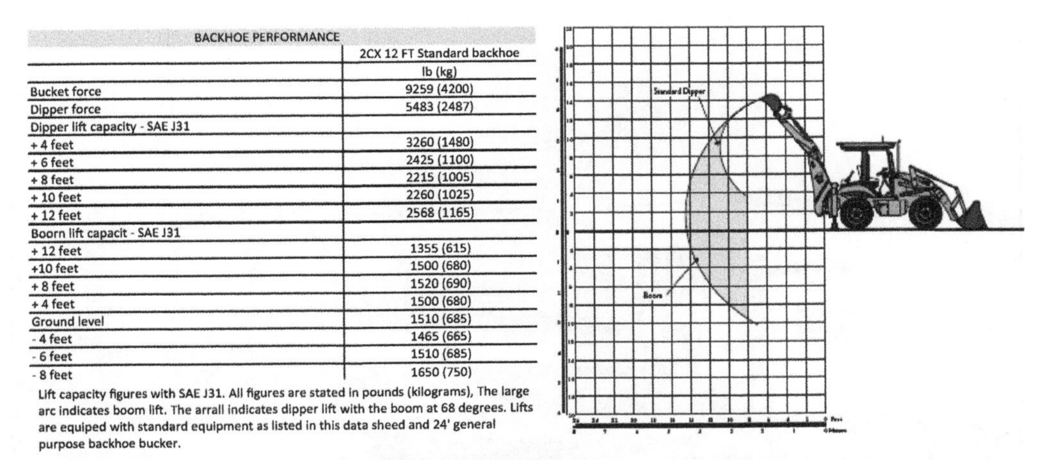

BACKHOE PERFORMANCE

	2CX 12 FT Standard backhoe
	lb (kg)
Bucket force	9259 (4200)
Dipper force	5483 (2487)
Dipper lift capacity - SAE J31	
+ 4 feet	3260 (1480)
+ 6 feet	2425 (1100)
+ 8 feet	2215 (1005)
+ 10 feet	2260 (1025)
+ 12 feet	2568 (1165)
Boorn lift capacit - SAE J31	
+ 12 feet	1355 (615)
+10 feet	1500 (680)
+ 8 feet	1520 (690)
+ 4 feet	1500 (680)
Ground level	1510 (685)
- 4 feet	1465 (665)
- 6 feet	1510 (685)
- 8 feet	1650 (750)

Lift capacity figures with SAE J31. All figures are stated in pounds (kilograms), The large arc indicates boom lift. The arrall indicates dipper lift with the boom at 68 degrees. Lifts are equiped with standard equipment as listed in this data sheed and 24' general purpose backhoe bucker.

FIGURE 4-23 Backhoe arm information showing that the amount of weight that can be lifted decreases as the extension increases
Source: JCB

Excavator

For larger and deeper construction, an excavator is used (Figure 4-24). An excavator operates similarly to the boom on the backhoe except it has a much larger reach and is normally on tracks. A track creates a significant amount of damage to pavement; thus protection of the pavement is required. As with the backhoe, the boom has some limitations on the length and distance for which weight can be held, but both the distances and the amount of material removed are significantly larger compared to a backhoe. Like a backhoe, it is easy to maneuver and operate. Unlike a backhoe, it does not have the versatility of having a front end bucket for covering up the pavement and laying bedding materials, it cannot ride across pavement (without damage), and it cannot fill the excavation. Excavators come in many sizes, including mini-excavators (Figure 4-25) and a rubber-tired version for light-weight work (Figure 4-26).

Front End Loader

Where an excavator is used, either a rubber tire backhoe or a front end loader is required. A front end loader carries much larger amounts of material than the front end of a backhoe. The front end loader is typically rubber tired and easily maneuverable for placing materials. It can place material relatively quickly or move it from place to place. The benefits are a limited amount of pavement damage, ease of operation, ability to travel the road, and ability to move large amounts of materials. The buckets can be huge (see Figure 4-27) or replaced by other equipment, such as a forklift (see Figure 4-28). The limitations are that it is a large piece of equipment, has limitations on where it can go, and has no ability to do any excavation.

FIGURE 4-24 Track excavator used for deep excavations

FIGURE 4-25 Small track excavator for small jobs

FIGURE 4-26 Rubber-tired wheel excavator for lightweight work

FIGURE 4-27 Front end loader bucket carries cubic yards of material

FIGURE 4-28 Front end loader bucket can be converted to other equipment

Trencher

A popular alternative to open excavation is a trencher. Trenchers come in a variety of sizes from a few horsepower to machines that have a cutting wheel that may be 50 ft (15 m) in diameter. Trenchers cut a deep, narrow trench that pipe can be laid in. It is important to ensure that the pipe trench is wide enough to allow maneuverability of the equipment that goes into the trench.

Trenchers create a limited amount of surface damage, have limited cost and good ease of operation, and can often travel the road or be easily hooked up to a trailer that travels the road. However, because they lay in a straight line on a blade, trenchers are not easy to maneuver. They are limited to relatively small pipes for municipal applications. Trenchers cannot lift any objects and cannot go very deep unless large trenching equipment is used. Large trenching equipment would be used for deep piping, river bank filtration, exfiltration, and simliar applications. Trenchers may be small and inexpensive, but their lack of maneuverability creates some limitations. At the same time, the fact that they create very limited damage when compared to an excavator or a backhoe is a significant benefit.

Where the surface damage from a trencher is not acceptable, boring machines can be used. These have become more popular as the technology has improved dramatically over the past 20 years. Boring machines allow the operator to drill underneath the ground and lay pipe without significant surface damage. The benefits are the lack of residual damage, lack of excavation, and the fact that backfill is not required because, in most cases, the piping is laid within a hole that is filled with mud materials similar to those of a drilling rig. However, boring machines are not easy to maneuver, nor are they easy to operate. They lift no object of significant size, cannot travel the road, and are limited to pipes that are generally less than about 24 in. (61 cm) in diameter. It also should be noted that there are limitations in the material to be bored. A small boring machine will not be able to drill through very hard material or alluvial types of deposits. As a result, the horsepower of the boring machine needs to be matched up with the job application. Unfortunately, in a 1,000-ft (305-m) bore, conditions can change and the boring contractor may find the machine being used is not appropriate for the conditions in the field.

Bulldozer

A bulldozer is a less common piece of equipment for utilities but very common on roadway projects (see Figure 4-29). Bulldozers move dirt. They are on tracks and can create significant damage to pavement, which is why they are not used in most utility applications. They are relatively easy to maneuver and operate and can move large amounts of dirt by pushing. The larger the bulldozer, the larger the amount of dirt that can be pushed; thus very large bulldozers are made.

The limitations of bulldozers have to do with pavement damage, slow travel, inability to travel on roads, and the fact that a bulldozer lifts nothing; it only pushes dirt around. So if the job is to move large amounts of dirt from one place to another, bulldozing is inefficient. However, if the job is to smooth rock out on a roadway bed or to repave a road after it was torn up by utility construction, a bulldozer is efficient.

FIGURE 4-29 Bulldozer for spreading rock and material

Road Grader

An alternative to a bulldozer is a road grader (Figure 4-30), commonly used for finishing roads. This is a useful piece of equipment, although not common to utility systems. A road grader is used to level road subgrade, dirt roads, and rock courses before paving. A good road grader operator can create very accurate grading of the road, the crown of the road, and ditching to the side of the road. Road graders are fairly heavy pieces of equipment that are on rubber tires and can move dirt very efficiently while not damaging pavement. They are fast and they can cut a very fine grade.

The limitations of road graders are that they cannot move large quantities of material at one time, so they are typically employed after a bulldozer or front end loader has been used. They move material only, although they can cut material on the side of the road for ditching. They cannot replace material and require a significant amount of skill to operate.

Asphalt Machine

In most cases, asphalt is renewed as opposed to thrown away. Asphalt is one of those materials that is easily recycled, and an asphalt grinder, used on most roadway projects to remove asphalt, can also be used on utility projects. The benefit of an asphalt grinder (see Figure 4-31) is that it quickly removes the original asphalt, and either leaves grooving or can completely remove the asphalt where necessary to allow for later pavement. They are connected up to a truck, move slowly, and are easy to operate. Asphalt grinders take up space and are noisy, thus are normally not used during rush hours. Because they are designed to damage pavement, if protecting pavement is important, then an asphalt grinder is *not* a useful piece of equipment.

FIGURE 4-30 Road grader used for spreading rock and material and for fine grading

FIGURE 4-31 Asphalt grinder

The opposite of the asphalt grinder is the asphalt machine (Figure 4-32). The asphalt machine reheats ground-up asphalt, and additional asphalt is added and the asphalt is relaid. The asphalt machine is tied to a truck full of hot asphalt. It is easy to remove or push the truck along. Asphalt machines are laser guided, relatively fast, and require only the rollers to compact the asphalt. The laser guidance must be set up properly. Asphalt machines come in a variety of sizes, including very small machines to allow for paving over utility cuts.

FIGURE 4-32 Machine for spreading asphalt to repave roads

Vibrators/Rollers

Rollers are used to compact asphalt and can be used to compact some materials in a trench. There are steel rollers that may use vibrations (to compact the surface prior to pavement) or not (see Figure 4-33) and rubber-tired rollers (see Figure 4-34) that are used after the initial rollers with a heavy steel roller. Benefits of this equipment are that it is easy to maneuver and operate and can travel on the road, although this is not a preferred option for steel rollers. Rollers come in a variety of sizes from very large to small machines that would be owned by a utility system. A danger with rollers is that over-rolling asphalt damages it; more than three or four passes will actually ruin the asphalt.

Cranes

Cranes are important for utilities. They are used to lift large loads or to extend the boom where an excavator or backhoe cannot. Cranes can be large truck-mounted units (see Figures 4-35 and 4-36), track mounted, or used to lift people (see Figure 4-37). Pieces of equipment within a treatment plant are often removed with a crane. The utility may not own the crane but needs to have access to one. Various sizes, boom lengths, and heights are available. The number of uses is infinite, but the load that can be picked up is specific to the crane. Access is easy as well with a crane. However, cranes are not easy to maneuver and it's important to carefully balance loads and weight or the crane will tip over. Cranes that fall at construction sites are front-page news in most areas and have the potential to damage property and injure people. The correct crane and operator are both needed for the job.

FIGURE 4-33 Steel roller to compact soil and roll asphalt

FIGURE 4-34 Rubber-tired roller

FIGURE 4-35 Crane for lifting and moving large pieces of equipment

Trucks

A variety of trucks will be needed for any construction or operating program. Trucks come in all shapes, sizes, and types, depending on the use. Typically utilities require dump trucks to move dirt and bring in rock and fill material (see Figure 4-38), as well as water trucks (see Figure 4-39) and a variety of pickup trucks with utility bodies to cart equipment. The benefits of trucks are they have many uses and are easy to maneuver. A commercial driver's license may be required for operation, which is an issue that the utility needs to address as a part of its ongoing program. Trucks have limitations for off-road access. The rubber tires can also be a problem if there are a lot of nails or debris at the

FIGURE 4-36 Truck-mounted crane for lifting and moving large pieces of equipment

construction site that may damage them. Trucks are easy to maintain as most garages and operations personnel are familiar with them.

A sweeper is another piece of equipment that may be useful when workers are ready to repave the road (Figure 4-40), and it is also road traveling.

REDUCING POWER USE

Utilities face many financial challenges in funding, including capital, operations, and power. Water pumping systems are typically designed to last 20 years, but may be in service much longer (Bloetscher et al., 2010a). Efficiency of old motors and pumps may be significantly less than of current pumps and motors. A properly designed installation with variable frequency drives (VFDs) and correctly sized pumps and motors is a start in reducing power consumption. Variable frequency drives will allow operators to adjust flows without throttling valves, which wastes energy. Ongoing preventive maintenance also will reduce wear on pumps and motors, which maintains efficiency longer and extends asset life (Moran and Barron, 2009).

Large water and wastewater treatment plants are often among the largest users on the grid (Bloetscher et al., 2010a). As can be seen in Table 4-2, power costs are significant, especially as more exotic treatment is employed (seawater desalination being the most costly). Because much of the power generated creates greenhouse gases, the need for more and higher quality water will increase the demands on the power grid exponentially. As an example, south Florida has discussed creating indirect potable reuse projects to recharge aquifers and increase water supplies through recycling. The cost to implement reuse at

FIGURE 4-37 Crane for lifting and moving large pieces of equipment

large wastewater facilities in southeast Florida (600 mgd) would be $6 billion in capital construction and would demand an estimated 1.8-GW power supply. To treat 250 mgd of brackish Floridan Aquifer water supplies with reverse osmosis is estimated at $4.5 billion in capital, plus about 1.25 GW in new power.

If wastewater plants can generate power on-site using sludge and digester gas, and water plants can modify practices to limit the demand for membranes, the power

FIGURE 4-38 Dump body on a truck—full

FIGURE 4-39 Truck with tanker on the back

FIGURE 4-40 Sweeper

TABLE 4-2 Power costs per million gal per day

Treatment Process	Power (MW/mgd)
Aeration/high-service pumps/wells	1.4
Lime softening	2.3
Nanofiltration, 125 psi	2.7
Low-pressure reverse osmosis > 200 psi	
Brackish water supply	3.3
Secondary wastewater/ pure oxygen	3.4
Reclaimed wastewater	3
Seawater desalination	13

demands for the water and wastewater sector could decrease. This would create additional capacity in the present power grid for growth and development and limit the need to add power capacity. It would also limit the need for new power with its associated greenhouse gas emissions. Many treatment plants can or do generate most of their power on-site, such as the Perth desalination facility in Perth, Australia, which creates all its own power with wind (Bloetscher et al., 2010a).

Example 1

Utility X is located in southeast Florida (Bloetscher et al., 2010a). It is concerned about the potential for severe tropical storms and hurricanes to impact water availability, and wants to use sustainable technologies to meet its future energy needs. The city proposed a nanofiltration facility to complement the current lime softening plant (see Figure 4-41).The purpose of the facility is to provide cleaner, clearer water to city customers. Nanofiltration is a process that works like reverse osmosis, only it removes hardness instead of salt.

As opposed to the current lime softening on the site, nanofiltration is very energy intensive. In this case, the proposed nanofiltration facilities were determined to require about 50 percent more power than is currently being consumed—roughly 235.6 kW per day (Table 4-3). The nanofiltration process will more than double current power demands, while only increasing capacity by 67 percent.

The building program involves on-site piping, pretreatment, nanofiltration skids, the building, chemical storage, cleaning system, electrical connections, parking improvements, connections to the clearwell, and other ancillary items. It is also the intention to provide operator office and laboratory space, and to connect all instrumentation to one central control center in the new operations building.

There are two especially innovative issues associated with the plant. The first is that pilot testing indicated that up to 95 percent recovery of water is possible, so that will be pursued. This will be the first plant to construct to that recovery level (Bloetscher et al., 2010b). Second, the new process building will be designed to be Leadership in Energy and Environmental Design (LEED®) certified. It will be the first water treatment plant building and the second building owned by Dania Beach to be LEED certified. LEED certification parameters were suggested as a bonus to the city.

FIGURE 4-41 Conceptual nanofiltration facility

TABLE 4-3 Power costs at treatment plant

Load	Number	Current (kW/day)	Future (kW/day)	Difference (kW/day)
High-service pumps	4	134.1	134.1	0.0
Backwash pump	1	55.9	55.9	0.0
Lime slurry pumps	2	7.5	7.5	0.0
Sludge pumps	2	4.5	4.5	0.0

(continued)

TABLE 4-3 Power costs at treatment plant

Load	Number	Current (kW/day)	Future (kW/day)	Difference (kW/day)
Accelerator drives	3	22.4	22.4	0.0
Nanofiltration membrane feed pumps	3	0.0	223.5	223.5
Degasifier blower	1	—	7.5	7.5
Chemical dosing pumps	10	3.7	1.9	–1.9
Membrane cleaning pumps	1	—	18.6	18.6
Lights, A/C, and controls	lot	7.5	18.6	11.2
Total		235.6	494.5	258.9

LEED is a process developed by the US Green Building Council to recognize buildings that encompass the concepts of building green into their process. Five major categories are used to evaluate the buildings. The first is sustainable sites. The intent is to use existing disturbed sites as opposed to virgin sites. In addition, the idea is to increase green space on the site. The nanofiltration facility will be located on the same site as the current lime softening plant, which will be refurbished and available to supplement the nanofiltered water. The improvements will integrate the new nanofiltration water treatment plant facilities with the current facilities and coordination of construction to minimize disruption to current activities. The final site plan proposes to increase the amount of perviousness on the site by reducing asphalt and removing unused structures and impervious pavement. To encourage employee use of alternative transportation, spaces will be provided for carpooling and vehicles that use alternative fuels. A bus stop is located at the boundary of the site. Showers and a bike rack will be provided.

The second category is water use. The plant will have a minimum recovery of 90 percent, which is greater than most membrane systems. The goal is to achieve 95 percent, which will improve plant efficiency by 10 percent. In addition, low-flow toilet fixtures, waterless urinals, and low-flow faucets and showerheads will be used. The project expects to use half the water of a comparable similar structure, or roughly 100,000 gal (3,785 L) per year, in addition to the increased water savings from process improvements (up to 36 mil gal per year). The utility will alter its irrigation ordinance to preclude the need to irrigate when Florida-friendly plants are used. Rainfall recapture of runoff (for use as irrigation) is planned.

The indoor air quality section (the third category) is intended to reduce air pollution, which means no materials, paints, or finishes that contain volatile organic chemicals (VOCs) will be employed; smoking will be prohibited in the building; and the HVAC system will be attuned to inside conditions and occupancy. Exterior lighting will be used to reduce power costs and improve indoor comfort.

The fourth category is materials. Recycled concrete and steel are available locally. All materials will meet the "Buy-American" clause and likely will be purchased within 500 mi of the site, which reduces transportation costs (and achieves a LEED point). All concrete and steel removed from the site will be recycled to the maximum extent possible, and construction debris will be separated and recycled to minimize landfill impacts. Crushed concrete and asphalt can easily be recycled, which is a goal of the project. The contractor should be able to achieve over 75 percent material recycling and recovery, thereby reducing landfill costs by 75 percent.

The fifth category is energy and atmosphere. Goals here will be difficult to accomplish because the plant process is power intensive. However, lights will be changed to compact fluorescent bulbs and turned off when not in use (automatically), and VFDs will be employed to increase energy efficiency. The project will evaluate energy recovery turbines on the permeate and concentrate streams. The building will have a white roof, which has been found to lower attic temperatures by 30°F (–1°C) (to under 10°F [–12°C] above ambient). Insulation will be provided to increase efficiency. HVAC will use high seasonal energy efficiency rating (SEER) equipment. Automatic systems to adjust temperatures will be employed. The goal is to reduce energy use by 30 percent over a similar building, in addition to making process improvements. Interior use is expected to be reduced by 2 kW•h, while the increased demands for the nanofiltration plant may save 20 to 30 kW•h.

As an academic exercise, the initial project was outlined to investigate the feasibility of operating a membrane facility with no additional power purchased from the local electric utility company. Furthermore, the initial project design was conceived to prohibit on-site fossil fuel and additional power from the grid. Adding to the complexity of the design was the fact that there is no real potential to create biogas-derived fuels, as is possible in wastewater applications, and membrane plants traditionally use far more power than conventional lime softening plants (see Table 4-2). Evaluation of the power options indicated that several kilowatts of power reduction can be gained with changes to existing electrical fixtures, such as use of compact fluorescent light bulbs and low-wattage fixtures, and by reducing chemical feed at the current lime softening plant. However, a net of an additional 258.9 kW was needed (Table 4-3).

The exercise evaluated the following options for developing power: (1) solar cells, (2) wind turbines, (3) pressure recovery, (4) fuel cells, and (5) compressed methane (derived from a local landfill and piped to the facility). An evaluation of the power options indicated that solar cells could generate up to 100 kW of the needed power on-site. An off-site parcel (10,000 ft^2 [929 m^2] owned by the utility could contribute another 100 to 125 kW, as needed. One solar panel vendor provided units with an estimated daily capacity of 385 kW•h, requiring a total of 415 solar panels on-site. Another option evaluated 3-ft × 5-ft (1-m × 1.5-m) solar panels without on-site battery storage, using an existing commercial buyback program to sell power back to the provider during the day in exchange for off-peak power at night. This system provided 612 kW based on 2,073 solar panels on-site. Large wind turbines were determined to be impractical. However, there are manufacturers of mini-wind turbines that have the potential to provide 40 kW and can be hung on towers to increase the power generated per square foot of ground area. Pressure recovery was also evaluated at a concentrate pressure of 72 psi and flow of 48 gpm. Using a recommended micro direct current output turbine/generator, the unit was capable of recuperating 0.8 kW of power from the concentrate line.

Fuel cells are expensive, and the technology is not currently developed enough to make them a useful option at this time. Compressed landfill gas needs to be pretreated with a scrubber before it can be used, but since there was a landfill within 4 mi of the site, a low-Btu methane-powered turbine was considered. The scrubber, pipeline, and turbines add significantly to the cost when a low-volume system is used. This option has more appeal to wastewater plants with on-site methane generation.

Figure 4-42 outlines where solar panels were proposed on the plant site. The intent was to cover virtually every surface of a building or covered tank with solar panels. Table 4-4 outlines the total power generated on-site—622 kW. Off-site, the utility also owned a 10,000-ft^2 site. This site could also be a solar field. Table 4-5 outlines the new site plus the plant site. The off-site location includes a water tank that could house six small wind turbines. As a result, the total power that could be generated at the two sites is over 800 kW, about 20 percent beyond that needed to operate the facility. However, it should be noted that solar power depends on sunlight and storage of power in batteries on the grid. Backup power is required. The Florida Power and Light (FPL) grid can be used for this purpose. Although initial costs to create the new power options were determined to add approximately 40 percent to the price of the plant, the benefit of green power is a long-term payback. By saving 236.5 kW per day, at current electricity prices, the present worth of power costs, and at 6 percent per year over 20 years, nearly $3.5 million is generated, which demonstrates that the conversion is actually cost-effective.

The utility has recently completed renovations to its high-service pumps and added a new ground storage tank, and it is in the process of rehabilitating existing components. At present, the utility is building a 2-mgd potable water plant. Table 4-6 outlines the estimated cost for the facility. The green power options are $4.25 million, or about one third the total costs. It should be noted that the utility bid the facility without the green power requirement. The low bid was $7.5 million, as much of the green power costs were not included. Figure 4-43 is the actual proposed facility. The new nanofiltration plant will include the following:

- Optimized nanofiltration system with areas for chemical storage, membrane cleaning facilities, pipes, pumps, and other appurtenances
- Operations center (400 ft^2 [37 m^2])
- Two offices (each 150 ft^2 [14 m^2])
- Maintenance area (200 ft^2 [18.5 m^2])
- Americans With Disabilities Act (ADA) public access restroom with safety shower (125 ft^2 [11.5 m^2])
- Water quality laboratory (400 ft^2 [37 m^2])

The design incorporates the LEED Gold elements of LEED certification for function, aesthetics, security, and safety. The current plant must remain in service at all times. Operator safety and plant security are important issues. Site vegetation will consist of a mixture of native low- and medium-growth natural vegetation.

Example 2

Example 2 is a regional facility that serves nearly half the population of Broward County, Fla. (Bloetscher et al., 2010a). Current flows are 84 mgd (460 m^3/s). The plant capacity is over 100 mgd. The plant is a secondary activated sludge process with disposal to injection wells and an ocean outfall. A small reuse facility is located on-site that adds filtration and high-level disinfection to the water. The plant site encompasses 160 acres and five treatment modules, plus digesters. The digester gas is flared to the atmosphere. It should be noted that methane has 22 times the greenhouse gas effects of carbon dioxide and is a useful fuel, if it can be captured.

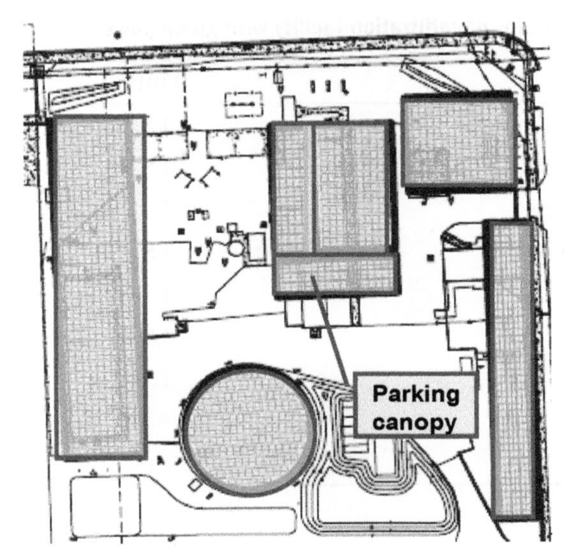

FIGURE 4-42 Location of solar panels on site

TABLE 4-4 On-site power generated

Location	Number of Solar Panels	Power (kW/day)
Building	198	59.4
CW/west	136	40.8
Tank	262	78.6
Pond/west	693	207.9
Parking	162	48.6
Mon. wells	270	81
East	352	105.6
On-site total	2,073	621.9

TABLE 4-5 Total power generated

Source	Area (ft^2)	Power (kW)
Solar field	10,000	125
Wind	10,000	60
Off-site subtotal		185
On-site		621.9
Total system		806.9

TABLE 4-6 Cost estimate for nanofiltration facility with green power

Item Description	Unit	Cost
Site preparation, mobilization	LS	$150,000
Building, equipment	LS	$4,000,000
Filtration, water process	LS	$3,125,000
Storm sewer, grading	LS	$250,000
Power, on-site	LS	$3,500,000
Power, off-site	LS	$750,000
Landscaping	LS	$50,000
Engineering	%	$2,000,000
Total		$13,825,000

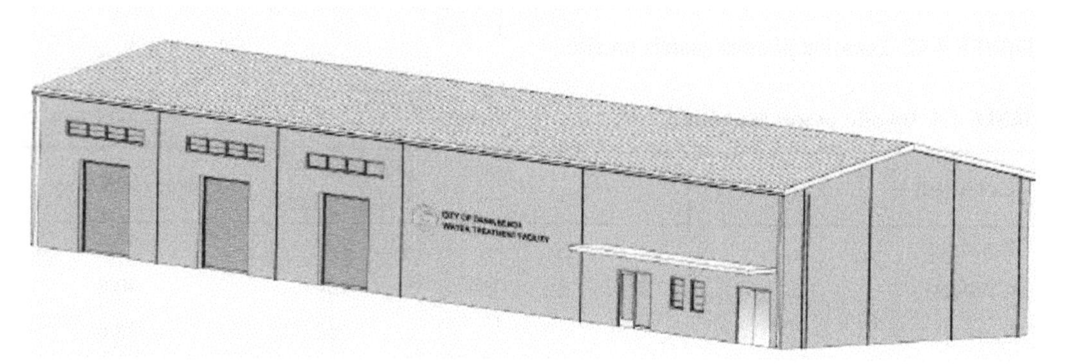

FIGURE 4-43 Actual preliminary design of proposed water plant

At an average a daily demand of 133,000 kW, the plant is the largest power user on the grid in Broward County. The aeration portion of the activated sludge process consumes half of the power needs, while the injection wells consume another 15 to 20 percent. Unlike the prior example, here the presence of on-site digester gas changes the analysis considerably.

A series of items were identified that would improve current power use. The aeration modules could be improved with the installation of fine-bubble diffusers, variable speed drives, and more efficient motors and blowers. The savings are 2 to 5 percent of the total costs. Slip power recovery would improve efficiency for the deep wells. Up to 5 percent improvement could be made with lighting, increased HVAC system upgrades (to 18 SEER plus insulation), and lighting sensors.

A small wind farm could be created on the site and on the buildings. However, south Florida is not in a zone that is conducive to wind energy production since the average wind speed is less than 10 mph. As a result, limited amounts of wind power can be generated. Sunny south Florida is much more conducive to photovoltaic power generation.

The plant produces 23,700 ft^3/s of methane from its anaerobic digesters, so the use of digester gas is a viable option (currently this greenhouse gas is flared). Digester gas options include microturbines and fuel cells. Microturbines operate like the generators that operators are familiar with. Characteristics of fuel cells are listed below:

- Work like a battery
- Require only fuel, air, and water
- Generate DC electricity that is converted to AC electricity through an inverter
- Are more efficient at generating electricity than the combustion process found in cars and power plants
- Have low environmental impact

A methane fuel cell was evaluated; running continuously, it can produce up to 68 percent of the power usage from 21,720 ft^3/s of methane. A present-worth analysis determined that the use of digester gas in fuel cells and microturbines could generate a third of the power demands for the plant. The microturbines generate 97 percent of the energy that fuel cells can, but cost 10 percent of the cost of fuel cells. Note that in both cases, costs are below those of the local power company. The microturbines require minimal maintenance and a limited capital investment. There are no major repair parts and the technology is well developed. Fuel cells require higher capital and higher maintenance costs. Both require "cleaning" the methane of impurities.

The addition of a solar array could increase on-site power generation by another 100,000 kW but the cost is triple the current power costs. Between the solar cells and microturbines, the full power of the plant could be generated on-site. This would not be palatable to the wastewater plant, just as the high cost was not palatable in the prior example. However, the blended (solar plus microturbine) costs are less than the local power costs, which provides an opportunity the utility could pursue.

ASSET MANAGEMENT PROGRAMS

During the past 150 years, the North American water and wastewater industry has grown to the point where extensive water and sewer systems serve most people in the United States and Canada. However, these systems are aging and the rate of replacement of the infrastructure has not kept pace. Pipeline infrastructure in North America, especially, has been built over 100 years, and much of it has passed its useful life (Stadnyckyj, 2010). While this infrastructure is not expected to fail at once, it does need attention.

An asset management decision-making framework is a business process that draws from economics, operations, and engineering to optimize decisions about a broad range of assets. A good asset management system incorporates items that would affect infrastructure reliability (such as quality of the equipment initially installed, quality of construction, adequacy of the design, assessment of ongoing condition, potential for disruption, and vulnerability of failure or public health risks) into an economic assessment of trade-offs among alternative investment options and uses this information to help make cost-effective investment decisions.

USEPA (2002) indicated that long-term maintenance and asset management principles can help the utility staff overcome the tendency to have a short-term, visible focus of most elected bodies, a "key challenge to asset management." Critical assets in visibly poor condition are obvious needs, but buried (hidden) equipment or piping that is near the end of its useful life would also be a priority, especially if failure could create widespread impacts. Equipment that costs more to operate and maintain (e.g., old pumps and motors) should be another area for investigation by the utility.

In addition to aging infrastructure, changes in federal and state clean water and drinking water programs will require additional upgrades to current plants and piping systems, which will require investments in the form of debt and increased rates. As time goes on and the states/provinces realize that infrastructure has deteriorated, there is likely to be an increase in regulatory enforcement, revenue required (meaning higher rates), and need for information to be developed about the assets.

Both the need to replace older infrastructure and these new demands will likely engender resistance from consumers who are facing economic distress, a poor economy, losses of service population, and instability in the debt markets. The result is that utility managers will deal with balancing the needs for system reliability, the need to meet necessary growth, regulatory optimization, and workforce objectives and skills through better management of system assets and by developing tools to enable priority issues to be resolved before to failure.

This is where asset management comes into play. (Note that asset management, capital management, and infrastructure management are all basically the same concept.) It requires that each infrastructure component be deemed a capital asset and allows the appropriate accounting methodologies to be applied to develop information about the potential condition of the system and the likelihood that failure or replacement is pending. Using asset management techniques may benefit utilities constrained by local decision makers reluctant to make investments.

The goals of asset management are to minimize the cost of asset ownership and improve maintenance practices by using the life cycle approach (discussed below). It is important to ensure that appropriate maintenance opportunities are in place, that the infrastructure provided meets the needs and objectives of the utility, and that the required service levels are met by ensuring that the assets are reliable and maintained appropriately.

For most utilities, maintenance activities typically fall into two categories—those that are planned and preventive, and those that are unplanned and reactive. Most daily activities are focused on emergencies and reactive maintenance, which leaves limited time for preventive tasks, such as procuring parts, equipment, or expertise (Gaha and Urquhart, 2005). Preventive maintenance and replacement is less costly than reactive emergency repairs, and better planning can be performed to minimize service disruptions. During emergencies, this is rarely possible. While there are no rules about how much time should be devoted to preventive maintenance, Gaha and Urquhart (2005) indicate that 60 to 80 percent of efforts should be planned, preventive activities, which rarely happens in practice. Budget cuts, loss of experienced staff, complex equipment, and fluctuations in maintenance needs may direct efforts toward reactive maintenance, which comes at a greater risk of system failure than preventive maintenance. As a result, there is limited time for planned, preventive maintenance activities.

A means to help with evaluating asset condition involves life cycle analysis, whereby assets beyond their useful life will increase in priority versus other asset needs. Life cycle analysis should include not only the initial cost of the asset, but the operating and maintenance costs, and the life and expected replacement design (Yarlott, 2010).

In using the life cycle approach, the balance between cost-effective maintenance and (usually higher) initial capital cost can optimize the reliability and minimize risk potential for utility managers. The utility must pursue replacement of the infrastructure in established or stable communities when the asset is no longer economical to operate, is deteriorated to a point where replacement is more cost-effective than repairs, or no longer serves its intended purpose or meets regulatory standards. These established utilities will be faced with many such replacement projects simply due to age, but the number and magnitude of the projects will increase if the utility has not maintained its systems adequately due to funding or other factors. The utility can control the timing of its own projects to a degree, but must establish a program to ensure that the proper projects are constructed to meet operational timelines.

In most cases, the maintenance costs exceed the cost of the infrastructure over the life of the asset. The use of comparative options and appropriate techniques from engineering economics is required to compare multiple options to find which one would provide the best life cycle costing from an economic perspective and then determine whether the least-cost option is the appropriate alternative from a management, supervision, finance, and operations perspective (it may not be since avoiding failure is the underlying goal of all utilities). Economic analysis can also be used to help identify when the asset should be taken out of service. As an example, most police departments have some means to evaluate when patrol cars should be taken off the road because the average annual maintenance costs of older cars will at some point exceed the cost of purchasing a new vehicle and paying for its lower annual maintenance.

Note that economics should be one of the drivers, but *never* the only one, when making decisions on utility assets. An example where return on investment is probably inappropriately used is looking at how quickly a pipe extension into a relatively undeveloped area would receive payback. The payback may be so distant that the present worth of the revenues would not justify the extension. However, dealing with a pipeline extension may give rise to a developer or private utility that the public system would need to acquire, at much higher cost than the extension.

Other examples of this can be found with respect to the future costs of acquiring right-of-way or trying to get down developed streets to provide service to an area. Had the utility installed the pipeline prior to development, those costs could have been avoided. Another example would be using inferior materials that will lead to premature replacement or failure. In this case, consider that digging up pipe often has significant costs even when total failure has not occurred.

Using proper asset management protocols, system components are evaluated and maintained regularly over long periods of time to repair or replace assets prior to failure, when consequences are minimal (Goldwater, 2010). However, it should be noted that even continued, proper maintenance may reach a point where the time and effort expended in the protection of an asset is higher than the cost of the asset (or identifying the fiscal issue) (see Figure 4-44). The most common example is the repair of an asset that no longer has value—adding a repair provides no significant new value. Another example would be finance personnel spending hours tracking down a few pennies of difference in the balance sheet. It is important to be fiscally responsible, but there is a point where the time and effort to track down those pennies exceed the value of the difference (Matchich, 2010).

Annual investments can sometimes be determined, but often they lack the detail to make the information useful. To create an appropriate asset management system, there are a number of tasks that need to be undertaken. The first is an inventory of assets, which involves site visits and review of maps and other records. This means that the initial cost, the initial installation

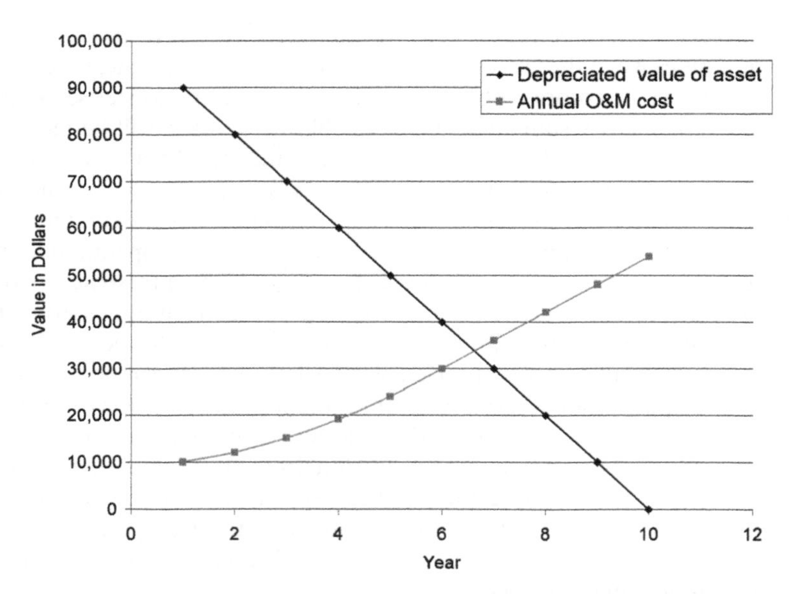

FIGURE 4-44 Graph showing the point where the cost of effort and time is higher than the cost of the asset

date, the quantity of the item, material, and other information would need to be developed. Unfortunately, the exact installation date of a lot of infrastructure cannot be determined easily. However, approximate dates can be determined based on the experience of existing staff. The installation dates can be approximate; often within a few years is sufficient. If the infrastructure is broken into many components, it is likely that true deviations will balance out. The concept of grouping infrastructure into periods to characterize condition and propensity for breaks to gain a perspective on condition is acceptable when more detailed information is lacking. This tool is useful where some idea of development patterns exist. Once estimates of current installation costs and date of installation are calculated, they can be devalued from current.

Most utilities can (and should) compile extensive lists of assets, from valves, meters, services, and hydrants, to piping, pumps, tanks, and treatment facilities. To assess the infrastructure of a utility, three tasks must be accomplished.

1) Identify the Asset

Real property

- Determine utility-owned property and existing facilities from maps and local property records.
- Verify utility-owned property records.
- Visit each site for field verification to determine condition and on-site facilities owned by the utility.
- Verify records of utility insurance documents.
- Compile inventory.

Water and sewer

- Update utility infrastructure (water and sewer) from facilities plan.
- Update mapping records of utility.
- Determine approximate installation date of all infrastructure components via discussion with utility staff and available drawings.
- Compile inventory.

Having a GIS system is useful to this end. Separate maps need to be generated for each asset type to improve efficiency.

2) Determine Condition of Asset

The second step is to determine the condition of the asset at the present time. Condition assessment is the process used by engineers and operators to evaluate the overall value or health of an asset. Visual observation can be used to derive condition data for some assets, such as roads and sidewalks, but buried infrastructure cannot be evaluated without excavation. Few utilities rigorously implement condition assessments or update assessments based on new conditions or knowledge. Condition evaluation is an ongoing issue that can only be developed from conducting field observation, recording the information on work orders, and entering it into some form of maintenance management system. Field conditions may alter the rate of depreciation or indicate that asset deterioration is accelerated over that anticipated from an accounting perspective. Maintenance entities gathering this information may have difficulty transitioning from the situation where they do not record data to a more paper- and information-intensive system.

For many utilities, asset age data are often used to estimate condition. Age may not reflect true conditions, e.g., different pipe materials and local conditions may cause differences in the pipe life cycle. As a result, the process requires judgment, experience, and attention, but any pattern used for condition assessment must be consistent and uniformly applied to ensure comparisons between competing priorities are appropriate.

3) Depreciation of Asset

Once life and condition are determined, the next issue that needs to be developed is a depreciation method. Depreciation does not provide actual value. For example, once a water line is installed, it has only functionality, not value, since no one would dig up the infrastructure. The value is functionality, but depreciation cannot capture this aspect. However, depreciation is performed to provide the utility with an idea of the remaining value of the infrastructure (hence requirements under Government Accounting Standards Board [GASB] 34). In theory, the average age of "properly maintained" utility infrastructure should never exceed half its useful life, meaning that each year investments in capital facilities should be made to replace those facilities whose useful lives have been reached. Rarely does this happen, as the trend seems to be large expenditures sporadically to catch up on obsolete facilities.

Straight-line depreciation is normally the method used for depreciating assets. It is not the only acceptable method; only the easiest to understand. As noted previously, assets tend to depreciate much more slowly in earlier years and then rapidly in the final few years of their lives.

One method of gaining a perspective on overall infrastructure depreciation is to create a table containing the inventory of assets, replacement values, and estimated installation costs using the inflation indexes. Annual inflation rates going back in time are required to determine the asset value at the time of installation, so records can be developed in the absence of full information. However, these data still provide limited information on the appropriate amounts for repair and replacement funding, especially where historical data are sparse. Replacement value of assets of the utility will be substantially more than many expect when compared to the net assets in audit statements, because the value of money has changed considerably with time (see Figure 4-45). For a cumulative perspective, see Figure 4-46. Obsolescence of assets affects the utility's ability to optimize operation because newer technology tends to be more efficient. Obsolete infrastructure should have a value of zero even if it has remaining life.

There are many thoughts on the life of different assets. General estimations by the author follow:

- Concrete structures—100 years, with repairs and refurbishment every 20 to 30 years, depending on the environment.
- Steel potable water tanks—100 years, with repairs and refurbishment every 10 to 15 years, depending on the environment.
- Steel treatment structures—50 years, with repairs and refurbishment every 10 to 20 years, depending on the environment.
- Water mains (except galvanized)—60 or more years (material and soil conditions may shorten the life of pipelines).
- Sewer lines (clay)—50 to 100 years, although slip-lining and other repairs will need to occur every 25 to 30 years.
- Personnel access openings—50 years, longer if ongoing maintenance and protection are pursued.
- Mechanical equipment should be assumed to be viable for 15 years or less.

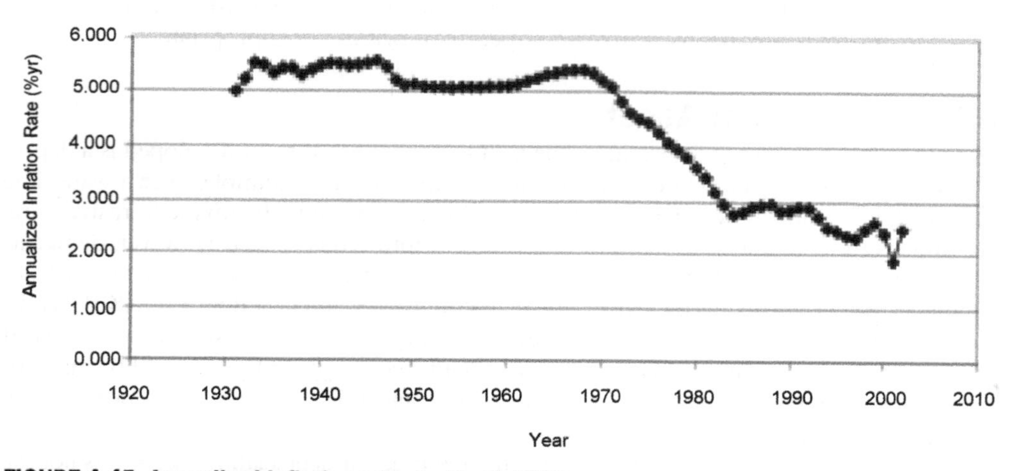

FIGURE 4-45 Annualized inflation rate starting in1930

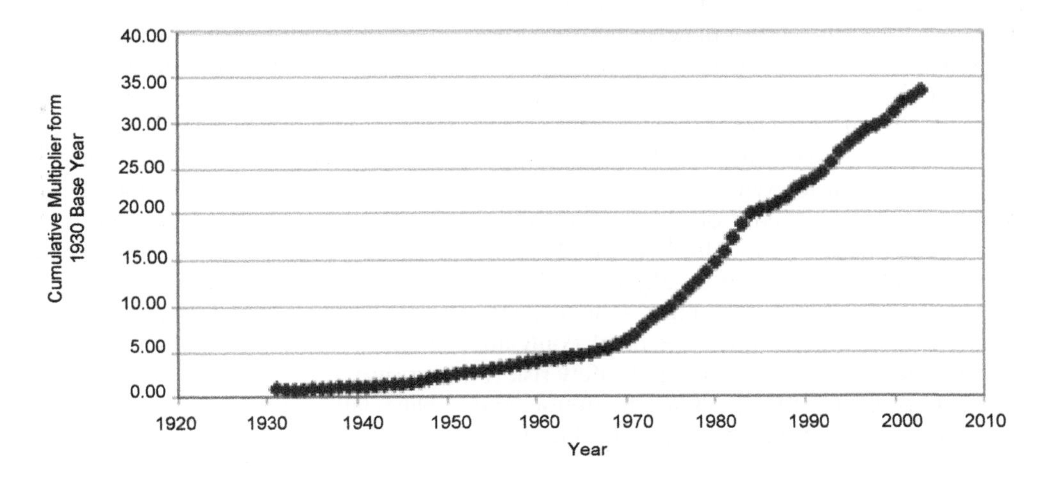

FIGURE 4-46 Cumulative inflation with time

While these values should be adjusted for local conditions, the age of the facilities is relevant as all pipe and facilities have a finite life. Keep in mind that it may be difficult to know whether the life is 10 years or 100 years, which makes a significant difference in a depreciation amount. Certain pipelines may last 100 years, but it is assumed that they will last as few as 60. Rarely would it be appropriate for pipelines to be assumed to last over 100 years, because there is no evidence to indicate that they would last that long given the materials that are commonly used in water and sewer piping systems. Judgment is required. Nonsensical values like pipe life of 250 years should be avoided as they are not helpful to infrastructure valuation.

Risk Assessment

Once condition data and an asset inventory are complete, minimizing the risk of failure from a regulatory reliability or public health perspective is the only real priority of utility managers. To improve asset management, the utility needs to assess the risks of system failure. Understanding failure modes is useful to determine these risks. With a public utility, the confidence of the public is damaged by utility failures, often with a political cost to elected officials and cost of employment to employees.

The techniques for identifying risk are often less robust than people think. There are two factors that need to be evaluated: the likelihood that failure might occur and the impact of that failure (Lovely, 2010). The asset's importance to the utility is crucial in assigning priority when evaluating between competing needs (Yarlott, 2010). Determining the likelihood of failure is difficult, and reflects opinions about the asset condition, age, capacity, and level of service (Lovely, 2010). The first two factors are physical, i.e., older assets and those assets with the most exposure or use may be more likely to fail than newer assets (although this is not always the case). The capacity issue relates to the asset's functionality. At the time it was installed, the asset may have been adequate, but the system has now grown. Level of service is generally related more to public perception of the asset, e.g., too noisy, too many odors, or lacking in safety. Any of these issues might accelerate the need to replace the asset, but none relate to failure unless safety is at issue. Safety may be the highest priority to the public and those whose responsibility is to maintain the system.

Once the utility staff has created an inventory of major assets and determined age, expected life, any extensions to that life, and asset condition, judgment needs to be applied to the likelihood of failure of the asset and the consequences of that failure. Quantifying the likelihood of failure is subjective. For example, galvanized pipe may be more likely to leak or fail than ductile iron even though the galvanized pipe may be newer than the ductile iron. Work order tracking data on the local conditions would help with these judgment calls. Mechanical assets are more likely to fail than pipe. Both examples lead to the consequence issue, i.e., galvanized pipe is usually smaller so the consequences of galvanized pipe failures are less than for ductile iron pipe. A high-service pump failure may shut down service for many customers, while most pipe breaks only affect a few. Consequence evaluation should take into account public health and safety risk (top priority), regulatory compliance, property damage, cost of repair, service disruption, and public relations (Lovely, 2010).

A matrix of consequences and likelihood of risks for any given utility will identify critical infrastructure. Vulnerability assessments are intended to provide this information, although their scope may be more limited to critical assets of the water system. A short example of this may be helpful. Table 4-7 shows five assets, each showing age, number of people served, likelihood of failure, and the score for consequences (1 being high and 5 being low). Figure 4-47 shows a matrix that is another means to analyze risk. Dots in the figure indicate various assets of the utility and the risk perceived to exist, which helps identify assets critical to utility functionality.

The outcomes of asset management should include (Goldwater 2010):

- Meeting goals of regulatory compliance, service reliability, public health, and customer service. This will likely require creation of a set of goals to guide utility managers in identifying asset maintenance and replacement priorities.
- Assembling an inventory of utility assets.
- Identifying threats to vulnerable areas of the utility.
- Scheduling maintenance or replacement of assets based on risk and consequences.
- Assessing effectiveness of the actions taken.

Decisions to replace an asset, contrary to some engineering economics books, should be based on more than just cost, although cost is an important consideration. An asset may be old, but continue to provide good service with a minimum of downtime. Operator comfort and minimizing disruption are worthy goals to consider along with economics. For instance, a cast-iron water pipeline with leaded joints installed during the 1930s will tend to leak under major roadways with truck traffic as a result of vibration of the trucks. If this roadway were the only major feed to the community from the water plant and its leak history indicated joint leaks were a problem, the utility staff could immediately identify this pipeline as a candidate for replacement in the near future as a result of its age (over 50 years), its material (cast iron with lead joints that leak), its history (leaks), and the potential risk of a failure (no water to the community). This is a simple example, but the thought process can be applied to virtually every asset the utility operates.

TABLE 4-7 Example of means to prioritize asset replacement

Asset	Age	Customers Served	Condition	Probability of Failure	Probability of Failure	Consequences of Failure	Consequences of Failure	Risk	Priority
Raw ductile iron water main	25	50,000	Good	Low	0.1	High	1	0.1	3
Local galva-nized pipe	15	150	Poor	High	0.5	Low	0.1	0.05	4
High-service pump	15	35,000	Fair	Medium	0.25	High	0.75	0.1875	1
Water tank	50	10,000	Poor	Medium	0.2	Low	0.1	0.02	5
Ductile iron pipe under major thoroughfare	40	15,000	Fair	Medium	0.2	High	0.9	0.18	2

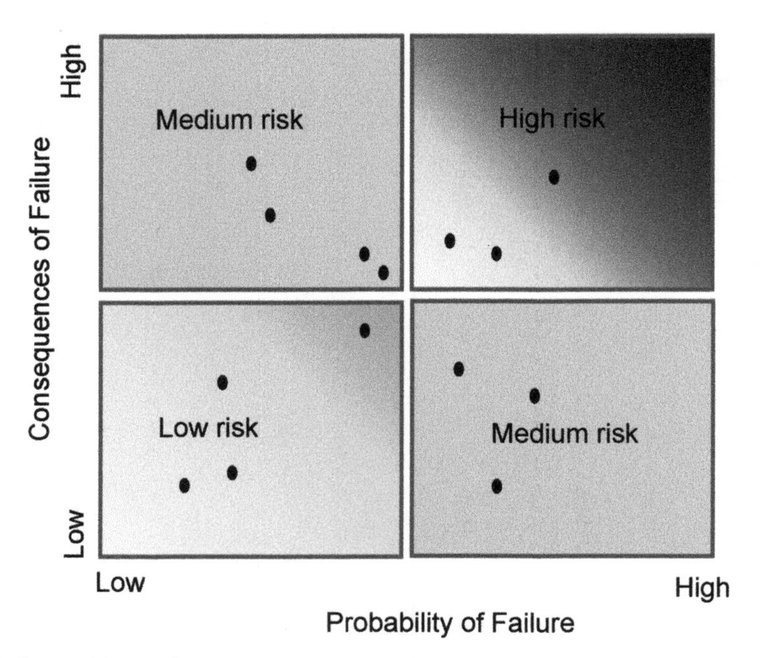

FIGURE 4-47 Asset risk matrix

A second example is shown in Figure 4-48. This figure shows the costs in the life of two assets—the original $550,000 machine and a replacement machine that costs $750,000 but can produce 20 percent more. What the graph shows is that after four years, the original machine should be replaced with the new machine as the annual value is higher for the new machine. The break-even point is seven years.

Initial capital costs are spread over time through borrowing, but throughout the life of the asset, the operations and maintenance costs will continue to climb. Asset management decisions must rely on the following:

- Work order tracking systems that note the location of breaks/repairs/leaks/reactive maintenance on pipes and equipment
- Mapping of pipeline breaks
- Operator knowledge

The decision to replace capital assets is made on the basis of one or more of the following:

- More economical to replace than repair
- A critical asset with a history of failure
- Obsolescence as a result of technology (i.e., more efficient to run with new technology)
- Repair parts unavailable
- Water quality goals are not met

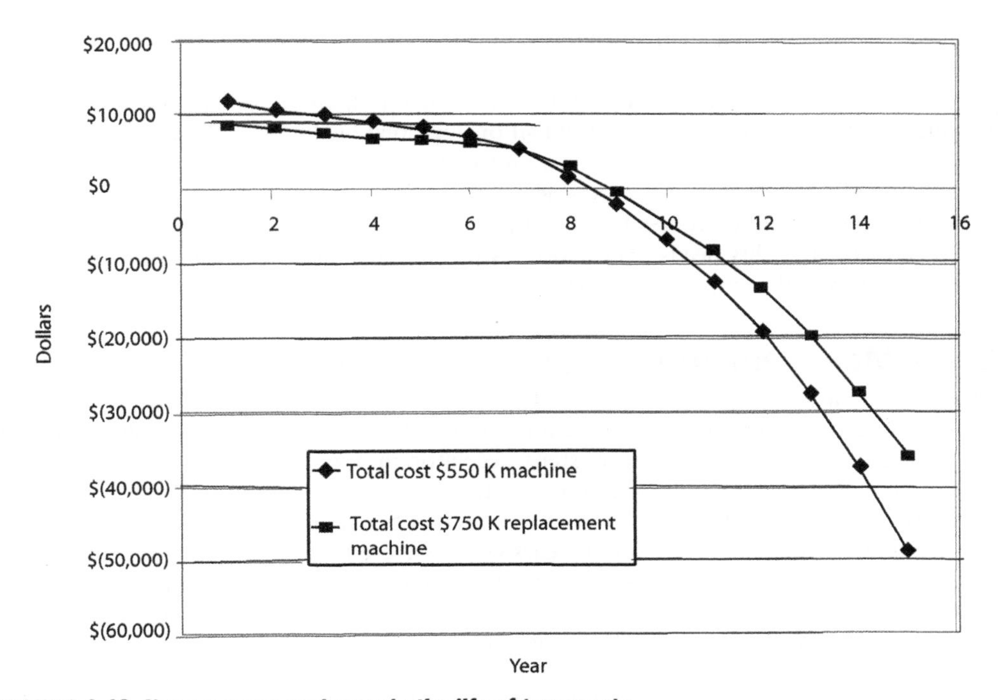

FIGURE 4-48 Net revenues each year in the life of two assets

In determining the appropriate amount of money to set aside for system development, repair, or replacements, historical data may not be representative, particularly if maintenance of the system has been neglected or deferred over time. Too often the reason given for an expenditure is as a response to new regulations, consent orders, or ongoing and persistent failure of the component, which indicates the asset is beyond its useful life. Proper asset management could avoid the magnitude of some of these borrowed expenditures and their associated rate impacts to customers.

Asset management involves finance, operations, planning, engineering, and the management and political will to ensure the appropriate infrastructure is in place. These functions will have to interface with advanced procurement and project delivery and affordability of costs to the customer (Monsma 2010). Depending on utility staffing concerns, it may be beneficial to replace the asset through borrowing. (In regard to borrowing, note that no asset should be acquired with borrowed funds and then retired prior to paying off the loan, with the one following exception: if an old utility system is acquired and merged into a larger utility, there may be significant cost efficiencies associated with retiring the old facilities. In most cases, the avoided costs, i.e., costs to upgrade the acquired system, are considered.)

For the asset shown in Figure 4-48, the break-even point is year 4, and a strictly economic decision could be made to replace the asset at this point (e.g., vehicles are commonly replaced early in their lives to avoid maintenance costs). However, the figure shows that the cost of the asset in years 3, 5, and even 6 are only marginally higher, and therefore potentially could be deferred if need be. The replacement of an asset should be included in a capital plan so the funding is available (year 4 in the asset in Figure 4-48). Delaying the replacement beyond year

4 may not increase the annual maintenance costs dramatically, but the potential for failure, the one event utilities strive to avoid most rigorously, increases significantly.

Ultimately, on a yearly basis, the utility staff should prepare the design and construction budgeting documents for major replacement or upgraded facilities for the coming year and should reprioritize the ensuing years as a part of the capital planning process. The capital improvement planning process is a product of the future directions, issues, and goals identified during the planning process for the utility system. It necessitates the identification of specific project or facility needs to develop existing or anticipated deficiencies on the system, maintain or more efficiently operate the system, and provide better service to the customers.

Asset Management Example

This example involves a small utility system that currently owns and operates two wells and one water treatment plant. The plant is a lime softening facility. The eastern accelator was constructed in 1952 and the western accelator in 1963. Piping for a third reactor appears to exist west of accelator number 2. Both accelators were refurbished in 1992 to extend their useful life. The condition of the plant is better than one would expect for its age. Straight-line depreciation over 50 years would be used for plant improvements.

Based on the assumptions described in this section, conditions are determined for each asset. As in most other cases, the age of the infrastructure is uncertain, but can be estimated. For example, the bulk of the older part of the water system was constructed in the early 1950s. These pipes have been assigned dates from 1951 to 1955. A later addition was made in the late 1950s/early 1960s. These areas were assigned dates from 1957 to 1962. Each corresponds with treatment plant construction (1952 and 1963). A complete update was undertaken for this exercise. Estimated dates were used for pipe installation. A reinventory of the water and sewer piping and updating of the system were conducted. The city has eliminated most of its 2-in. lines over the past 10 years by replacing them with 6-in. lines. Where water lines were constructed, it was assumed that the sewer lines and storm drains were constructed 10 years later. Therefore the same dates for installation are used. This matches with an orderly development pattern for the utility.

Next, actions were taken to:

- Define the replacement value of the asset based on current construction data
- Define the installation cost by deflating the replacement value to the years estimated for installation
- Determine the life of the asset
- Use straight-line depreciation to depreciate the asset from the installation date to current date

The same inflation values were used for water and sewer piping. The life of this infrastructure can be estimated from 15 to 50 or more years. Based on experience in other communities and engineering judgment, an estimate of 15, 30, or 50 years was used for this infrastructure. Tables 4-8 through 4-12 outline portions of the asset tables developed for the city. Each increment of water and sewer systems (block to block) was separated and noted on the tables. Table 4-13 summarizes the infrastructure of the utility.

The utility's current situation presents two problems—lack of records on contracts, including their value, and lack of dating of improvements. Many improvements are over 30 years old, yet belong in the asset inventory, such as piping, parks, and buildings. The dates and original construction values of certain assets are unknown. Example 1, presented earlier in this chapter, is not the preferred way to accomplish this. In going forward, the actual cost should be used.

GASB 34

Several methods can be used in conducting condition assessments. The intent is to create a hierarchy that will lend itself to prioritization of repair and replacement needs, especially in comparison with new facilities or growth-related requirements. To do a condition assessment, some idea of the age and the physical condition needs to be generated. Materials and information on materials throughout the system are also needed. Specific information on condition is often hard to judge, so many utilities pursuing asset management will use some sort of priority scale (such as 1 to 5, mentioned earlier) or something similar that would indicate a relative condition compared to infrastructure.

Asset management has become a much larger player in local utilities as a result of GASB 34. The Governmental Accounting Standards Board, or GASB, is an independent, private-sector, not-for-profit organization that establishes and improves standards of financial accounting and reporting for US state and local governments. GASB standards aid in determining the ability of governmental entities to provide services and repay their debt. Governments and the accounting industry recognize the GASB as the official source of generally accepted accounting principles (GAAP) for state and local governments. In line with this mission, GASB issues standards that

- Result in useful information for users of financial reports (e.g., owners of municipal bonds and members of citizen groups) and
- Guide and educate the public, including issuers, auditors, and users about the implications of those financial reports (refer to www.gasb.org).

GASB 34 set requirements for asset valuation and depreciation. Prior to the adoption of GASB 34, there were no specific requirements on how assets were to be added to the balance sheet. GASB 34 is intended to create a consistent reporting mechanism. While GASB 34 is relatively new in the United States, Australians have been using it for some time in the form of the AAS 27 requirement. GASB 34 is modeled after the Australian requirement. The American Public Works Association, AWWA, and other organizations are in support of the GASB 34 reporting requirements. GASB 41 addresses a few issues that were not fully addressed in GASB 34.

GASB 34 works in two ways. It forces utilities to acknowledge the assets on their balance sheets as well as the depreciation of assets with time. In a properly run utility, the assets should continue to increase with time as investments are made in additional infrastructure, so that the ongoing audits that are conducted each year should show a general increase in net asset value with time. Obviously, if significant borrowing goes on at the same time, the liabilities associated with the assets through borrowing are also recorded.

TABLE 4-8 Partial listing of water line infrastructure for a small utility

ID Link	Installation Date	Inflation %	Life	Cost New (in dollars)	Lin ft	Size	Cost/lin ft	Annual Amount Installed	Cost Installed (in dollars)
207B	1947	5.461	50	22,050	490	12	45	490	1,123
207D	1951	5.106	50	7,875	315	6	25	315	591
130	1952	5.079	50	50,400	2,520	2	20		4,028
34	1952	5.079	50	11,000	500	4	22		879
207E	1952	5.079	50	6,160	280	4	22		492
36	1952	5.079	50	14,000	560	6	25		1,119
37	1952	5.079	50	117,250	4,690	6	25		9,371
38	1952	5.079	50	84,000	3,360	6	25		6,714
40	1952	5.079	50	70,875	2,835	6	25		5,665
42	1952	5.079	50	17,500	700	6	25		1,399
43	1952	5.079	50	17,500	700	6	25		1,399
44	1952	5.079	50	18,375	735	6	25		1,469
46	1952	5.079	50	17,500	700	6	25		1,399
47	1952	5.079	50	17,500	700	6	25		1,399
207C	1952	5.079	50	37,800	840	10	45		3,021
255	1952	5.079	50	45,000	1,000	10	45	18,720	3,597

TABLE 4-9 Partial listing of fire hydrant assets

ID	Installation Date	Inflation %	Life	Cost New (in dollars)	Number	Installation Cost (in dollars)
Dixie Highway	1966	5.399	50	2,500	1	357
SW 2nd Ave	1966	5.399	50	7,500	3	1,072
SW 3rd Ave	1959	5.108	50	10,000	4	1,117
US 1	1959	5.108	50	12,500	5	1,396
SW 4th Ave	1952	5.079	50	17,500	7	1,399
SW 2nd Ave	1952	5.079	50	22,500	9	1,798
US 1	1952	5.079	50	22,500	9	1,798
SW 1st Ave	1952	5.079	50	2,500	1	200
JA Ely	1968	5.412	50	17,500	7	2,766
Phippin Waiters	1968	5.412	50	25,000	10	3,952
SW 3rd Street	1969	5.35	50	2,500	1	425
SW 7th Ave	1969	5.35	50	5,000	2	850
Alley SW 2 T / DBB	1969	5.35	50	2,500	1	425
SE 4th Ave	1984	2.74	50	22,500	9	13,463
SE 3rd Ave	1977	4.067	50	32,500	13	11,528
SE 2nd Ave	1954	5.085	50	32,500	13	2,860
SE 5th Ave	1975	4.428	50	27,500	11	8,174
SE 11th Terr	1979	3.801	50	10,000	4	4,085
SE 10th St	1979	3.801	50	12,500	5	5,106
SE 1st Alley	1955	5.086	50	2,500	1	231

TABLE 4-10 Partial listing of sanitary sewer piping for a small utility

Street	From	To	Estimated Year Built	Inflation %	Sewer Lines	Sewer Size	Cost per lin ft	Cost per Pipeline	Sanitary Sewer Personnel Access Openings	Cost New (in dollars)	Cost Installed	2005 Value
Alley SW 15th/16th	SW 4th Ave	US 1	1981	3.428	1,903	8	38	72,314	7	81,064	38,618	20,081
Alley SW 14th/15th	SW 4th Ave	US 1	1981	3.428	1,911	8	38	72,618	7	81,368	38,763	20,157
Alley SW 13th/14th	SW 4th Ave	US 1	1981	3.428	1,978	8	38	75,164	7	83,914	39,976	20,787
Alley SW 12th/13th	SW 4th Ave	US 1	1981	3.428	1,011	8	38	38,418	7	47,168	22,470	11,685
			1981	3.428	800	10	44	35,200		35,200	16,769	8,720
Alley SW 11th/12th	Dixie	US 1	1981	3.428	1,328	8	38	50,464	5	56,714	27,018	14,049
Alley SW 11th/10th	Dixie	US 1	1981	3.428	947	8	38	35,986	4	40,986	19,525	10,153
Dixie Highway	SW 4th Ave	US 1	1973	4.62	1,112	8	38	42,256	7	51,006	13,158	4,737
SW 2nd Ave	Sheridan	Dixie	1973	4.62	2,413	8	38	91,694	0	91,694	23,654	8,515
US 1	Sheridan	Dixie	1973	4.62	50	10	44	2,200	0	2,200	568	204
Alley SW 9th/Cemetery	SW 4th Ave	SW 2nd Ave	1973	4.62	1,320	8	38	50,160	6	57,660	14,874	5,355
Alley SW 8th/9th	SW 4th Ave	SW 2nd Ave	1973	4.62	1,102	8	38	41,876	5	48,126	12,415	4,469
Alley SW 7th/8th	SW 4th Ave	SW 2nd Ave	1973	4.62	1,603	8	38	60,914	6	68,414	17,648	6,353
SW 7th St	SW 4th Ave	SW 2nd Ave	1973	4.62	1,554	8	38	59,052	6	66,552	17,168	6,180
SW 6th St	SW 4th Ave	US 1	1973	4.62	1,452	8	38	55,176	6	62,676	16,168	5,821
SW 5th St	SW 4th Ave	US 1	1973	4.62	1,558	8	38	59,204	6	66,704	17,207	6,195

TABLE 4-11 Partial listing of sanitary sewer laterals for a small utility

ID	Installation Date	Inflation %	Life (years)	Cost New (in dollars)	Number	Cost Installed (in dollars)	2005 Value (in dollars)
	Service Lines						
1963 SS Services	1963	5.317	50	105,999	303	13,345	2,135
1964 SS Services	1964	5.355	50	8,337	24	1,090	196
1965 SS Services	1965	5.399	50	57,774	165	7,833	1,567
1966 SS Services	1966	5.412	50	33,265	95	4,732	1,041
1967 SS Services	1967	5.412	50	—	0	—	—
1968 SS Services	1968	5.350	50	15,929	46	2,570	668
1969 SS Services	1969	5.216	50	—	0	—	—
1970 SS Services	1970	5.105	50	—	0	—	—
1971 SS Services	1971	4.816	50	—	0	—	—
1972 SS Services	1972	4.620	50	137,958	394	34,018	11,566
1973 SS Services	1973	4.502	50	147,838	422	39,451	14,203
1974 SS Services	1974	4.428	50	30,430	87	8,662	3,292
1975 SS Services	1975	4.244	50	87,364	250	27,281	10,912
1976 SS Services	1976	4.067	50	6,428	18	2,191	920

TABLE 4-12 Listing of real property

Label on Map	Account Number Property Appraiser	Location	Lot Size in Records	Impervious Area (ft^2)	Use of Property	Initial Land Value 1980 (in dollars)	Initial Bldg Value 1980 (in dollars)
					Restrooms		147.9
J	514201028250	101 N BEACH ROAD	62,850	62,000	Building	1,255.1	
J	514201022920	101 N BEACH ROAD	78,705	6,000	Intracoastal Waterway	65.9	
J	504236000031	101 N BEACH ROAD	70,000	0	Beach	42.0	
J	504236000032	101 N BEACH ROAD	91,000	0	Beach/marina/boat slips	68.9	11.8
J	504236000050	101 N BEACH ROAD	9,700	1,739	Dania Beach Grill ?	18.6	46.8
J	504236000040	101 N BEACH ROAD	197,000	56,700	S end of beach	366.8	
J	504236000041	101 N BEACH ROAD	41,300	31,100	End of FAU bldg?	203.7	
K	504234012810	24 SW 5 AVE esmt	1,250	0	Esmt	0.9	
L	504234230011	NW 7th Ave	2,533	0	Res lot	1.5	
M	504235160020	MARINA	54,600	16,476	Harbourtown Marina	63.6	
					IT Parker		
M	504235160010	1000 NE 3 ST MARINA	809,000	374,000	Harbourtown Marina	1,161.9	546.0
N	514203070120	SE 2nd Ave	5,000	2,250	SE 2nd Ave ROW	8.8	
N	514203070340	SE 2nd Ave	10,000	4,500	SE 2nd Ave ROW	8.8	
N	514203070680	SE 2nd Ave	50,000	2,250	SE 2nd Ave ROW	8.8	
P	514203360101	ROW-PS?	100	0	Not clear—tiny area side bldg	0.1	
Q	514203360172	ROW	750	0	Alley rear of 733 SW 4 ST	0.2	
R	No number	Park	24,500	7,500	Mullikin Park	22.1	

TABLE 4-13 Summary of infrastructure assets for a small utility

Item	Amount	Units	Value New (in dollars)	Value Installed (in dollars)
2-in. and under water main	15,085	lin ft		
4-in. water main	8,760	lin ft		
6-in. water main	213,032	lin ft		
8-in. water main	80,220	lin ft		
10-in. water main	5,025	lin ft		
12-in. water main	34,860	lin ft		
14-in. water main	2,345	lin ft		
16-in. water main	910	lin ft		
TOTAL MAINS	360,237	lin ft	10,644,635	4,926,963
Elevated water tank	1	ea	350,000	17,821
Water treatment plant	3	mgd	6,000,000	956,046
Land		ac	500,000	220,976
Shallow Wells	2	ea	682,650	388,466
⅝- or ¾-in. services	4,352	ea		
1-in. services	142	ea		
1 ½-in. services	80	ea		
2-in. services	77	ea		
3-in. services	4	ea		
4-in. services	8	ea		
TOTAL services	4,664	ea	1,632,400	732,417
Meters	4,664	ea	287,805	187,402
Fire hydrants	402	ea	1,005,000	414,220
TOTAL WATER	68.2	mi		7,844,312

mi = miles of pipe

To meet the GASB 34 requirements, the following assumptions must be made:

- All assets included in the inventory are of significant value and long life.
- With regard to landscaping, only landscaping installed and maintained by the city was included.
  - The inventory captured all assets, despite age, as well as possible, with some estimate of age, assuming they were initially installed as having significant value (e.g., utility piping that may be as much as 50 years old).
  - The asset value should tie to the "Net Plant Assets" in the annual audit.
  - Disposable infrastructure, such as signs, were deemed expenses, not capital assets, regardless of cost.
  - Since most infrastructure age cannot be accurately determined, averaging was employed where necessary.
  - Property records were used to value land or real property, as discussed previously.
  - Pipeline and utility infrastructure was based on replacement value, deflated, and depreciated for estimated installation dates using bid information and engineering judgment for new improvements.
  - Roadway infrastructure will be based on condition surveys and devalued from replacement values determined from bid information for new improvements, discussion with engineers, and engineering judgment for consistency among municipalities.
  - Depreciation will be straight line for all utility improvements.
  - Depreciated value for roadways will be based on the following formula: right-of-way value plus the remainder of pavement and base multiplied by present cost of installation.
  - Sidewalks will be assumed to have an average of 70 percent of current value remaining given that over half are more than 30 years old, and the remainder are less than 10. If a sidewalk has a life of 50 years, this seems to be a reasonable approximation (Bloetscher, 2003).

The valuation is based on using the depreciated value of the replacement value for infrastructure, unless otherwise noted. The results of the valuation methods use the condition assessment for roadways and the age of assets for utility infrastructure.

VULNERABILITY ASSESSMENTS

Water systems were required to conduct vulnerability assessments of their utility systems per the Public Health Security and Bio-terrorism Preparedness and Response Act of 2002 (PL 107-188), approved by Congress in the wake of the Sept. 11, 2001, incidents. The intent of the act was to force water systems to identify key system components that are vulnerable to threats and create plans to protect the assets and guard against those threats. The act requires the following to be reviewed:

- Piping and conveyance systems
- Physical barriers to access of facilities

- Raw water collection systems
- Treatment facilities
- Storage facilities
- Electronic or computerized systems, including SCADA
- Chemical handling

Systems serving between 3,300 and 10,000 customers completed this vulnerability assessment by June 30, 2004, and then completed the associated emergency response plan by Dec. 31, 2004. Self-assessment tools have been created by a number of agencies to facilitate the vulnerability assessment. The goals of a water and sewer system to be used in developing the vulnerability assessment include:

- Provide safe water quality
- Provide sufficient water quantity to meet community needs
- Provide adequate pressure for fire protection

The methods and resources used in this assessment of facility vulnerability involved the following:

- Current staff experience with the utility's water and sewer system
- Consultant experience with water and sewer systems
- Neighborhood and location of facilities
- Engineering judgment

Vulnerability Example

The utility in this example has a series of critical assets. The Sandia RAM-W software system was used to help with identifying severity of risk. For the most part, since the sanitary sewer system is collection only, potential impacts to the sewer system are limited to hazardous materials entering into the system that might explode, causing damage to houses and the street. The water system is different. Table 4-14 outlines the critical assets of the utility system with associated priority. The high-service pumps are the highest-priority items as a result of the interconnect with a neighboring system that delivers water to the site. While the interconnect permits water supplies to the city to continue, the high-service pumps are required to meet demand and fire flow goals. Loss of the high-service pumps would appear to pose a threat to meeting the appropriate water quantity and fire pressure goals. The next priority would be the current treatment system.

Table 4-15 outlines the current protections offered for the critical assets identified in Table 4-14 and likelihood of damage if attacked. Since most of these facilities are located at the water plant site, the current security measures are noted. It should be noted that the water plant site is gated with controlled access at all hours. However, the current system is not restricted to water plant employees only—any city employee can gain access. This has been identified as a concern. Facilities located outside are more vulnerable than those that are not; however, building damage may have collateral damage.

TABLE 4-14 Critical asset priority

Facility	Priority	Value	Factor
High service pump #1	High	3	0.9
High service pump #2	High	3	0.9
High service pump #3	High	3	0.9
High service pump #4	High	3	0.9
Reactor #1	High	3	0.8
Reactor #2	High	3	0.8
Well G	Medium	2	0.4
Well H	Medium	2	0.4
Well I	Low	1	0.1
Bulk water line	Medium	2	0.4
Raw water line	Medium	2	0.4
Generator	High	3	0.8
FPL service	High	3	0.8
Cl regulators	High	3	0.8
Cl tanks	High	3	0.8
Sodium hypochlorite tanks	Low	1	0.1
HSiF6 tank	Low	1	0.1
Phosphate tanks	Low	1	0.1
Polymer tank	Low	1	0.1
Lime silo	High	3	0.8
Acid storage	Medium	2	0.4
Lab chemicals	Low	1	0.1
170,000 clearwell	High	3	0.8
200,000 clearwell	High	3	0.8
250,000 elevated tank	Low	1	0.1
Water distribution system	Low	1	0.1
Backwash pond	High	3	0.8
Monitoring well 1	Medium	2	0.4
Monitoring well 2	Medium	2	0.4
Monitoring well 3	Medium	2	0.4
Water plant building	High	3	0.9
Administration building	Medium	2	0.4
Fuel pumps	Medium	2	0.4
Chlorine building	High	3	0.8
SCADA	Low	1	0.1

There are four classes of individuals who pose a threat to the facility. These are outlined in Table 4-16 based on discussions with officials in other water systems and on Sandia data. Vandalism and insider sabotage are the most-studied threats to the plant. The former is typically the result of the activities of bored teenagers and creates cosmetic damage. Insiders would sabotage the system but the likely point would be revenge against supervisors or maintenance issues they feel have been unaddressed. In either case, it is unlikely that the damage would be more than an inconvenience.

TABLE 4-15 Current site protection and likelihood for damage

Facility	Protection Offered	Vulnerability	Vulnerability Likelihood
High service pump #1	Secured WTP site	Exposed	Accessible outside to damage
High service pump #2	Secured WTP site	Exposed	Accessible outside to damage
High service pump #3	Secured WTP site	Exposed	Accessible outside to damage
High service pump #4	Secured WTP site	Exposed	Accessible outside to damage
Reactor #1	Secured WTP site	Exposed	Contamination threat
Reactor #2	Secured WTP site	Exposed	Contamination threat
Well G	In fenced offsite building	Exposed	Contamination threat
Well H	In fenced offsite building	Exposed	Contamination threat
Well I	Does not exist now	Exposed	None—offline
Bulk water line	Secured WTP site	Buried	Inaccessible
Raw water line	Offsite1	Buried	Inaccessible
Generator	Secured WTP site	Inside WTP building	Limited mech disruption
FPL service	Secured WTP site	Exposed	Accessible outside to damage
Chlorine regulators	Secured WTP site	Inside chlorine building	Limited mech disruption
Chlorine tanks	Secured WTP site	Inside chlorine building	High chemical impact
Sodium hypochlo-rite tanks	Secured WTP site	Exposed	No real disruption opportunity
HSiF6 tank	Secured WTP site	Exposed	No real disruption opportunity
Phosphate tanks	Secured WTP site	Exposed	No real disruption opportunity
Polymer tank	Secured WTP site	Exposed	No real disruption opportunity
Lime silo	Secured WTP site	Exposed	Accessible outside to damage
Acid storage	Secured WTP site	Exposed	None at present
Lab chemicals	Secured WTP site	Inside WTP building	No real disruption opportunity
170,000 clearwell	Secured WTP site	Buried	Contamination threat
200,000 clearwell	Secured WTP site	Buried	Contamination threat

Note: Water treatment plant (WTP) site has fencing and gate access and therefore has been termed "secured." However, access to site is readily available to non-water-plant employees, a situation to be addressed with the plant upgrades.

TABLE 4-15 Current site protection and likelihood for damage (Continued)

Facility	Protection Offered	Vulnerability	Vulnerability Likelihood
250,000 elevated tank	Secured WTP site	Exposed	None—Offline
Water distribution system	Secured WTP site	Buried	Inaccessible
Backwash pond	Secured WTP Site	Exposed	High potential for contamination
Monitoring well 1	Secured WTP site	Exposed	Contamination threat
Monitoring well 2	Secured WTP site	Exposed	Contamination threat
Monitoring well 3	Secured WTP site	Exposed	Contamination threat
Water plant building	Secured WTP site	Exposed	Collateral damage
Admin building	Secured WTP site	Exposed	Personal injury
Fuel pumps	Secured WTP site	Exposed	Collateral damage
Chlorine building	Secured WTP site	Exposed	Collateral damage
SCADA	Secured WTP site	Inside WTP building	Limited mech disruption

Note: Water treatment plant (WTP) site has fencing and gate access and therefore has been termed "secured." However, access to site is readily available to non-water-plant employees, a situation to be addressed with the plant upgrades.

TABLE 4-16 Classifications of threat

Threat Group	Group Included	Motive	Damage Effect	Likelihood
Insider	Employees	Revenge, political damage	Minimal	Slight
	Ex-employees	Revenge	Minimal	Slight
Outsiders	Political activitsts	Political change	Limited	Minimal
	Vandals	None	Minimal	Minimal
	Resident terrorists	Political change	Limited	Minimal
Cyber	Cyber-terrorists (Hackers)	Games, political change	Limited	Minimal
Terrorist	Foreign terrorists	Acts against the United States	Potentially significant	Minimal

Source: Sandia RAM-W use in other water systems

Terrorists, whether cyber, political, or international, are unlikely to select a small system as a target as they desire more publicity than would be generated. To create much impact on the system, the compound would need to be taken over, which would likely be noticed by neighbors, frequent patrols, or passersby on the road out front. Downing a water tank, which would generate publicity but have little more than cosmetic impact, is the most likely target. Damaging facilities would cause the city to buy water from a neighboring utility, which would obfuscate the damage.

Table 4-17 outlines the likelihood of a threat to the plant. As noted, vandalism is the most likely event. The cost of the damage would be minimal. Given there have been no existing reports of vandalism or other incidents at the plant in the past three years, based on engineering judgment, the likelihood of a threat has been assigned the value of 10 percent. Outsiders, hackers, and terrorists would be magnitudes smaller in size. Foreign terrorist threats were estimated at 1 in 1 million. Table 4-18 outlines the definition of the potential loss on the system. This is required to define the consequences or risk of an impact. The same factors were used by other neighboring cities.

Likewise, the probability of a threat of entry to the site was evaluated depending on current security on hand. These were divided into fenced areas, restricted-access areas, buildings, and SCADA monitoring. These factors ($1–P_e$) are noted in Table 4-19. At the present time, the utility has very limited SCADA capabilities. As a result, the potential for risk via SCADA or computer hacking is minimal and will not be considered further in the threat analysis.

Table 4-20 identifies the total risk calculation based on the prior data collected and analyzed with priorities. The bold items include everything that has a likelihood exceeding a 1 percent potential for incident. Table 4-21 identifies the relative priority of potential countermeasures based on the facility in question. The high-service pumps are the highest risk priority. Without them the city will fail to meet most of its objectives for water supply, fire flow, and water quality. As a result, this area should be protected with backup parts and equipment. Table 4-22 outlines how each of these facilities might be affected and the objectives that might be frustrated.

Based on the results of the prior tables, Table 4-23 outlines measures that should be undertaken to improve protection of the facilities based on the likelihood of damage to the facility and its importance in continued operations. Specific projects are as follows:

- Security cameras, motion detectors, and SCADA should be provided at the water plant site and remote well sites. The video feed should be directed to the police department as well as the water plant during nonbusiness hours. These modifications should be made as a part of the water plant upgrades.
- The water plant site should be cordoned off from the rest of the public works compound. This is planned as a part of the water plant upgrades.
- Storage is required per the Broward County Public Health Unit. Ten state standards recommend that storage equal average daily flows. The city currently has only 380,000 gal of storage on-site and none in the distribution system. A 1.5-mil-gal ground storage tank is proposed to meet this need. Having storage will provide help during short-term disruptions and provide a place where Hollywood can discharge its incoming treated water.
- Gas chlorine is currently used on the site. The system is over 20 years old and given security and risk issues, the city desires to move to an on-site generator system.
- Fuel tanks are in the middle of the water plant compound. The tanks should be moved with the rest of the public works traffic and cordoned off for protection.

Most of this work can be done at relatively minor cost increases from the current planned improvements. All of the suggested security measures add to less than $250,000, with the exception of parallel high-service pumping, which adds a similar amount to the total.

Table 4-24 lists a number of security suggestions from plant staff, some of which have already been implemented by the city.

TABLE 4-17 Threat probability (P_A)

Threat Grouping	Likelihood	Probability
Insider	Slight	0.1
Outsiders	Minimal	0.005
Cyber	Minimal	0.001
Terrorist	Minimal	0.000001

TABLE 4-18 Consequence identification (*C*)

Loss		High Loss	Medium Loss	Low Loss
Economic loss		$250,000	$100,000	$25,000
User health		500 people impacted	5	0
Derivation of loss		48 hrs +	24–48 hr	0–24 hr
Value	High	0.9 or 0.8		
	Medium		0.4	
	Low			0.1

TABLE 4-19 Protection factor (P_e)

Prevention Measure	Likelihood of Success	Likelihood of Detection	Speed of Response	$1–P^e$ Factor
Fencing/gate	0.5	0.1	0.2	0.5
Enclosed building	0.3	0.1	0.2	0.7
Buried entry	0.1	0.1	0.1	0.9
Equip. damaged	0.2	0.5	0.1	0.7
Security gate	0.2	1.0	0.8	0.8
SCADA	0.3	0.9	0.8	0.9

TABLE 4-20 Total risk calculation

Facility	*C* Factor	P^e Factor	Insiders	Outsiders	Cyber	Terrorist	Total Risk	Priority
High-service pump #1	0.9	0.5	0.045	0.002	0.00045	0.0000005	**0.048**	2
High-service pump #2	0.9	0.5	0.045	0.002	0.00045	0.0000005	**0.048**	2
High-service Pump #3	0.9	0.5	0.045	0.002	0.00045	0.0000005	**0.048**	2
High-service pump #4	0.9	0.5	0.045	0.002	0.00045	0.0000005	**0.048**	2
Reactor #1	0.8	0.5	0.040	0.002	0.00040	0.0000004	**0.042**	3B
Reactor #2	0.8	0.5	0.040	0.002	0.00040	0.0000004	**0.042**	3B
Well G	0.4	0.7	0.028	0.001	0.00028	0.0000003	**0.030**	4
Well H	0.4	0.7	0.028	0.001	0.00028	0.0000003	**0.030**	4

TABLE 4-20 Total risk calculation (Continued)

Facility	C Factor	P^e Factor	Insiders	Outsiders	Cyber	Terrorist	Total Risk	Priority
Well I	0.1	0	0.000	0.000	0.00000	0.0000000	0.000	
Bulk water line	0.4	0.9	0.036	0.002	0.00036	0.0000004	0.038	
Raw water line	0.4	0.9	0.036	0.002	0.00036	0.0000004	0.038	
Generator	0.8	0.7	0.056	0.003	0.00056	0.0000006	**0.059**	1A
FPL service	0.8	0.5	0.040	0.002	0.00040	0.0000004	**0.042**	3
Cl regulators	0.8	0.5	0.040	0.002	0.00040	0.0000004	**0.042**	3A
Cl tanks	0.8	0.5	0.040	0.002	0.00040	0.0000004	**0.042**	3A
Sodium Hypo tanks	0.1	0.5	0.005	0.000	0.00005	0.0000001	0.005	
HSiF6 tank	0.1	0.5	0.005	0.000	0.00005	0.0000001	0.005	
Phosphate tanks	0.1	0.5	0.005	0.000	0.00005	0.0000001	0.005	
Polymer tank	0.1	0.5	0.005	0.000	0.00005	0.0000001	0.005	
Lime silo	0.8	0.5	0.040	0.002	0.00040	0.0000004	**0.042**	3B
Acid storage	0.4	0	0.000	0.000	0.00000	0.0000000	0.000	n/a
Lab chemicals	0.1	0.7	0.007	0.000	0.00007	0.0000001	0.007	
170,000 clear-well	0.8	0.1	0.008	0.000	0.00008	0.0000001	0.008	
200,000 clear-well	0.8	0.1	0.008	0.000	0.00008	0.0000001	0.008	
250,000 EST	0.1	0	0.000	0.000	0.00000	0.0000000	0.000	
Water distribution system	0.1	0.1	0.001	0.000	0.00001	0.0000000	0.001	
Backwash pond	0.8	0.5	0.040	0.002	0.00040	0.0000004	**0.042**	3C
Monitoring Well 1	0.4	0.5	0.020	0.001	0.00020	0.0000002	**0.021**	7
Monitoring Well 2	0.4	0.5	0.020	0.001	0.00020	0.0000002	**0.021**	7
Monitoring Well 3	0.4	0.5	0.020	0.001	0.00020	0.0000002	**0.021**	7
WTP Building	0.9	0.7	0.063	0.003	0.00063	0.0000006	**0.067**	1
Admin Building	0.4	0.7	0.028	0.001	0.00028	0.0000003	**0.030**	5
Fuel pumps	0.4	0.5	0.020	0.001	0.00020	0.0000002	**0.021**	6
Chlorine building	0.8	0.5	0.040	0.002	0.00040	0.0000004	**0.042**	3
SCADA	0.1	0	0.000	0.000	0.00000	0.0000000	0.000	

TABLE 4-21 Potential effect on facilities and objectives not met

Facility	Total Risk	Priority	Consequence
High-service pump #1	0.048	2	Quality, quantity, fire protection
High-service pump #2	0.048	2	Quality, quantity, fire protection
High-service pump #3	0.048	2	Quality, quantity, fire protection
High-service pump #4	0.048	2	Quality, quantity, fire protection
Reactor #1	0.042	3B	Quality, quantity
Reactor #2	0.042	3B	Quality, quantity
Well G	0.030	4	Quality, quantity
Well H	0.030	4	Quality, quantity
Well I	0.000		None
Bulk water line	0.038		Quality, quantity, fire protection
Raw water line	0.038		Quality, quantity
Generator	0.059	1A	Quantity, fire protection
FPL service	0.042	3	Quality, quantity, fire protection
Cl regulators	0.042	3A	Quality, quantity, fire protection
Cl tanks	0.042	3A	Quality
Sodium hypo tanks	0.005		Quality
HSiF6 tank	0.005		Quality
Phosphate tanks	0.005		Quality
Polymer tank	0.005		Quality
Lime silo	0.042	3B	Quality
Acid storage	0.000	n/a	None
Lab chemicals	0.007		None
170,000 clearwell	0.008		Quality, quantity, fire protection
200,000 clearwell	0.008		Quality, quantity, fire protection
250,000 EST	0.000		None
Water distribution system	0.001		Quality, quantity, fire protection
Backwash pond	0.042	3C	Quality, quantity, fire protection
Monitoring well 1	0.021	7	Quality
Monitoring well 2	0.021	7	Quality
Monitoring well 3	0.021	7	Quality
WTP building	0.067	1	Quality, quantity, fire protection
Admin building	0.030	5	None
Fuel pumps	0.021	6	Quality, quantity, fire protection
Chlorine building	0.042	3	Quality, quantity, fire protection
SCADA	0.000		None

TABLE 4-22 Proposed countermeasures

Facility	Priority	Countermeasures	Cost
WTP building	1	Increase SCADA and secutiry; separate WTP area from PW compound	
Generator	1A	As part of WTP upgrade; spare generator parts	
High-service pump #1	2	Have pumps/parts on hand to replace, provide backup transfer pump,	<$100,000
High-service pump #2	2	additional camera security wired to police department, SCADA, separate	inc in CIP
High-service pump #3	2	WTP compound from access of rest of PW compound	
High-service pump #4	2		
FPL service	3	Rehab and wall off FPL transformer, currently exposed	
Cl regulators	3A	Complete hypochlorite system and remove gas cylinders	<$35,000
Cl tanks	3A		
Chlorine building	3A		
Reactor #1	3B	Additional camera security wired to police department; separate	<$75,000 incl
Reactor #2	3B	WTP compound from access of rest of PW compound	w/WTP upgrade
Lime silo	3B		
Backwash pond	3C	Need fencing to secure site from public access, cameras, control of return	$25,000
Well G	4	SCADA	$25,000 incl
Well H	4	SCADA	w/WTP upgrade
Admin building	5	Additional camera security wired to police department, access control	w/WTP upgrade
Fuel pumps	6	Move from WTP building proximity	w/WTP upgrade
Monitoring well 1	7	Locking caps on wells/test frequently	$5,000
Monitoring well 2	7		
Monitoring well 3	7		

TABLE 4-23 Proposed projects to improve security

Project	Location	Cost (in dollars)
Motion sensors/SCADA	Well sites	25,000
Redundant parts for HSP	WTP site	100,000
Transfer pump/backup HSP	WTP site	250,000
Security cameras	WTP site	10,000
Connection to police dept	WTP site	25,000
Fencing at WTP site	WTP site	25,000
Motion sensors/SCADA	WTP site	10,000
Pressure sensors/SCADA	Distribution	15,000
Hypochlorite system	WTP site	35,000
Total		495,000

TABLE 4-24 Staff security suggestions

Staff Checkpoint	Implemented	Comments
Written Emergency Response Plan	No	In process
Restricted access to WTP through gates/fences	Yes	
Procedures for inspecting incoming vehicles	Yes	Limited
External lighting	Yes	
Posted warnings about access/tampering	No	
Regular patrols	Yes	
Alarm systems	Yes	Limited
Chemicals delivered in presence of plant staff	Yes	
Proper storage of chemicals	Yes	
Secured wellheads	Yes	2003
Secured tank ladders	Yes	
Staff background checks	Yes	
Photo identification for staff	Yes	
Protected computer access	n/a	None exists
Secured maps	No	
Neighborhood watch	Yes	Limited
Community education/relations program	No	

REFERENCES

AWWA. 2005. AWWA Policy Statement on Cross-Connections. www.awwa.org/policystatements. Denver, Colo.: American Water Works Association.

Bloetscher, F. 2003. *Vulnerability Assessment for Utility.*

———. 2009. Water Basics for Decision Makers: What Local Officials Need to Know About Water and Wastewater Systems. Denver, Colo.: American Water Works Association.

Bloetscher, F., D.M. Meeroff, L. Hess, G. Beck, L. Simons, and M. Arockiasamy. 2010a. FAU Explores Green Energy for Water Treatment. In *Proceedings of the International Conference on Environmental Sustainability with Green Building Technology, March 15–17, 2010, Tumil, India.* Tumil, India: Meenakshi Sundarajan Engineering College.

Bloetscher, F., D.E. Meeroff, and A. Toro. 2010b. Concentrating the Concentrate—Pointing to Solutions for Concentrate Management. *IWA Journal*, 1(1):39–51.

Bohm, M., J.F. Devinny, and G. Rosen. 1998. On a Moving-Boundary System Modeling Corrosion in Sewer Pipes. *Applied Mathematics and Computation*, 92:247–269.

Brilliant, M., and K. Mains. 2010. Beware of SCADA System Vulnerability. *Opflow*, 36(5):6, 8.

Estoup, J.M., and R. Cabrillac. 1997. Corrosion of Biological Origin Observed on Concrete Digestors. *Construction and Building Materials*, 11(4):225–232.

Gaha, J., and T. Urquhart. 2005. Breaking the Cycle of Reactive Work: How Utilities Can Shift the Focus from Asset Failure to Strategic, Planned Maintenance. *WE&T*, 17(4):23–26.

Goldwater, D.A. 2010. Rehabilitation—An Important Tool for Asset Management. *Water Utility Infrastructure Management*, 3(1):22–24.

Idriss, A., S.C. Negi, J.C. Jofriet, and G.L. Hayward. 2001a. Corrosion of Steel Reinforcement in Mortar Exposed to Hydrogen Sulfide, Part 1: Impressed Voltage and Electrochemical Potential Tests. *J. Agric. Engng. Res.*, 79(2):223–230.

———. 2001b. Corrosion of Steel Reinforcement in Mortar Exposed to Hydrogen Sulfide, Part 2: Diffusion Tests. *J. Agric. Engng. Res.*, 79(3):341–348.

Keresztes, Z., I. Felhosi, and E. Kalman. 2001. Role of Redox Properties of Biofilms in Corrosion Processes. *Electrochimica Acta*, 46:3841–3849.

Lovely, R. 2010. Risky Business: Quantifying Risk Is a Fundamental to Any Physical Asset Management Program. *Florida Water Resources Journal*, 62(5):62–65.

Ma, H., X. Cheng, G. Li, S. Chen, Z. Quan, S. Zhao, and L. Niu. 2000. The Influence of Hydrogen Sulfide on Corrosion of Iron Under Different Conditions. *Corrosion Science*, 42:1669–1683.

Matichich, M. 2010. How Much Is Your Time and Effort Worth. *Public Works*, v. 4, pp 23–27.

Molak, V. 1997. *Fundamentals of Risk Analysis and Management.* Boca Raton, Fla.: Lewis Publishers.

Monsma, D. 2010. Redefining the US Infrastructure Challenge. *Opflow*, 36(2):22–25.

Moran, D., and C. Barron. 2009. Low-Cost Strategies Optimize Energy Use. *Opflow,* 35(12):10–14.

Mori, T., T. Nonaka, K. Tazaki, M. Koga, Y. Hikosaka, and S. Noda. 1992. Interactions of Nutrients, Moisture and pH on Microbial Corrosion of Concrete Sewer Pipes. *Water Research*, 26(1):29–37.

Owens, E. 2010. Valve Maintenance: Use Existing Resources to Create an Effective Exercising Program. *Opflow*, 36(5):20–22.

Pourbaix, A., L.E. Aguiar, and A.M. Clarnival. 1993. Local Corrosion Processes in the Presence of Sulphate-Reducing Bacteria: Measurements Under Biofilms. *Corrosion Science*, 35(1–4):693–698.

Quist, G.M., D. Drake, C. Davisson, and D. Wasko. 2009. Alarming Trends, Remote Monitoring at Multiple Locations Helps Avert SSOs. *WE&T*, 21(3):63–67.

Stadnyckyj. 2010. Condition Assessment: Bridging the Gap Between Pipeline Investments and Risk Reduction. *Water Utility Infrastructure Management*, 2:25–27.

Tekin, T., N. Boyabat, and Bayramoglu. 1999. Kinetics and Mechanism of Aqueous Oxidation of H_2S by Fe^{+3}. *Int. J. Chem. Kinet.*, 31:331–335.

USEPA. 2002. *The Clean Water and Drinking Water Infrastructure Gaps Analysis.* USEPA Report EPA-816-R-02-020. Washington, D.C.: USEPA.

Wang, A.M., W. Zhang, H.D. Crisp, and J. Garland. 2009. Conservation versus Water Quality—Beware of the Side Effects. *Opflow*, 35(7):22–26.

Werner, S.E., C.A. Johnson, N.J. Laycock, P.T. Wilson, and B.J. Webster. 1998. Pitting of Type 304 Stainless Steel in the Presence of a Biofilm Containing Sulphate-Reducing Bacteria. *Corrosion Science*, 40(2/3):465–480.

Yamanaka, T., I. Aso, S. Togashi, T. Minoru, K. Shoji, T. Watanabe, N. Watanabe, K. Maki, and H. Suzuki. 2002. Corrosion by Bacteria of Concrete in Sewerage Systems and Inhibitory Effects of Formates on Their Growth. *Water Research*, 36:2636–2642.

Yarlott, M. 2010. Factoring Condition Assessment and Asset Management into Capital Planning. *WE&T*, 22(4):42–48.

CHAPTER 5

Employment Rules and Managing Employees

POLICIES AND REGULATIONS

As noted in chapter 2, there are many regulations affecting water quality and operations associated with water and wastewater utility systems. In chapter 3, the discussion focussed on the capital assets of the utility system, while chapter 4 reviewed maintenance of those capital assets. Another significant asset of a utility is its employees. Sustainability of the utility involves maintaining its infrastructure and its personnel assets (Ralston and Ginley, 2010).

Just as the capital assets must be protected and maintained, the personnel assets must be protected and nurtured. This involves acquisition, training, and development so employees contribute positively to the utility mission. Personnel have knowledge of the utility, an attribute that can be saved by developing an appropriate work order tracking system that transfers information discovered during breaks and other maintenance activities to paper, and ultimately some form of electronic analysis.

Just as there are many regulations in the operation of the utility assets, there are even more associated with the human resource aspects of utilties. Some are rules and regulations, while others arepolicies. Enforcement differs between them. The ability to change, alter, or ignore them varies as well. Regulations have legal consequences, while policy implications do not. Most personnel rules are regulatory in nature. However, purchasing is usually associated with policy implications.

A policy is a guideline that employees are expected to follow. Decisions are made on policies that are generally defined by a governing board or executive management. The goal is to ensure that decisions regarding a given situation are consistently made (e.g., the utility always turns the water off if people fail to pay their bill in a timely fashion, as opposed to turning certain people off, which would be unfair.) An inconsistent policy is an area where the utility could be challenged about its policy-making ability.

Policies should provide some capacity for interpretation, however, which is why they are deemed to be guidelines as opposed to rules. Policies allow the employees and, at times, residents to understand what is expected and what reasonable implementation is to occur. Hence, most utilities have a purchasing policy that requires a certain number of quotes or bids to ensure the utility gets materials and work performed at reasonable prices. Yet, if it is an emergency, they do not have to follow those rules. An emergency would allow the utility to take care of the situation without following the policy, especially when life, safety, and welfare of the public might be concerned.

Policies Affecting Supervisors

There are three important policies a supervisor should be aware of. The first one is budgeting policy. The supervisor is likely the person who prepares the initial budget for the areas under his/her control. Money must be available to pay for the work that is planned; thus a sufficient amount of money needs to exist for this purpose. The supervisor knows the work better than anyone else in the organization and may need to explain what is needed. During the year, the supervisor is responsible for the judicious allocation of funds in order to accomplish the tasks at hand. This includes both labor and materials.

Materials lead to the second policy; there needs to be a written policy for how to purchase materials, tools, and so forth. All are routine items that employees will require to accomplish their work. If there are no purchasing rules, the potential exists for the organization to pay more for materials than it would otherwise need to. In the public sector, there are usually state laws that outline how municipalities are to purchase tools, equipment, and other services. Even for private-sector utilities, there should be a program to ensure that multiple sources are contacted and that the lowest prices are accepted as long as they meet the requirements to be responsive and responsible. These terms mean that the bidder has the equipment or services to adequately provide the work or that the materials meet certain specifications or qualify within the guidelines the utility needs, i.e., they can provide them on the date needed and with the appropriately sized equipment.

The supervisor also needs to understand personnel. There is a protocol for hiring personnel, monitoring their efforts, and supervising them. Also, there may be union rules regarding hiring or utilization of the personnel during certain hours. Most labor rules are based on laws and statutes approved by elected bodies and must be complied with. The supervisor should keep in mind that there may be regulations associated with the work. It is the responsibility of the supervisor to ensure compliance with the regulations. Failure to comply with regulations falls on the supervisor, even if it is the employees who are violating them. He/she needs to take action to ensure policy compliance but, ultimately, if the employees are not trained or not advised of the situation, then it is the supervisor's responsibility. Some discussion of the basic employment laws follows.

EMPLOYMENT REGULATIONS

Employment Laws

Employers and employees need to be aware of regulations associated with personnel. Employers are required to post notices to all employees advising them of their rights under the law, including Equal Employment Opportunity (EEO) regulations. Such notices must be accessible to the employees regardless of disabilities. There are several acts of particular importance from a federal perspective. These include the Civil Rights Act of 1964, the Equal Pay Act of 1963, Age Discrimination and Employment Act of 1967, Americans with Disabilities Act of 1990, the Rehabilitation Act of 1973, and the Civil Rights Act of 1991.

Civil Rights Act of 1964

Title 7 of the Civil Rights Act of 1964 is the major issue associated with this act. Title 7 prohibits employment discrimination based on race, color, religion, sex, or national origin (USDOL, 2008a). It has been amended a number of times and each time, more prohibitions were added relating to discrimination in hiring practices, promotions, pay, discharge from employment, fringe benefits, job training, classifications, referrals, and other aspects of the employment process. Such discrimination includes sexual harassment and issues associated with pregnancy. National-origin issues prohibit discrimination against individuals because of their birthplace, ancestry, culture, or linguistic characteristics common to an ethnic group (USDOL, 2008a). Note that these rules may apply where employees are required to speak English only on the job, unless it can be shown that the ability to speak English is a requirement necessary for conducting business. Such examples would be dealing with the public, customer service, or dealing with the media. In such cases, the employee must be notified that English is required and the consequences of failure to provide appropriate English speaking or writing skills. Likewise, religious accommodation is required for employees, unless it would impose undue hardship. Time off for legitimate religious holidays associated with recognized religions is an issue that must be planned for by the supervisors.

Sexual discrimination and harassment are major issues associated with employment law and are subject to a significant number of lawsuits. Sexual harassment can range from direct requests for sexual favors to conditions that create what is termed a *hostile work environment*. Hostile work environments apply to persons of either sex and extend to harassment on the basis of race, color, national origin, age, religion, and disabilities (USDOL, 2008a). If an employee complains about a violation of the Civil Rights Act, retaliation by the employer or the supervisor is prohibited by law.

Equal Pay Act of 1963

The Equal Pay Act of 1963 requires that men and women who perform substantially the same work in the same establishment be compensated based on the same principles (USDOL, 2008b). This assumes that they are hired for a given task, and that their qualifications, starting date, and experience are the same. It does not mean that someone who has been in place for 20 years should be making the same money as someone who has been there for one year, which is often an issue that people are confused about. It does not mean that periodic raises must be granted to both employees if their performances are different. Employees may be confused about this as well.

Age Discrimination and Employment Act of 1967

The Age Discrimination and Employment Act of 1967 prohibits and protects individuals 40 to 70 years old from discrimination in hiring based on their age, or from discrimination relating to promotion, discharge, compensation, conditions, or privilege of employment (USDOL, 2008c). The intent was to prevent employers from hiring only young people to keep salaries low or from acting on the misperception that older people were incapable of learning to do the tasks at hand. The act does not prohibit the employee from needing to demonstrate their capabilities with regard to doing the job or that they have physical or mental tools required (USDOL, 2008c).

Equal Opportunity Act of 1996

The Equal Opportunity Act of 1996 prohibits discrimination and preferential treatment on the basis of race, color, national origin, or sex with regard to federal employment and federal contracts and programs. As a result, if the utility is a recipient of federal funds, not only do the Civil Rights Act of 1964 issues need to be complied with, but those of the Equal Opportunity Act of 1996.

Americans with Disabilities Act of 1990

Title I and Title IV of the Americans with Disabilities Act of 1990 (ADA) prohibit discrimination against qualified individuals with disabilities, in both the private sector and state and local governments. This act protects qualified applicants from discrimination in hiring, promotion, pay, discharge, job training, fringe benefits, classifications, and other aspects of employment on the basis of any disability (USDOL, 2008d). A person with a disability is defined as any individual who has a physical or mental impairment that limits one or more major life activities (USDOL, 2008d). A person must have a record of having this impairment, or be regarded as having the impairment, which may include impairments such as walking, breathing, seeing, hearing, speaking, learning, and working. A qualified applicant with a disability is someone who satisfies the skill, experience, education, and other job requirements of the position, not someone who cannot meet those requirements due to their disability. This person, with a reasonable accommodation, must be able to perform the functions of a person without a disability (USDOL, 2008d).

The act requires that reasonable accommodation be provided unless the accommodation would impose an undue hardship on the employer (USDOL, 2008d). Reasonable accommodation may include such activities as making facilities handicapped accessible, job restructuring, modification of work schedules, providing additional unpaid leave, modifying or adjusting training materials or examinations, and modifying policies associated with job placement and promotion. The reasonable accommodation must be necessary to apply for the job or to enjoy the benefits and privileges of employment. The employer is not required to lower production standards as an accommodation (USDOL, 2008d). Examples would be if the employee is required to lift certain objects and move them, or to add fluoride to a tank. In this case a person in a wheelchair might not be qualified for that job because accommodation would impose an undue hardship to the employer. The ADA also requires that employees be provided parking spaces, wheelchair ramps, and so forth for jobs where an undue hardship does not exist (USDOL, 2008d). There are a variety of places where Title I and Title IV apply.

Title VII of the ADA of 1990 extends the prohibition on discrimination to recruitment, testing, use of company facilities, training, and other issues (USDOL, 2008d). The act reinforces the fact that harassment against disabled people as well as all others based on race, color, religion, sex, national origin, or age is prohibited and prevents the retaliation against any individual who files a complaint, participates in an investigation, or who vocally opposes discriminatory practices (USDOL, 2008d). If employment decisions can be shown to be based on stereotypes or assumptions about the person's abilities, traits, or performance, or because they are married to or associated with someone with any of the protected traits, then denying employment opportunities to this person is an actionable offense.

An employer may not ask a job applicant about the existence, nature, or severity of any disability, although applicants may be asked about their ability to perform certain job functions in order to determine whether or not they are capable of fulfilling them (USDOL, 2008d). This applies to any prospective employee, not just those with disabilities. Job offers may be conditional based on the results of a medical examination of their ability to perform the duties involved. Most employers offer a trial period to determine employees' capabilities. Medical examinations of an applicant must be job related and consistent with necessities of the position in question, not with issues that are unrelated to job performance.

Civil Rights Act of 1991

The Civil Rights Act of 1991 made major changes to certain prohibited discriminations. The intent of this act was to modify certain Supreme Court decisions that limited the rights of people protected by civil rights laws. The act authorizes monetary and punitive damages when it can be demonstrated that the employer intentionally discriminated against an employee, and provides remedies of how those damages may be claimed (USDOL, 2008e). Sections 501 and 505 of the Rehabilitation Act protect against discrimination against individuals with disabilities who work for the federal government (USDOL, 2008f). The Rehabilitation Act does not apply to local governments, but, as noted earlier, other legislation with the federal government has extended these protections to those who receive federal funds or contracts with the federal government (USDOL, 2008e).

Family and Medical Leave Act of 1993

The Family and Medical Leave Act of 1993 (FMLA) requires that employers grant up to 12 workweeks of unpaid leave during any 12-month period for the birth and care of a newborn child of the employee, the placement to the employee of a child for adoption or foster care, the care of an immediate family member (defined as a spouse, child, or parent) with a serious health condition, or because the employee is unable to work because of a serious health condition (USDOL, 2008g). During the time of leave, the employer must maintain the employee's health coverage under any group plan that is normally offered to the employees and in which the employee previously participated. The intent is that illness should not deny the employee the same rights as he or she would have while employed.

On return from leave, the employee must be restored to his or her original position with equivalent pay, benefits, and employment terms (USDOL, 2008g). The use of leave under this act cannot result in the loss of any employee benefits or positions, pay, or status the employee would have had had he or she worked continuously through the entire period.

The employee cannot use the provisions of the FMLA without providing the proper advance notice to the employer and providing information on the medical condition for which the leave is requested. If the employee just takes the leave, the provisions of the act do not apply (USDOL, 2008g). The employee must provide at least 30 days advance notice if the leave is foreseeable (USDOL, 2008g). Obviously, for emergency conditions such as a heart attack or traffic accident, there is some expectation for accommodation on the part of the employer. The employer may require a medical opinion to support the request for leave for serious health conditions and may require second or third opinions at the employer's expense to determine if the employee is fit to return to work after a serious health condition. If the employer does not believe that the employee is physically fit to return to work after the incident, the employee may be required to return to a different position more acclimated toward the health condition (USDOL, 2008g).

Not all employees are subject to the FMLA. Eligible employees must have been employed for at least 12 months, with 1,250 hours of service during the 12-month period prior to when the leave is requested (USDOL, 2008g). For private employers, the regulations apply where there are 50 or more employees at the work site and within 75 mi (120 km) of the work site.

There are a variety of violations for employers to watch out for. The first is that the act makes it unlawful for any employer to interfere with, restrain, or deny exercise of any right by the employee under the FMLA or to discriminate against any person for any practice covered under it. For example, since the act specifically requires or permits up to 12 weeks unpaid leave due to pregnancy or child adoption, a discrimination complaint would result should the employer deny this leave under the provision (USDOL, 2008g).

Employee Polygraph Protection Act of 1988

During the 1980s, some employers required that their employees take lie-detector tests, either for preemployment screening or while employed. The intent of the Employee Polygraph Protection Act of 1988 was to prohibit the use of lie-detector tests to collect information about other employees or to base employment decisions on the results of the tests because polygraph tests have limitations in their accuracy (USDOL, 2008h). Where tests are administered, employers cannot require or request any employee or job applicant to take a lie-detector test or to discharge, discriminate against, or discipline employees for refusing to do so (USDOL, 2008h). Employers may not use or inquire about the results of a lie-detector test that has been administered, or discharge or discriminate against an employee or a job applicant based on the results of the test.

The act prohibits most private employers from using lie-detector tests for any reason. Remedies are the same as for civil rights complaints (USDOL, 2008h). When the employer is covered by a specific exemption that permits a polygraph test, the employer must maintain records that set forth the specific activity or incident for which testing was performed.

The act specifically does not cover federal, state, and local governments because they may have situations, such as law enforcement or the use of security clearance, where there is a need for ensuring that employees tell the truth (USDOL, 2008h). Law enforcement routinely uses polygraph testing for employee screening and to help defend public safety personnel. From the water plant perspective, it should be noted that, because of the types of activities that occur at water plants, there may be a belief that new employees coming on board need to meet certain criteria from an honesty perspective. However, employers should judiciously exercise care in the implementation of such testing.

Fair Labor Standards Act

The Fair Labor Standards Act (FLSA) establishes minimum wage, overtime, record keeping, and youth employment standards affecting employees in both the private and public sectors. As of July 24, 2009, the minimum wage was raised to $7.25 an hour (USDOL, 2009). The act specifically suggests that overtime pay will be paid at one and a half times the regular rate of pay after 40 hours of work in a work week.

There are certain employees that are exempt from the provisions of the FLSA. Employees in management or professional positions, due to the nature of the work, may require more or less work in a given period (USDOL, 2009). These employees are not covered under the act and do not have to be paid overtime. Likewise, certain public safety officers (e.g., fire fighters) are not required to be paid overtime as long as routine shifts do not exceed 48 hours in any given week over a two-week period. This is usually done because fire fighters are on one day and off the next and over a two-week period, they would work 48 hours in one week and 36 hours the next for an average of just over 40 hours. Most water plant operators, utility employees, and field employees are covered by the act and can be expected to receive time and a half for any hours over 40 a week.

The FLSA does not limit the number of hours that can be worked (over 40) in a week by any employee. It does not require that overtime be paid on weekends, holidays, or regular days of rest, unless more than 40 hours per week are worked during those periods. It only requires that they be paid time and a half. Penalties to the employer for violating the provisions of the FLSA are $11,000 per violation plus recovery of any back wages withheld from the employee (USDOL, 2009).

The Fair Labor Standards Act defines child labor as any worker under the age of 18. The limitations on child labor include a requirement to pay the minimum wage, overtime pay, and record keeping of full- and part-time workers in both state and local governments. The limitations are 2 hours per day and 18 hours per week during the school year, 8 hours on a nonschool day, or 40 hours per week during summer or nonschool periods. Work cannot begin before 7:00 a.m. nor continue after 7:00 p.m., except in the summer when it can continue until 9:00 p.m. (USDOL, 2009). In other words, persons under the age of 18 cannot be working late nights or early mornings. Once the employee is 18, the child labor laws do not apply, regardless of whether or not the person is in school. Federal child labor rules do not require a work permit as they once did; however, if some form of permit is required, the Department of Labor will issue age certificates if the state does not do so. All states have child labor standards and the most stringent ones apply.

Uniform Services, Employment and Reemployment Rights Act

The Uniform Services, Employment and Reemployment Rights Act (USERR) is specifically addressed to veterans who have been required to leave their civilian employment to serve in the armed forces (USDOL, 2008i). These tours of duty may extend for a number of months or even years and are beyond the control of the employee. The only thing needed from employees is that they give notice to the employer that they are being required to proceed to active duty and that they apply in a timely manner at the conclusion of service (USDOL, 2008i). There is an assumption that they have not been discharged from the military or, if discharged, it was not less than honorable. The intent is that an employee who leaves a job for military service (the war in the Middle East, for example), must be restored to the job and benefits that would have been attained had they not been required for active duty. In other words, the employee should not suffer because the federal government requires his or her military service (USDOL, 2008i).

The act applies to both active-duty military service requirements and training of the individuals covered under the act. If the person is a member of the National Guard and required to go to training two weeks per year and one weekend a month, then the employer is required to accommodate this and not to engage in any discriminatory actions or denial of benefits as a result (USDOL, 2008i). The law is intended to encourage people to participate in the reserves and National Guard. The act specifically does not apply to those who have a full-time career in the military and are serving in active-duty situations.

Drug Free Workplace Act

The Drug Free Workplace Act (DFWA) of 1988 was passed by Congress and allows employers to use random drug tests as a basis of employment. The act details how they are applied and what privacy rules are required. The act permits random testing during employment but the results of these tests and the solutions to the problems must remain private between the employer and the employee (USDOL, 2008j). Violations of privacy can subject the employer to damages from the employee.

Employees and applicants currently engaging in the illegal use of drugs are not covered by the DFWA if the employer acts on the basis of such use. While certain examinations and inquiries are prohibited, tests for illegal use of drugs are not considered medical exams and are, therefore, not subject to legal restrictions. Employers may hold employees who are illegally using drugs or individuals who are abusing alcohol or using alcohol while on the job to the same standards of performance as other employees. There is an automatic right by the employer to terminate employees when their use of these substances may impact the health, safety, and welfare of other employees.

Any individual who believes that his or her employment rights have been violated may file a discrimination complaint with the Equal Opportunity Office. Another person or agency can also file a complaint against an employer for the same charge if they believe that exposing the employee's identity would subject the employee to retaliation by the employer or could compromise his or her ability to do their job.

There are a variety of remedies that are available for employment discrimination, whether caused by intentional acts or by practices that have the effect of being discriminatory. These may include forced hiring of the employee. If an employee is already on board and feels that discrimination has impacted his or her ability to gain a promotion or raise, the relief remedy may be to require the employer to promote that employee and to provide the appropriate raise. Employees who have been terminated for what they believe are discriminatory practices may file suit to be reinstated with back pay, with any promotions or raises that they would have received had they not been terminated (or other actions that would make the individual whole). In addition, in all cases, the courts will award attorney's fees, expert witness fees, and other court costs. Punitive damages are not recoverable against federal, state, or local governments, but are recoverable against all private-sector employers.

The US Equal Employment Opportunity Commission (EEOC) enforces all these laws. They provide oversight and coordination of all the federal rules and create regulations to clarify the provisions of these rules and regulations. The EEOC acts on behalf of the employees in trying to reach a satisfactory solution to apparent or actual violations of these rules and attempts to help employers ensure that they are compliant with the requirements of the rules so that violations do not occur and penalties are not incurred.

In summary, the requirements of all these rules are that no employee may be discriminated against as a result of race, color, national origin, disability, military service, etc. There is an attempt to help both employees and employers meet the requirements of these acts to reduce potential problems. EEOC will step in in cases where blatant problems exist and may attempt to help both parties where perceived problems may exist.

Workplace Safety

There are a variety of rules for employers regarding the safety of workers and the workplace. This is because for many years, unsafe working conditions caused numerous injuries to employees. As a result, productivity of the employees and businesses diminished and additional costs for making the employee whole were significant. In some cases, employees were killed or permanently disabled as a result of workplace activities.

The Occupational Safety and Health Administration (OSHA) was created to help try to minimize the number of workplace accidents and injuries and to require employers to provide safe conditions and appropriate training. It is recognized that not all workplaces can be made completely safe. An example would be a water plant or a construction site. By their nature, accidents can occur and working with equipment can lead to accidents.

Because operations personnel come into contact with construction, road work, chemicals, and sewage, these workers should be cognizant of the need for proper use of hard hats, safety vests, eye protection, hearing protection, safety shoes, and safety gloves. Respiration devices are required for entry into many confined spaces. Confined spaces require entry training, multiple person crews, and airspace testing. Ongoing training is required to reduce workers' compensation claims, lost time, and legal liabilities.

Also required are a variety of rules involving chemical usage and handling, hazardous material safety data sheets (MSDS), regular training of staff in the proper use of chemicals, chemical cleanup procedures, and emergency leak response plans. Chlorine is often the most dangerous chemical used by utilities, but safety training is often readily available through chlorine dealers.

One person should be assigned safety responsibilities in larger utility systems. It is best if this person reports to the utility manager as opposed to a field supervisor for two reasons: safety is elevated in stature when management is involved, and there is less potential for conflict between field activities and safety procedures.

It is vital to minimize the frequency of incidents that occur by ensuring that the employees are not subjected to undue hazards and that appropriate training is provided. A failure to comply with proper procedures, either by the employer or the employee, may bring OSHA into action. Employees have a right to notify the employer or OSHA about workplace hazards. The reporting employee's name can be kept confidential. Employees have the right to request that OSHA provide an inspection of a workplace if it is believed that unsafe or unhealthy conditions exist. In addition, employees have the right to file a complaint with OSHA if, as a result of a complaint, retaliation or discrimination has occurred. Employees have the right to see any OSHA citations that may be issued against the employer for unsafe conditions. In part, this is permitted to allow the employee to determine if there is a recurring pattern or if an accident is an isolated incident. OSHA treats these differently. If a citation has been issued, the employer is required to post those citations near the alleged violation site. The employer is required to correct workplace hazards as identified and in compliance with the date on the citation, and must certify that the corrections have been completed. If the employees have been exposed to harmful conditions or toxic substances during the period of the violation(s), then they have the right to have copies of this information provided to them and to their physicians.

Worker's compensation

Prior to the enactment of the workmen's compensation laws (back in the mid-twentieth century), the only option that employees had was to file suit against the employer for negligence. Such actions had significant drawbacks for the workers, especially since there was no protection against discrimination or discharge as a result of the suit. Hence, retaliation was a real threat and the source of current employee protections.

The initial workmen's compensation laws were enacted as a result of increases in litigation and to mitigate the requirement that workers prove that injuries were the fault of the employer (USDOL, 2008k). If hazardous conditions exist, somebody should pay if workers are injured, and, as a result, workmen's compensation rules and the insurance associated with it were developed. Employers pay into workers' compensation insurance pools and the workers' compensation insurance pool pays for the worker's injuries. The first such workers' compensation laws were approved by the state of Maryland in 1902 and federal employees were covered in 1906. By 1949, all states had some sort of workers' compensation system.

Workers' compensation rules have developed over time to protect injured workers so that they will receive appropriate medical care, not lose wages due to on-the-job injuries, and, where necessary, to retrain and rehabilitate them so they can return to the workplace in another position if the injuries are severe enough to prohibit them from continuing in the job they had when injured. When workers are killed on the job, members of the worker's families may receive workers' compensation benefits.

There are certain exclusions to workers' compensation, depending on the jurisdiction. Employees may be restricted from collecting benefits if injuries are a result of willful misconduct or intoxication. The employer's workers' compensation insurance will not pay if the injury or death is the responsibility of the employee as a result of his or her negligence or impairment due to use of illegal or alcoholic substances. Willful misconduct includes instances where the employee does not follow the training guidelines provided by the employer.

Most employers are required to participate in some form of a workers' compensation insurance program. Failure to do so subjects the employer to financial penalties. Usually very small companies with fewer than five employees or sole proprietorships with fewer than five employees may be exempt from workers' compensation insurance requirements (USDOL, 2008k). This does not exempt them from the requirements for training, or litigation on behalf of injured employees. Policies are available from a variety of sources, including commercial insurance companies. The costs are based on the perceived risks. Construction and public works activities are typically assigned high risk. Since the risk is higher than for most occupations, most governmental entities self-insure for workers' compensation or provide compensation through a governmental pool. For high-risk employers, coverage is available through assigned risk programs or pools.

Most workers injured on the job recover quickly because the injuries are minor. However, all injuries should be reported to a workers' compensation system in order to qualify for benefits because, at the time of the injury, the extent of the injury and how it may impair the employee's ability to work in the future may not be clear. Failure to notify the employer about an injury provides a justification for the workers' compensation carrier to deny benefits, even if the injury is serious. Failure to provide a timely notification (i.e., taking a long time to inform the employer) also may impact the ability to recover benefits, even if the injury is serious. The employee is required to provide the following information about any injury that occurs: names of all witnesses; a description of how, when, and where the injury happened; the type of injury; the activity that was occurring; and any other pertinent information (USDOL, 2008k). More serious injuries are more obvious and the workers' compensation rules permit litigation via an administrative setting to recover monetary damages.

It is particularly important that employers be notified if the employee has been exposed to toxic substances such as pesticides or asbestos. These may cause long-term issues that do not appear immediately; therefore, if 10 or 15 years later the employee finds out that exposure to these toxins or asbestos may create a health condition, then the claim information will be on file. This is an example of where it is likely to have the employee denied benefits if exposure is not reported in a timely fashion, since it's difficult to go back years to determine when an incident occurred.

In most states, when a worker qualifies for workers' compensation due to injury, the employee is required to return to work as soon as he or she is able. This does not necessarily mean returning to work to perform his/her prior task. For instance, field personnel who are required to do physical labor may be able to work, but must do so from a desk. This is termed *light duty.* The worker does not get fully restored to his/her prior job until medically cleared to do so. Ultimately the employer must provide suitable work and return the worker to the job he or she left, or an equivalent job, assuming such a position is available. Workers' compensation benefits may be required to make up the difference between the salary of that position and the salary the employee was previously making.

Workers' compensation pays two thirds of the salary or wages the injured employee was making when the injury occurred. Taxes are not paid on workers' compensation benefits until the employee is released to return to work or it has been determined that the employee's injury is a permanent disability. A lump sum is paid for a permanent incapacity.

If there is a dispute over whether or not a compensable claim has occurred, the employer may dispute the claim. The workers' compensation carrier will investigate and ultimately the matter may go to some form of state workers' compensation commission or other agency for a determination. In some situations the employer may believe the employee is not actually disabled. There are numerous claims of fraud, where employees claim an injury and then are pursuing other employment opportunities. These subject the employee to criminal prosecution if it is determined that fraud has occurred with workers' compensation. The rules are there to help people, not for people to take unfair advantage.

Workers' compensation claims only apply if the injury occurred at work and is related to work. Additionally, the employer has no responsibility to comply with the accommodation requirements for an employee injured outside the workplace.

In most states, the employer will immediately begin to pick up the costs for medical expenses as well as pay a percent of income once injured. Cost containment is approached by looking holistically at the workers' compensation program. Employers do not focus on one specific concern, but look at the entire regime of potential risks that may injure employees on the job. In this manner, employers can attempt to train employees on procedures within the control of the employees and the employer, as opposed to things that lie outside their control. Delegating responsibility to the workers to take charge of those things within their control is an important component of this issue. If the workers recognize the hazards, they can adjust to them and try to avoid problems.

A workers' compensation risk reduction program will have four major areas: assessment of the problems and recommendations to solve them; design and development of a protocol to minimize potential for ongoing injuries; implementation of the protocol; and rollout and evaluation. What is typically found is that an ongoing training program that involves the employees will often lead to a significant decrease in the number of claims made for workers' compensation. This is a benefit for both the employer and the employee. For the employer, it reduces lost work time costs; for employees, it reduces the potential for injury.

HIRING, PERFORMANCE EVALUATION, AND TERMINATION PRACTICES

A major issue associated with supervision and management is staffing. Human resources managers cite a number of reasons why it is difficult to recruit employees, including the following, with the percentage of times the issue was reported in parentheses (Olstein et al., 2005):

- Small pool of applicants (60 percent)
- Compensation limitations (51 percent)
- Certification/skill set (44 percent)
- Budget constraints (39 percent)
- Hiring process/unions (30 percent)
- Residency in some jurisdictions (9 percent)

Once the rules regulating human resource activity are defined, as presented in earlier sections, the next step is to acquire the right human assets. There are four main situations that will indicate when an employee or additional employees need to be hired. The first is when the organization does not have the skills or expertise on hand to do the job. The second is when it will cost more to do the job (such as with overtime) than to hire additional full-time (nonovertime) staff to do it. Third would be when one person alone cannot do the job. Fourth would be when people generally dislike the job or the task that must be done, so leave it as quickly as possible.

In looking at each of these situations, the skill and expertise issues are problematic and engender themselves for some sort of supplemental help. If this is a permanent, long-term, routine task, having employees in-house is probably of more benefit. Where certain tasks are not done regularly, cannot be done in-house, or when the work loads have increased, it is likely time to increase staffing. The increase in staff should improve productivity because more can be done, and less stress and demands may exist with current staff. This will improve their productivity as a result of improving their attitude.

Finding the right employee with the right skill sets is one of the major goals of the staffing process and can be a full-time job in large organizations. People with special skills in human resources are often used to do this. The staffing process includes finding the people with the correct skill sets and attitudes, training them, and providing them with evaluations and guidance for their long-term success. The key is long-term success and productivity of the organization, thereby ensuring the success of the employee, as well.

A common problem is that supervisors and those hiring don't really know the employees, their talents, or the impact of new technology or new methods on meeting organizational goals. As a result, the talents may be misapplied (e.g., a person may be pushed into a position for which his/her skills don't exactly line up, and that employee may not be successful). Seniority has often been used in unionized organizations as a means to determine which employees should be promoted; however, if one thinks of sports examples, the most senior person may not be appropriate person to actually provide the solution. Younger people with new knowledge and a different perspective may be more capable of addressing changing circumstances.

There are six steps in the staffing process. The first is that someone needs to identify the skill set needed and the position job description. This is normally something a human resources person and the line supervisor will do together. Other management personnel may be involved, but it is important to end up with a skill set and job description that match the work that is needed and can match up with employees. Human resources will then initiate recruitment, which includes advertising and soliciting in a variety of sectors. Job applications will be filled out, which will need to be screened. Some applicants will not meet the basic qualifications and can be ruled out immediately. However, there may be a large number of applicants who meet the minimum criteria, along with some who far exceed them. Selecting appropriate candidates is the next step and usually is done in conjunction with a human resources manager with experience in evaluating resumes. The selection process continues with an interview or some discussion whereby the applicant is actually talked to, and is evaluated. Once the appropriate employee is selected, an orientation and training program along with subsequent follow-up on job training will be required to help improve the odds of success for the employee. While not all employees will be successful, the lack of training and lack of communicating expectations of goals and objectives for the employee are often issues cited where difficulties exist between supervisors and employees. A survey of salaries should be conducted periodically to determine whether or not the salaries are competitive within the organization. If they are not, adjustments should be made, because good, highly qualified and motivated employees are the ones most likely to leave, not the poor performers who cannot find jobs elsewhere.

Filling a position involves a significant amount of paperwork and discussions with individuals. This may not be perceived as a particularly productive time, but it is time for which decisions are being made about how efficiently the work gets accomplished over a 6-month to 10-year period. Investing some time and effort into ensuring that the job descriptions, advertisements, and candidate selection procedures maximize the potential for success is an important factor. A background check should be included. It is surprising how many employers fail to check background information, such as whether or not the applicant has a driver's license, a claimed college degree, prior work experience, or a license to practice work. Some relevant information, such as applicants' graduation from college or possession of licenses, is easy to check, but when ignored is a common source of issues that arise well after the employee has been hired. Engineering licensing boards regularly discipline people who claim licensure but do not have it. These are statutory violations, meaning the offenders could face jail time. An employer can be subjected to potential risks and liabilities from negligence to follow up on these simple issues.

Recruiting Good Employees

Good organizations can fail relatively quickly when positions are not filled properly. Good recruitment involves a good job description, competitive pay and salary, and a wide base from which job applicants can be secured. One must also understand the job skills in the marketplace and the employee market to determine what the employees are doing or where potential employees might exist (who are not reached through traditional methods). Nontraditional methods include everything except for walk-ins. Utilities may participate in job fairs, use employment agencies or professional recruiters, or may solicit people they work with in the field to find potential job applicants.

The objectives of recruiting are to determine if there is a candidate for the job, if that candidate has the necessary qualifications, and if he or she is capable of actually performing the work. To do this, the following is needed:

- A good job description,
- Competitive benefits and salary,
- A broad base from which to draw employees,
- An understanding of job skills in the market, and
- An understanding of the employee market.

The key is to determine if the job advertisement and description actually reach the pool of potentially qualified applicants, and if so, whether the job offers enough opportunity to make it competitive enough that they would consider leaving their current positions.

There are four barriers to filling positions. One is a highly competitive environment. If there are a lot of people hiring the same position, it will be difficult to secure the best people. Private entities have a little better opportunity to deal with that problem in the short term because they are not required to go through elected bodies to increase salaries (something that is generally avoided during the fiscal year). The second issue is inadequate human resource support. If the job descriptions are poor, recruiting is limited, or salaries are inappropriate, then good employees will not apply for the job. The lack of training opportunities, educational opportunities, or a skilled labor pool will limit options. What options exist may be expensive or require extensive training. Inadequate compensation will not appeal to better workers or those currently with jobs.

Job Descriptions

The applicant will apply for a job based on the duties and expectations listed in the written job description. This job description should also indicate the tools needed to successfully complete the work along with the knowledge and education required and any special licenses. It may be determined later that this was not exactly the skill set required, and that employee may move on from the position relatively quickly as a result of this being a poor match. Salary and other compensation may also limit the potential for employees to grow within the job. It is important that job descriptions be reviewed on a regular basis.

Job applications cannot contain questions about any of the following: pregnancy or related health conditions, arrests that did not result in convictions, the existence of disabilities, or military history, unless the job requires a background check for these items (these are typically law-enforcement-oriented positions and would not apply to water and sewer utilities). Further, job applications cannot contain questions about the following: the applicant's height or age, unless it's specifically pertinent (such as in law enforcement, where various guidelines have been established by national organizations). Questions regarding organizational affiliations outside of professional memberships are not permitted, nor are those relating to the lowest acceptable salary for a position or the status as a high school graduate. Applications can ask about arrest records that resulted in convictions or time served and an explanation and disposition of same. It is permissible to request an applicant's details of educational history, however, not necessarily about specific degrees unless those are required for the job.

Points to consider when hiring include consideration of whether or not the employee has the potential to move upward in the organization or is likely to be a part of high turnover due to unforeseen issues of the job or the fact that the job is undesirable. Developing a list of questions for applicants should be aimed at getting answers to these questions. The reason the applicant left a prior position may be important. If it was a particularly difficult departure, the employee may not want to discuss it. However, just because an applicant may have been discharged may not be an indication that it was completely the employee's fault. It may be that the supervisor had unrealistic expectations or was incapable of properly communicating what was expected or appreciating the complexities involved. In hiring any employee, references and past experiences should be verified. This is especially important when licensure is at stake. Employees should also be asked about short- and long-range personal and professional goals. Personal goals should be oriented toward training and issues associated with the job; they cannot be issues associated with their personal lives.

Legal Issues in Hiring

Once a shortlist of potential candidates has been made, the focus migrates to whether or not the applicants are capable of working within the organizational structure with the current employees and existing resources. From a long-range perspective, utilities should evaluate the potential for employees to grow and develop within the organization. It is often obvious which applicants have limited upsides or have skills that are useful in the immediate future but may not have a value in the long term. Such applicants should not be brought on staff unless there is the potential to retrain them for long-term objectives. It is important for organizations to develop a vision and a long-range set of goals, so that the resources acquired at the present time, including personnel, will permit the attainment of that vision and the long-term goals.

Assuming that an applicant gets to the point of being a finalist for a position, they are typically interviewed. Interviews can have more potential for litigation than the employment application, mostly because of uninformed interviewers asking seemingly harmless questions that violate federal statutes. There are a number of questions that are strictly prohibited. First is anything to do with any of the potential federal codes dealing with employee relations, such as age, sex, religion, race, and national origin. No discussion of these items is permitted. The only age question allowed is whether or not the applicant is over 18 years of age. It is allowed to ask if the applicant has a legal right to work in the country either through citizenship or status as a resident alien. Proof of both can be required. Employers may not ask about marital status, maiden names, children, dependents, or other personal issues. No questions are permitted about the employment of the spouse or child care arrangements. These issues especially come to bear when hiring women. For example, the interviewer may not ask a 32-year-old female applicant if she anticipates having a family. It does not matter what the answer is; it does not matter what her plans are; the laws are specific—this question cannot be asked.

The goals of the interview are to gain information. To do so, ask open-ended questions about

- Performing the work
- Applicant's experience

- Applicant's approach to job tasks/challenges
- Interest areas
- Knowledge/depth of knowledge
- Job skills
- How the applicant will react to job situations

In regard to performing the work, you want to get a feel for whether or not applicants have an understanding of what the work is and if they have performed similar work in the past. If they do not understand the work, they probably do not understand the position, and much more training will be required. It is also possible that they are not being completely honest about their experience. Experience is important and usually verifiable, so it's vital to determine that applicants were truthful in regard to employment history. Past experience provides an indication of whether or not they are capable of handling the job proposed.

The approach to jobs and challenges is more oriented to understanding how the employee will react to certain situations. This should be an open-ended question that allows the applicant to explain his/her thoughts on how to approach a given situation. Applicants who have done the work before, or have seen the issue before, they will provide a fairly detailed and straightforward response. If not, this is an indication that, if hired for the position, additional training may be required or that their experience may be limited. Questions about interest areas have more to do with what the long-term potential of the applicant may be with the organization, rather than immediate issues. If the applicant is interested in a career as a water treatment plant operator, then that bodes well for the long term of a water plant. If the applicant is interested in getting a job because he or she does not have one, then this does not. Interest areas will tell you whether or not they have spent any time learning things about water, science, treatment operations, mechanics, or other areas that will be useful to the utility. If their interests lie outside that, then they are not relevant and further depth and explanation of these areas should not be pursued.

In addition to experience and approach, knowledge of situations should be evaluated. The point here is to understand the value of classes applicants have taken, what educational levels might contribute to their long-term potential, and whether or not they have retained knowledge from similar situations in the past. This is a link with the prior issues that will lead directly to what job skills the applicant actually has.

Training/Acclimation

Once selected, new employees need to be acclimated properly. Unfortunately, many organizations do not do a good job in this regard; they send a new hire out to do the job, and the employee succeeds or fails on his or her own. This does not build a long-term support structure, nor does it indicate a long-term commitment by management for the success of the employees. Instead, it is better to assign the employee to a veteran of the department so that he or she can learn the organizational goals, work guidelines, and objectives (i.e., mentoring). New employees should be mentored to help them adapt to the organization; thus they should not be overloaded with work.

Meanwhile the supervisor needs to determine what the new hire can do, i.e., what his or her strengths and weaknesses are, so training can be provided. In most cases it is the responsibility of the supervisor to ensure that employees maximize their strengths and receive training and guidance in their weaker areas, which are probably not primary functions of the job. To make these primary functions would only ensure that the employee will fail in the long term.

Initially, new hires should have a routine follow-up with the supervisor to see how they perceive things are progressing, what issues they might have, and what ideas they may have to help the work effort. In this way, the employee understands that he or she is part of a team and that the team has a real interest in his/her thoughts, actions, and insights.

Performance Appraisals

Employees are normally subjected to periodic evaluations or appraisals of their performance because they need feedback on their work. No feedback means tacit approval of the employee's performance and no correction in performance or attitudes is required. So, unless feedback is provided, whether positive or negative, nothing changes. Performance appraisals are a formal tool used for this purpose. The appraisal is worthless, though, unless the employee knows what is expected of him/her. Performance appraisals should involve the following:

- Assess the employee's work performance, preferably against preestablished objectives
- Provide a justification for salary actions
- Establish new goals and objectives for the next review period
- Identify and deal with work-related problems
- Serve as a basis for career discussions

The performance appraisal should evaluate the strengths and weaknesses of the employee and determine what training or resources might be useful to help the employee improve or develop job skills. Weaknesses should be addressed with additional training. At the same time, it is important to ensure that in a potentially changing workplace, the operation provides support to:

- Develop current job skills
- Measure achievement of utility goals
- Ensure goals are measurable
- Provide solutions when goals are not met

Most employees know that their raises are tied to performance appraisals. For a performance appraisal to be useful, it requires periodic written and verbal communication about the employee's performance and strengths and weaknesses. Employees should be provided ample opportunity to address weaknesses within their skill set and their performance. The case of the typical city manager shows the importance of appraisals.

The city manager works for the governing body. It is they who are his or her supervisors, not the citizenry. The governing body provides direction on policy issues, budgets, and agreements. It is the manager's responsibility to implement them. The manager's performance should be based on his/her successful implementation of governing body directives. The fact that employees may be unhappy might relate more to actions taken by the governing body and implemented by the manager, e.g., no raises, benefit cuts, and so forth, than to the manager. At the same time, the governing body lays out how agendas, action items, budgets, and so forth are to be brought to them, and what reporting they need to monitor performance for the city government.

A new board rarely has had any significant interaction with the manager and certainly has not been in a position to provide expectations. Because the city manager reports to the governing body, the new governing body should first have a discussion to provide the manager with its expectations, realizing these may be significantly different than those of the prior governing body. If this discussion does not happen and the manager does not know that expectations have changed, then who is at fault? Until the city manager knows what is expected, it is difficult to comply. The new governing body firing the city manager may be very unfair unless he/she demonstrates an inability to comply with the board's expectations for directives. The same applies to line supervisors and their charges.

Employees need to know what is expected of them. Expectations must be measurable, definable, and clear. The employee can then be judged based on meeting or failing to meet these objectives. The employee's raises, future promotions, and even employment may be related to meeting expectations. Through the performance appraisal process, the supervisor must identify and deal with any work-related problems and counsel the employees on how they can meet the expectations. If an employee fails, how much responsibility is on the supervisor? Employee failures should be a part of the supervisor's appraisal. Failure can occur on a number of levels, and the supervisor may be responsible for some of them.

Discussions resulting from performance appraisals should include the topics of training and access to training resources. Supervisors should also take care to encourage employees to pursue opportunities. An issue that supervisors need to understand is that employees will generally sort themselves out between high performers and those who are less so. Because raises are provided in this regard, it puts the supervisor in a difficult situation, especially if they have come up through the organization and the people working for them have been their friends. Not everyone gets the big raise. A common problem is supervisors who try to make all their charges happy by recommending maximum raises for everyone. Neither management nor the employees will deem this to be acceptable.

Supervisors' performance is measured as sort of a combination between that of managers and employees. As a result, the effectiveness of the employees is important to the supervisor's appraisal. If employees fail, it is the supervisor's responsibility and can impact his/her raises, future promotions, and employment. At the same time there is an expectation that the supervisor will provide the appropriate tools to help the employees succeed in accomplishing their goals. Failure to do this is seen as a failure by the supervisor, i.e., he/she is not putting his/her employees in a position to succeed.

Managerial performance is measured by overall project management effectiveness, organization, direction and leadership, and team performance. Performance management is ultimately how the organization and the employees will be evaluated. The focus is on establishing certain goals and criteria by which the employee is expected to perform. The work is then measured as successful or not in meeting those criteria, which often leads to whether or not the employee is successful in the job. Performance management is a tool to determine where weaknesses may lie versus strengths within the organization and within the utility as a whole. Performance management includes the development of knowledge through training and observation, the utilization of skills, and the ability to apply skills learned through training or other education. Ultimately questions in four areas arise:

- Skill set. If the employee does not have the requisite skills, why was he/she not trained to get them? Is the employee in the right position?
- Qualifications. If the employee was not qualified to do the job, why was he/she hired?
- Attitude. If the employee shows poor attitude, was he/she counseled by the supervisor?
- Is the supervisor using the correct approach to maximize the employee's potential?

Problem Employees

The employee is responsible for his/her actions, so drugs, alcohol, and reckless endangerment are problems that should not be tolerated and action should be taken immediately. However, these are typically not reflected in a supervisor's appraisal. There are a number of reasons for failure of the employee, including the following:

- Attitude was wrong for the job
- Skill set wrong for the job
- Education wrong for the job
- Did not respond to supervision approach

Where an employee does not meet the performance appraisal expectations, there are a variety of processes that can be employed. Counseling about performance or behaviors is first. If this is unsuccessful, an oral reprimand with documentation in the employee file should be completed. These may be removed later if deemed appropriate. If the verbal reprimand is not effective, a written reprimand describing the problem, corrective actions, timelines for corrections, and consequences for failure to complete corrections may be the next step. This would be followed by suspension and termination. In most cases, employees have an opportunity to appeal a termination.

Other alternatives to termination may be employee assistance or counseling, a change in the job duties to more closely match the employee's skill sets, paid leave of absence (if there are conflicts in the personal life of the employee), or intervention for substance abuse or other issues. In any case, the intent is to build some trust between the supervisor and the employee so that the employee will listen and allow the supervisor to guide him or her to better performance.

Terminations

Termination of an employee should be the last resort (OWBC, 2008). There are a number of issues to consider when terminating an employee. The first is determining if the problem relates to ability or attitude. Ability problems mean that the employee was not appropriate for the position in the first place and probably should be transferred somewhere else. It may be the fault of both the supervisor and the employee when the latter fails. When an attitude problem or a lack of desire to conform is the issue, there needs to be a personal interview and written report signed by management, the supervisor, and the employee documenting the issues. The report should be dated and included in the personnel file. Employees have the right to comprehend what the consequences or actions may be, just as supervisors have a right to know what the employees are doing. The two-way communication here is important to ensure that, if a termination needs to occur, it will be done in the appropriate fashion and not come back to the supervisor in a litigious situation.

Firing of employees should only occur after the employee has been given a clear indication that he or she has deviated from the expectations, i.e., from a written job description provided to the employee or in an appraisal of performance. It also requires that there be a clearly written personnel policy that specifies the conditions and protocols for firing employees and that the employee has acknowledged receipt of the policy. This way, employees cannot argue that they were never given this information or did not know the consequences of their actions.

Employees also must be warned in successive and dated documents about the degradation of their performance over time and provided a record that demonstrates that the supervisor has, on multiple occasions, attempted to provide guidance that would help them avoid termination. The number of memos or counseling sessions that may be needed in the file may vary, depending on policy (but there should not be more than three or four), after which the employee must continue to demonstrate the adverse behavior or performance. The only exception to this is if employees are terminated within a probationary period, which is specified in the personnel policies and for which the employee must acknowledge understanding.

Dismissing an employee is a somewhat personal decision, although based on business parameters. As a result, employees should never be terminated on the spot. Termination requires deliberate action, consideration, and consultation with human resource personnel and others. A letter of termination to the employee should be developed (OWBC, 2008). As with previous letters, it must be clear what the problematic behavior or performance is, what guidance the employee was given in the past, the indication of the consequences for failure to respond, and the fact that corrective action has not been taken and the consequences must be enacted.

Once the termination decision is made, discussion with the employee and others, as necessary, should be set up so that procrastination and the potential for damage to the organization is minimized. Computer access via password should be revoked immediately. Providing the employee with the letter of dismissal at a meeting is the next step. An explanation of the termination process is also required. This would include explanations of when the termination is effective; what materials, such as keys, other access devices and documents, need to be turned over; and that someone will guide the employee to their office to remove personal items. Employees should not be permitted to do this without supervision as they may sabotage computer systems or remove items that are not theirs. In all cases, a set of notes of the meetings and conversations, what was said in exchange, and witnesses should be included. Witnesses are usually a good idea for terminations.

Terminating an employee does subject the organization to difficulty relating to securing both noncompete and proprietary information. Both may create damage to an organization should valuable company information be removed from the premises. This is why employees should not be permitted access to their computers or allowed a significant amount of time with co-workers.

Improperly handled employee terminations generate a significant number of lawsuits against corporations and municipalities each year. The only defense against them is the appropriate documentation of the behavioral issues, notifications of actions taken, recommendations, and opportunities on multiple occasions for the employee to correct the problem(s). When these exist, the employer usually will not suffer any significant adverse impact. However, firing any employee creates stress with the employee, the supervisor, and other employees with whom the terminated employee had worked.

Before firing an employee (OWBC 2008):

- Be sure the action is approved by top management and conforms to written company policy.
- Seek legal advice regarding severance conditions for higher-level employees.
- Note the employee is entitled to a documented, concise explanation of the reasons for his or her dismissal.
- Plan the interview carefully to anticipate responses and defuse reactions.

Consider the possibility of an irrational response by the dismissed employee. Take the necessary precautions to change security codes, access codes to computers, and entry to the corporate premises.

A major concern of those terminated is what will be said about them behind their back. It is a good policy to reassure workers that, except for the management team involved in the termination or others on a need-to-know basis, the issue will not be discussed with employees.

The direct supervisor should terminate an employee. The direct supervisor should prepare what is to be said, reserve a private meeting place, and set a meeting tone of cordiality and empathy. It is useful to thank employees for the good they have done, but this discussion should be kept short lest the person somehow thinks he or she is being called into the office to be commended. The bad news can be given next. This needs to be done calmly and with empathy, without gloating. In handling the temination, the supervisor should be honest and clear about the reasons for discharge, and avoid personal statements that might degrade or humiliate the individual or vague statements that might suggest that the situation is reversible. It is customary to have another individual, such as a professional from human resources, present as a witness and a support for the employee, particularly if emotional reactions are anticipated. Information about severance pay procedures, benefits continuation forms, pension or profit-sharing payouts, and other available assistance, such as outplacement counseling, should also be provided. The employee should not be fired first thing in the day unless it is before the start of the work day.

The employee should be allowed to leave quietly with his/her personal things without confrontations or interaction with other employees. Evenings work best because people can plan on being around to help. The supervisor needs to make a list of organizational belongings and tools, i.e., a checklist of company property that should be accounted for, including keys, credit cards, ID cards, and computer disks.

When an employee is given the choice to resign or be terminated, it is considered a case of "constructive discharge" and is no different than a termination. This may help employees feel better about themselves but does nothing positive for the employer. It may subject the employer to legal issues about job performance, thus the prior advice about documentation hold. The only benefit here is for argument in an unemployment hearing, i.e., if there is a resignation letter, then it is harder for the employee to argue that he or she deserves unemployment compensation. However, if the employee argues he/she was not given a choice, and this can be documented, then there is no difference between resignation under pressure and termination.

Two common mistakes at this stage are when the supervisor is vague to the point that the employee does not know he or she has been terminated and/or the supervisor talks too much. Silence can make interpersonal situations uncomfortable, and in an effort to fill this silence, the supervisor is likely to say more than what should be said.

Employers have certain obligations even after an employee is fired. Human resource personnel should be involved to deal with pension, health and other insurance, COBRA, payment of vacation time earned, and so forth, that are beyond the purview of the supervisor. Any termination should be fully documented with human resources and a "company line" developed on why the person is no longer with the organization. Talking badly about the person is inappropriate and subjects the supervisor to slander suits.

Justifiable reasons for employee terminations are usually outlined in an employee handbook or personnel manual and include:

- Incompetence or failure to respond to training
- Gross insubordination
- Repeated unexcused absences or lateness
- Sexual harassment
- Verbal abuse
- Physical violence
- Falsification of records
- Theft
- Drunkenness on the job

Court rulings have determined that an employee cannot be fired for:

- Whistle-blowing in regard to employer policies or violations of laws
- Complaints or testimony regarding violations of employee rights
- Lawful union activities
- Filing claims for workmen's compensation
- Filing charges of unfair labor practices
- Reporting OSHA violations
- Garnishment for indebtedness

Any of these actions would be considered retaliation and subject the employer to severe penalties. This right protects the employee from "termination at will," the previously popular employer practice of discharging an individual for virtually any reason. Employees who have obtained a job and perform it competently have a right to expect that they will continue to be employed as long as the organization is solvent and the work they do continues to be required. This rule was enacted to limit at-will firings and retaliation by supervisors against employees they did not like.

If an employee is fired for cause, no recommendations of any type should be given. Never give a bad reference as it can only lead to legal difficulties for the employer. Employers only have the obligation to confirm the person worked there.

Layoffs

There is a difference between terminating an employee and layoffs. The latter have to do with the fiscal strength of the organization or the need to continue a certain job function. Termination has to do with performance of the employee personally. Layoffs may create panic among both employees and customers, so this needs to be handled delicately. Federal law requires 60-day advance notification to employees affected by layoffs and plant or office closings. Prematurely early notification may significantly affect production and possibly invite undesirable reactions. In the event of layoffs, as with termination, employers have obligations to review vacation time, COBRA health insurance, and other benefits, as well as unemployment insurance. Employees who have questions and concerns should be encouraged to meet with specified representatives of the company privately.

People experiencing job loss may go through some predictable emotional stages, including lowered self-esteem, despair, shame, anger, and feelings of rejection. The greater the positive feelings the employee held toward the supervisor, and the longer the period of employment within the operation, the more poignant these feelings may be.

ORGANIZATIONS

Typically, larger organizations that are effective in producing products or services do not have strict hierarchies or many levels of management. The tendency is for their organizational structure to be flat and for employees at all levels to be assigned and empowered to take responsibility for their efforts. At the lowest level, this means that the employee is responsible for whether or not his/her efforts succeed or fail, and action can be taken where repeated failure occurs. As mentioned previously, supervisors fail personally when the goals set by the organization are not met by the staff being supervised.

The way this relates to utilities is that, in most scenarios in small systems, one person may be responsible for all tasks and as a result, the success or failure of the organization is dependent on one person. In larger organizations, the number of tasks accomplished is the same, but there are more repetitions (e.g., more pumps to grease and more packing to be replaced). As a result, staff is hired to do these repetitive tasks and trained accordingly. The supervisor must ensure that the tasks are accomplished and retains responsibility if they are not. Few supervisors or managers can operate and maintain their water system, make repairs, operate the backhoe, purchase the materials, supervise construction, and do all reporting, bill paying, and rate collections, all by themselves. Understanding strengths and limitations is key to productivity.

REFERENCES

Olstein, M.A, D.L. Marden, J.G. Voeller, J.D. Jennings, P.H. Hannan, and D. Brinkman. 2005. *Succession Planning for a Vital Workforce in the Information Age.* Denver, Colo.: AWWARF.

OWBC. 2008. Online Women's Business Center. Preparing for a Termination section. "Parting Ways: Effective Termination Techniques." http://iwantmoresales.com/TerminateEmployees.htm. Accessed 8/28/08.

Ralston, S., and J. Ginley. 2010. In Deep Water. *American City and County*, p. 30–43.

USDOL. 2008a. Civil Rights Act of 1964. http://finduslaw.com/civil_rights_act_of_1964_cra_title_vii_equal_employment_opportunities_42_us_code_chapter_21. Accessed 5/10/2008.

———. 2008b. Equal Pay Act of 1963. http://www.dotcr.ost.dot.gov/Documents/ycr/EQUALPAY.HTM. Accessed 5/10/2008.

———. 2008c. Age Discrimination and Employment Act of 1967. http://www.eeoc.gov/laws/statutes/adea.cfm. Accessed 5/9/2008.

———. 2008d. Americans with Disabilities Act of 1990. http://www.ada.gov/pubs/ada.htm. Accessed 5/9/2008.

———. 2008e. Civil Rights Act of 1991. http://www.eeoc.gov/laws/statutes/cra-1991.cfm. Accessed 5/9/2008.

———. 2008f. Rehabilitation Act of 1973. http://www.dotcr.ost.dot.gov/Documents/ycr/REHABACT.HTM. Accessed 5/9/2008.

———. 2008g. The Family and Medical Leave Act of 1993. http://www.opm.gov/oca/leave/html/fmlafac2.asp. Accessed 5/9/2008.

———. 2008h. Employee Polygraph Protection Act of 1988. http://www.dol.gov/compliance/laws/comp-eppa.htm. Accessed 5/9/2008.

———. 2008i. Uniform Services, Employment and Reemployment Rights Act. http://www.dol.gov/vets/usc/vpl/usc38.htm. Accessed 5/9/2008.

———. 2008j. Drug Free Workplace Act. http://frwebgate.access.gpo.gov/cgi-bin/getdoc.cgi?dbname=browse_usc&docid=Cite:+41USC701. Accessed 5/9/2008.

———, 2008k. Workers' Compensation. http://www.dol.gov/dol/topic/workcomp/index.htm. Accessed 5/9/2008.

———. 2009. Fair Labor Standards Act. http://www.dol.gov/whd/flsa/index.htm. Accessed 5/9/2009.

CHAPTER 6

Supervision of Operations

DEFINING SUPERVISION

A supervisor is responsible for the efficient direction and use of human and material resources in such a manner as to effectively produce goods and services to meet customer needs. Some other terms associated with supervision would be *optimal*, *efficient*, and *low cost*, which further increase responsibility and expectations on the supervisor.

Supervision could be defined as using available resources, human capital, materials, equipment, and funding, to meet the organizational goals. The proper allocation of these resources, i.e., knowing when to use them and when not to, is one facet of successful supervision. The supervisor is responsible for developing goals for the employees under his/her control, in keeping with the goals of the organization, and to direct and allocate the necessary resources to permit the employees to achieve those goals. If the efforts do not appear to be working toward achieving organizational goals, then it is the supervisor's responsibility to make necessary changes to improve the likelihood that the goals are met.

Without supervision, basic direction, and an orientation toward achieving organizational goals, chaos would result. Everyone would do what they felt was most beneficial to them, which would not provide any cohesive effort or include any organizational goals. Frederick Taylor, the "Father of Scientific Supervision," tried to convince organizations to do the following four things (Taylor, 1911; FWPCOA, 1997):

1. Increase physical output
2. Decrease production costs
3. Lessen employee fatigue
4. Increase wages

The reasons for doing these are simple. At the time when these recommendations were made, the average wage was minimal and employees worked many hours to acquire those wages. Modern manufacturing processes were not invented until the twentieth century when Henry Ford invented the assembly line. There was limited mechanical power, so the employees did everything by hand. Most employees likely believed there was little or no hope of improving their lot in life. As a result, there were no incentives to work harder, because working harder simply meant doing more work, a disincentive to employees. Automation improvements in manufacturing processes improved profits, productivity, and efficiency, and provided less-expensive products that spurred other economic activity.

Part of supervision is getting the most out of people, which means letting them develop to their fullest potential (which may mean them becoming your supervisor). The advent of new ideas implemented to improve manufacturing productivity created the need to maintain equipment to ensure continued productivity and the need to improve labor skills. The way to improve labor skills was to improve education through training and public schools, and invest in technology that would minimize adverse work conditions and lower the potential for injury to employees. It was realized that trained employees were assets (human capital) just as much as equipment, buildings, and products. As a result, the investments in these assets needed to be protected because training new employees often required hours of downtime. Having skills also became more of an issue in hiring employees because profits were tied to efficiency.

To understand the change that occurred during this period one needs to look at the specifics of the evolution of technology and productivity. Effective training needs increase as technology advances. Technology is one of the major factors that has changed the workplace, starting with electricity, and moving to assembly lines, transportation changes, computers, and robotics. At each point, the intent of the change was to improve productivity while reducing human error and interface. One could view this as a way to cut jobs, but the opposite is true. What technology does is reduce the demand for unskilled-labor jobs, which are much more efficiently performed by machines.

An obvious utility example of this is ditchdigging. A four-man work crew is common for water distribution pipeline crews. How productive would these people be if they had to lay all piping by hand? How much pipe could they lay in a day? Working diligently, it is doubtful that they would lay more than 50 ft (80 m) of pipe, let alone backfill it properly. They can lay the pipe—the easy part—but it is the digging that is the problem. So backhoes were developed through advances in technology. The backhoe eliminates a substantial number of unskilled ditchdigging jobs, as it, with the appropriate operator, can dig as much as 1,000 ft (1,600 m) of trench a day, and backfill it. To achieve similar results, it would take a crew of 100 or more. Yet, the backhoe cannot put the pipe together. Putting the pipe together requires manpower to chain it for lowering and placement. The backhoe can help by lowering the pipe, and perhaps pushing it together, but the backhoe cannot guide the pipe. The number of pipe layers remain the same, but technology has put a lot of ditchdiggers out of work. But now people are required to operate the technology and service/repair it, which are jobs that did not exist before the technology was invented. These are higher-paying jobs, for which training is needed.

Technology and training help utility personnel do their jobs more effectively and, in most cases, more efficiently. When training, technology, and work conditions improve, higher effectiveness will follow. More highly skilled workers with proper training can do their work more efficiently, which improves morale. Greater efficiency benefits the organization as a whole because unforeseen costs should decrease.

The fact that there is less physical labor involved in the field means more inside work is needed to maintain complex technology (more training), which improves work environments. Less time is required outside in the weather, while more time is spent inside in the air conditioning using more SCADA and more computers. Better working conditions and higher-training-level requirements will attract different types of workers, who will seek higher pay and better benefits and may accomplish more as a result of new technology. Higher output means more can be paid for skilled workers. Overall work conditions improve as a result.

Organizational health is an important driver, especially for good employees. As Collins notes in *Good to Great: Why Some Companies Make the Leap…and Others Don't*, great organizations strive to place the best-qualified people in the proper positions (Collins, 2001).

The concept is that people who have the skills and drive to do their work will do it well and be happy. They will work as a team, demonstrating respect for one another, while solving organizational issues. The key is making the organization great, not individuals. The latter comes with recognition of the organization.

Organizations that just "fill positions" typically are more hostile. Hostile environments are not conducive to effective or efficient operation. In addition organizations that attempt to resolve discourse through replacement with "yes people" fail to adapt to changes in the field, which impinges on their ability to remain effective or efficient. "Good and great" supervisors are looking for people who strive to be great, who have the requisite skills, and who can provide perspective on their areas of responsibility (Collins, 2001). Having differing opinions is good, assuming some consensus can be reached, as opposed to infighting. This is where the supervisor rises to achieve effectiveness in the organization, or fails personally and organizationally. It is the latter issue that most supervisors do not understand; i.e., their responsibility is to the organization.

Great supervisors surround themselves with great employees (Collins, 2001). Great supervisors will tell you that what to look for in employees is people who can do more than the job they are being hired for. Great supervisors often look for employees with more skills and potential than they themselves have. The reason is that such people will be successful. They will ultimately leave the position, but they likely move upward in the organization.

There are a variety of factors that impact the supervisor's ability to accomplish his/her goals effectively. One is economics. If the economy is poor, pay increases or bonuses are difficult to provide. As a result, economic turmoil, recession, and depression will adversely affect the supervisor–employee relationship. Employees have little option for movement, so perceive little bargaining power with employers. Societal adjustments are also a confounding factor. Younger employees have different work ethics and different incentives than older employees. They value free time and less rigid work environments. Older employees accept rigid work schedules and may be more motivated by pay and security.

Legal, political, and regulatory issues are often related to one another. In the early twentieth century, major inroads were made to improve working conditions, especially for textile workers. Child labor was eliminated and education was encouraged for all children. Today the Fair Labor Standards Act limits work hours and requires overtime pay beyond a 40-hour workweek. Equal-opportunity laws prohibit discrimination in hiring employees and OSHA requires employers to provide safe work environments and train employees on workplace hazards. Supervisors must understand the implication of all of these types of laws and policies to the organization and to themselves personally (see chapter 5). Supervisors who do not comply with these rules can be held personally liable.

Given the responsibilities and restrictions, and based on the reasons why a person would want to be a supervisor, problems can occur when operations personnel are promoted from within. Line personnel often have relationships with one another, often close ones. This is normally useful in the field, but it is not appropriate between supervisors and those under their supervision, and remains a common problem for those who are promoted from within. One cannot maintain the same relationships with one's old "buddies" and supervise them effectively. A common reason for failure among supervisors promoted from within organizations, and a major reason that management does not often do it, is that operators who are promoted from line positions to supervisors cannot overcome their roots.

A part of supervision is the ability to adjust to a dynamic work environment, something that line employees are less comfortable with. A criticism often levied on people who have been in supervisory positions with utilities for many years is that they still do things the "old way" and have not updated their skill set. Dynamic supervision requires that the supervisor understand that different circumstances and different people will require different approaches. The "one-size-fits-all" view rarely applies in the workplace anymore. Supervisors who have dealt with this changed axiom are more successful than those who have not. Continuous improvement is an ongoing, never completely achieved goal that provides incentives to supervisors. This view also implies that the direction of staff and use of resources will be allocated to meet these goals.

SUPERVISORY SKILLS

The supervisor has a distinct role in the organization and many roles with employees internal and external to the utility. Supervisors (in theory) know more about the job and the organizational goals than line employees. They also, in theory, know how to motivate people. As discussed in the following paragraphs, the supervisor should function as a leader, decision maker, motivator, coordinator, counselor, communicator, and trainer of those under his/her charge (FWPCOA, 1997). He/she is responsible for them and responsible to them. The actions of the supervisor affect the employees directly and affect the perception of the employees, which in turn affect the employees' perception of the supervisor. A poor supervisor can do significant damage to an organization and thus should receive some training on how to implement each of these ideas correctly in their organization. Employees who are promoted from within may not understand how to convey leadership to their new subordinates, because they may not understand how to separate themselves from their co-workers.

From a leadership perspective, the supervisor is expected to demonstrate the ability to plan and coordinate work and assign work appropriate to the skills of his/her charges (FWPCOA, 1997). As a result, leaders must be willing and able to make decisions, so these roles are intertwined. The leadership role includes the ability to create a plan, evaluate alternatives, make decisions about those alternatives in a timely fashion, and take responsibility or ownership of those decisions. Not all decisions will be good ones and not all will work out well, but this is part of the risk taking that occurs in the leadership role. Some people are unwilling to take risks. For example, a person who has never made a decision is not a better supervisor than a person who is only right half the time. The person who has made no decisions has a 0 percent success rate and has accomplished nothing, whereas a person who has made 20 decisions, only 10 of them good, still has a 50 percent success rate and has likely accomplished something. However, decisions should not be made recklessly; this shows a lack of leadership, not decisive leadership.

The leadership role blends directly with the motivator role. The motivator needs to understand how his/her behavior as a supervisor affects the employees and how the behavior of the employees affects other employees and the organization (FWPCOA, 1997). The supervisor's responsibility is to protect the organization and to use the employees to accomplish the goals of the organization. To do this, he/she needs to be able to motivate the employees to buy into the goals of the organization and the tasks that need to be undertaken to accomplish those goals.

Workers need to be satisfied with their work and feel appreciated. They also need to understand why the tasks they are doing are important. As a result, the motivator must understand individual needs, needs for satisfaction, and how to deal with differing human behaviors. Many "old-school" supervisors believe that all people must be treated the same way. Younger supervisors realize that this is impossible to do; all it does is create difficulty for a large number of employees who may not respond to the same motivational techniques that others do. People respond to different stimuli. Some employees need rigid repetition to be successful. Others, especially younger people, are motivated by accomplishing tasks using their own methods. In some cases, their ideas may save time and effort. In other cases, they must accomplish tasks through more traditional methods.

An effective supervisor can discern what motivates individuals so as to get the best out of them. Because employees have changed, supervisors need to determine how their charges respond to stimuli and direction. Some people need and want direction. Others may need only limited input—they see the issues and address them without direction. Still others may need direction, but resent it. How each of these is approached will be telling in terms of the supervisor's effectiveness.

The coordinator role links the motivation and leadership roles together. The coordinator role is to ensure that reasonable work tasks are coordinated together among the crews as needed to accomplish the goals of the organization. The goal is to ensure that all parts of the operation work smoothly together and efficiently so that, for example, when parts or repairs are needed, they are scheduled in a timely fashion so that the equipment or systems are back up and running in as short a time as possible.

The coordinator role requires that the supervisor understand the skills and limitations of workers so that tasks are delegated to those who have the capabilities to accomplish them efficiently, as opposed to those who may fail due to a lack of ability. For example, if the task is to dig up and repair a water main, then a group of operators who are good at working with chemicals, SCADA, and jar tests would not be the appropriate people to assign. Instead, field maintenance crews with experience with backhoes, jackhammers, and pipe cutting would seem more appropriate. Assigning the right staff minimizes confusion and misunderstandings among the workforce, and allows the workers to understand what their responsibilities are.

The counselor role is designed to eliminate problems sometimes created with interpersonal relationships on the job. This includes the elimination of emotional issues, friction, stress, and frustrations (FWPCOA, 1997). Employees need the opportunity to talk to somebody about issues on the job that are creating difficulties in performing the tasks at hand. It may be that the employee is not suited to do the tasks being assigned, and that being assigned different tasks might prove beneficial to the organization.

Good communication enhances performance. It is the most important thing in ensuring that the goals and objectives of the organization are met. In the counselor role, the supervisor needs to have the ability to communicate with the employees to determine what the real causes of problems are and to convey to them what the need for certain projects or policies may be. The supervisor must communicate effectively both vertically and horizontally with other departments and with management (FWPCOA, 1997). The supervisor needs to know which issues are applicable to a given level of the organization.

The supervisor should never talk down to any employee or any other person in the organization. All people deserve and require respect. The supervisor should minimize his or her own emotions and avoid rash acts. At the same time, some employees just need to get something off their chest. Once that is done, the issue is actually solved. Trying to take action to solve the issue after the fact actually creates more of a problem than what existed initially. This is an extraordinarily difficult issue for young supervisors to understand.

While increased compensation, respect, influence, a better working environment, and more freedom are often cited as reasons people want to become supervisors, many new supervisors do not realize that with these things come the following:

- Responsibility
- Accountability
- Tough decisions
- The need to listen and to try to help your people with their problems
- Limited authority and more rules

The first of the issues that come with supervision is responsibility. A supervisor must take responsibility for the problems that occur with his/her crews. If the supervisor does not do that, he/she risks losing the respect of his/her employees. Since the supervisor is the one assigning resources and personnel, obtaining tools, and providing input for budgets, his/her failure to adequately plan any of these activities will create difficulty with both management and the staff. Supervisors must make tough decisions, often ones that are especially difficult if he/she has been promoted within a group. These may include making decisions that adversely affect old friends or co-workers as a part of the department. Such decisions can create a significant amount of friction and add to the stress on a supervisor. In order to avoid friction and stress, the new supervisor needs to listen and try to help the people with their problems while realizing that he or she has limited authority to make changes.

Choosing a supervisor is important to the success of the organization. When hiring a supervisor, managers usually look for a number of skills, among them the ability to comprehend both abstract and general ideas; the ability to investigate and identify the true problems that exist with equipment, operations, and people; technical training; the ability to work with people on an interpersonal basis through communication, motivation, and accommodation; and some political acumen. But remember, not everyone can be a successful supervisor.

SUPERVISORY FUNCTIONS

Not only does a supervisor need to have a variety of skills that must be judiciously implemented with the resources available, but he/she must also understand the supervisory functions. These include planning, organizing, staffing, directing, controlling, budgeting, and decision making. At various times the supervisor must implement each of these to optimize performance of the organization and to accomplish organizational goals (FWPCOA, 1997).

From a planning perspective, it is the responsibility of the supervisor to plan the work that is to be accomplished each day and over short- and long-term horizons. It is necessary to have a series of long-term goals for the organization to achieve so that the supervisor can make decisions and plan work so that all components necessary to achieve these goals occur at appropriate times. Planning requires not only understanding the goals, but understanding what is necessary to accomplish them in terms of proper utilization of employees, equipment, materials, and other resources.

Evaluation of how well planning occurs on the part of the supervisor is indicated by whether or not the goals are met and whether or not there is a minimal amount of wasted time, effort, or other resources. For example, you can have a great crew, highly motivated, efficient, and well equipped, but if they go to a job site to do a job, then find out that someone forgot to order all the parts, then this shows a lack of supervisory planning.

The planning horizons are short and long term. In many cases, decisions made in the short term affect the long-term ability to accomplish certain tasks or to complete certain types of work. Short-term goals may involve days, weeks, or months. For a line supervisor, many of the decisions are day-to-day implementation and utilization of resources. Longer-term planning horizons may be goals to complete by the end of the month or by the end of the fiscal year, whereas short-term goals might be completing preventive maintenance work, performing monthly operating report gathering, and daily sampling. Longer-term actions might include replacement of pumps or the shutdown and repair of certain treatment plant components. Long-term actions might be oriented toward plant expansions or new water sources that might be as many as 20 to 50 years out, but these are primarily management issues, not line supervisor responsibilities.

A fourth area of planning involves understanding what emergencies or conditions might occur, what breakdowns could occur, and where pipes and equipment might fail. This will help with creating contingency plans that are part of the supervisor's responsibility. The line supervisors are typically responsible for implementing the emergency plans and taking actions associated with vulnerable infrastructure (see chapter 4). In order to create a proper plan, whether this is done on paper or on a day-to-day basis within the supervisor's head, the plan requires that goals be established (e.g., what is to be accomplished today, tomorrow, this week, this month, and this year) with the understanding of the current situation.

Once the goals are established, it becomes clear what tasks need to occur in order to get from the current situation to the hoped-for future condition. Understanding barriers or issues that might prevent accomplishment of those goals in the future will become more obvious as more planning and discussion among those completing the work occurs. What this shows is that communication is required, which is why it is necessary to write things down versus, for example, permitting planning in the supervisor's head.

Once the goals, present condition, and barriers to accomplishing the future are understood, an action plan can be developed where resources are allocated, materials ordered, and staff and equipment provided to accomplish the task. A budget will be developed so that it can be determined what the various tasks may cost.

Looking at an example, operators take water samples on a daily basis to monitor the treatment process, measure chlorine residuals, and take biological samples. They also monitor equipment. Thus, on a daily basis, there should be a plan laid out so that everyone knows when these tasks are to occur and what results are expected. In a treatment plant, it might be that a jar test is conducted on an hourly basis to determine whether or not the coagulation process is functioning properly. Or, on an hourly basis, a chlorine residual test might be taken. This can be written out on a schedule so that everyone knows that every hour this test must be completed (and appropriate employees assigned to do the work). In certain months the operating data and daily operating data must be recorded, hence, once a sample is taken, data must be entered into some database to indicate that the work was actually completed. This should also be scheduled.

In doing an annual budget, it is clear how many tests are going to occur and how much effort it actually takes to accomplish them, so those resources can be allocated. Therefore, the short-term goal of taking samples hourly, recording the results, and determining what the long-term set of results will look like provides data so budget costs can be assigned.

A repair, such as to a pump, is an action that would occur rarely. The supervisor would have to determine how much downtime there might be, what resources are needed, and what contingencies could occur. In this case, contingencies could include taking longer than expected to get the motor rewound, bolts that do not allow the pump to be moved as efficiently as anticipated, bolts that break off, and/or a pump that actually cannot be repaired. All are potential issues that would keep the staff from accomplishing their goal in a timely fashion. Necessary resources that might need to be scheduled include equipment like a crane, an extra person or two to ensure that this gets accomplished, and electricians or others to connect or disconnect the pump. As a result, one can figure out what the cost of these factors would be, and schedule everything accordingly.

Once the plan is created, the action is taken and results are verified. If the supervisor does not verify that the results actually occur and that things occur as anticipated with regard to resources and budgets, then the planning process never has a balance to determine how effective actions may be. Hence, work order tracking and maintenance management systems have significant value in determining the frequency and cost to make the repair for budget purposes.

A lot of work on the part of the supervisor goes into planning and it may be one of the most important aspects of the supervisor's job. Planning work and budgets and making sure materials are on hand are all activities that will take some thought and time to accomplish properly. Meetings to discuss what is expected and how it is to occur are generally required even if they are brief. There is also the supervision to ensure that the work is actually done.

Organizing for the work means the following need to be defined: the task to accomplish, the human resources required, the skill set required, materials, and the work environmental where it will be accomplished. Most of these are obvious, but the work environment may be more difficult to define. The work environment may be a normal site during the day or it may be scheduled at a specific time and place. For example, a pump may need to be changed at midnight when flows are low because it needs to remain in service during the day. If the utility is doing work at night, extra lighting may be required, people may need to be brought in after hours, and so forth. Therefore the environment affects all those decisions in regard to resources required.

The supervisor must have authority from management. If the staff does not believe that there is authority from management to the supervisor to direct the work, then the work may not get done efficiently. In this case, authority means delegation of responsibility from management to the supervisor, which means that there is an expectation of certain work being accomplished. The supervisor should delegate the responsibility for task completion to the line personnel. The supervisor should not have to watch to make sure that every task gets accomplished; he or she delegates that to the staff and holds them responsible for a task's completion. For example, if operators fail to do the hourly sampling tests mentioned earlier, then they need to be questioned as to why and perhaps disciplined for failure to accomplish this task. Ultimately, this is a team-building exercise in which the team is only as effective as its weakest member.

Supervisors want to build an effective team, which requires that the team understand the work they will do and what is expected at the conclusion. Productive, happy employees are enthusiastic about what they are going to accomplish and are typically more productive and efficient in accomplishing tasks than those who are not. This fosters teamwork because everybody appears to be pulling their own weight and the employees believe they can count on one another. Teamwork requires that the supervisor not micromanage. Tasks are delegated so employees feel some empowerment and have the ability to make choices and decisions that affect what they are doing. In many cases, supervisors, especially young ones, do not understand that many decisions have results that are inconsequential to the accomplishment of the organizational goals but may have a very significant impact on the person who is actually doing the task. Choice of equipment is a good example. A group of employees may be much more confident in getting work done using one piece of equipment than another. As a result, if it is critical to get this work done, the issue may be minimal to the supervisor, but important to those employees.

Direction is defined as the issuance of orders, instructions, assignments, and the necessary guidance and oversight/follow-up. Direction is useless without the following:

- Planning (discussed in chapter 4)
- Evaluation of talents (chapter 5)
- Acquisition of tools and materials

To use a football team as an example of the process, it is accepted that the goal is to score touchdowns and win the game. The quarterback on the field is directing the action, but he has a supervisor, the coach, who is directing him (calling plays and sending in the personnel to enact those plays). The coach instructs all the players and tells them their responsibilities. If everyone accomplishes their responsibilities, they will score touchdowns. The more effective they are, the sooner this will occur. Once on the field though, the quarterback assumes the supervision of what happens. He may change the play; he may change direction; he may change receivers to whom the ball is thrown based on the conditions on the field. But they have to work as a team. If they do not, the quarterback gets sacked or throws an interception and progress is stopped.

If one observes team sports, it is clear that groups that do not act as a team are less effective than they would be otherwise. As a result, everyone understands what they are going to do, what is expected, and, as in sports, those who cannot comply with the activities are usually cut. If they are winning, everybody is happy. Productivity is winning.

Supervision requires a lot of planning, effective use and evaluation of talent, and acquisition of tools and materials, which takes a substantial amount of time and thought. Some employees do not understand how much time this actually takes and some supervisors, when they take the job, do not understand it either. If insufficient time is spent on planning, the organization will suffer.

There a few steps in the process of direction. In assigning responsibility for the work to be done, it must be clear who is responsible (i.e., which team members are responsible for it) and ensure that everyone understands what is to be accomplished. There must be a timetable for when the work is done or at the intervals at which the work is done. The team must accept the job responsibilities, which means that they understand the tasks at hand and will perform them (Diamond, 2007). If there are employees who refuse to accept responsibility, then they ultimately are probably not beneficial to the organization. But most employees, if they believe that their opinions matter and their skills are used, will accept responsibility for their piece of the operation.

Directions must be clear and reasonable. For example, "You must lay a 10,000-ft water line in three days," is clear but not reasonable. Unreasonable direction will not be conducive to the organizational goals or team building (Diamond, 2007). In addition, assigning work that has no relationship to organizational goals will cause the supervisor to lose the respect of his/her crews, and the work is unlikely to be accomplished (employees realize that failure to accomplish goals falls on the supervisor). It is the supervisor's responsibility that the work expected to be done is understood.

Miscommunication is the most common problem in the workplace. The common response to work not being done as the supervisor intended is to have lengthy discussions with the employee or employees responsible. Unfortunately, it may be that the supervisor failed to convey the information properly, not the employees' failure to do the job as intended. Failure to communicate only leads to hard feelings on the part of the employees and frustration on the part of the supervisor. Hence, it is important up front to make sure that everyone understands what is expected.

Teamwork skills are required in any organization. Individuals who go off on their own are rarely successful, although they may be successful in accomplishing personal goals. In part, good interpersonal relationships build an opportunity for positive progress and goals. For example, sports teams with divisive locker rooms are often not successful (i.e., not a winning team). The issue is directly akin to that in virtually every other workplace. It is necessary that appropriate expertise and resources exist within the team and that team members understand who is responsible for which resources and where to go for the necessary expertise. The goals for the team must be clearly defined and obtainable. Unrealistic goals will hurt teamwork because the team will believe it is doomed to fail. There is a need for management at all levels to be supportive of the goals of the team, provide leadership, and communicate to the team (Diamond, 2007). The expectation is that the team will be able to communicate among themselves. This does not mean that they need to be friends, but that they need to have the ability to work together to accomplish the greater good. If each employee does his/her share, the combined efforts are of greater value than the same individual efforts. Typically teams that work together will show a limited amount of interpersonal conflict or behaviors.

Control is another of the steps required to ensure that the work gets accomplished and that progress toward the goals of the organization is ongoing. Control is mostly about standards. Some questions that might arise following an activity are:

- Was the time to accomplish the work reasonable? Was it too long or too short?
- Did the team work effectively?
- Did the personnel perform satisfactorily?
- Did they all do their jobs as they were expected to?
- Was the cost reasonable or did they end up spending a lot more time and effort than was needed, or was additional equipment required that was not anticipated?
- Was the job done satisfactorily or were there issues?

In order to determine whether or not these goals are met, the supervisor needs some form of feedback, which is why tracking time, money, and resources on a work order is required as a part of the process. Once the work is complete, an evaluation can be made to see if the plan and the results coincide. It may be that certain things occurred that were not planned, the budget was overspent, or the time required exceeded expectations. The supervisor needs to be reasonable in his/her evaluation. His/her estimates of time may not have been correct or estimates of budgets may not have been correct. The process is to plan it, do it, determine whether or not it met the plan, and, if it did not, the next time or for the next portion of the plan, react and modify it.

Time management is one of the components that must be addressed to properly allocate resources. Often this involves delegation. Many supervisors have difficulty in delegation or, conversely, delegate too easily. Delegation involves assigning responsibility to someone else to accomplish a task on time, on budget, and with relatively quick decisions. There are other decisions that are delegated as a part of a task, which may include spending money, ordering materials, or determining how work is to progress. Expectations must be set.

Other issues with regard to time management include deciding simple things quickly; deciding who should attend what meetings; learning to avoid taking on issues that consume large amounts of time but are not within the core mission; moving forward immediately on tasks, including doing the difficult parts first so that the easy parts come later; and traveling lightly. Time management also includes avoiding tasks such as creating useless e-mails or memoranda as opposed to addressing or solving a problem. Avoiding unimportant tasks, looking forward to and anticipating coming tasks or workloads, and minimizing meeting and telephone time are important. Managers also need to ensure that their staff does not procrastinate in doing either work or decision making.

To make decisions, it is necessary to determine what the situation actually is. In many cases, a situation is poorly understood due to lack of data, experience, or time to respond. If the situation is not understood, decisions made will usually be in error. If the situation is understood, data can be generated and analyzed. The analysis should provide some direction about the choices that can be made. For example, in the treatment plant, a funny noise is heard in the pump room. It could be a variety of things, including motors, pumps, or valves. It is necessary to determine exactly what the situation is, as opposed to immediately assuming there is a bad bearing in a pump somewhere and starting to take pumps apart. If it is a clattering check valve, then that is a very different repair than replacing pump bearings. Once the action is taken, the choice needs to be reviewed and, if further action is required, that action can be taken. Making decisions means determining a direction from multiple options, and it is the supervisor's responsibility to choose the best option.

Decision Types

Different decision modes exist for any given issue. Decision modes can be summarized as follows: simple versus complex, short-term versus long-term, personal versus organizational, urgent versus nonurgent, and program versus unanticipated. Most of these are self-explanatory and examples for each will likely occur to the supervisor relatively quickly. The type of decision indicates the amount of time and effort that should be spent on the decision-making process. For simple decisions, little time or effort should be or is expected to be spent. Experienced supervisors make routine decisions much faster and easier in many cases than new supervisors because they know what routine decisions are.

Supervisors should understand that many decisions are inconsequential and have limited impact. These decisions need to be made quickly or delegated to the people doing the work. If a work crew needs a shovel, then they should get a shovel; that's not a decision that needs extensive thought. If the crew needs to fix a leak today as opposed to tomorrow because the road is washing away, then this is also not a decision that needs to be thought about. Failure to make simple decisions costs the supervisor respect, power, and authority very quickly. A supervisor who appears to be indecisive or cannot make decisions (and there are many people who struggle with making any type of decision) are misapplied in a supervisory role. He or she will not be effective in the long term and is rarely effective even in the short term.

Complex decisions are just the opposite. Short-term decisions may be important and may be complex when evaluated as to their long-term effects. As a result, short-term decisions may or may not be decisions that should be made quickly. Their intended or unintended long-term consequences need to be part of the evaluation process.

Urgent decisions are those that need to be made immediately. These apply to events such as water main breaks. For an urgent decision, time is of the essence and will supersede long-term goals, but will be made in compliance with organizational goals. These are emergencies for which one cannot know the full potential situation.

Types of supervisors

There are four categories of supervisor: risk takers, self-disciplined, flexible, and self-centered. Every supervisor should have some degree of each type, but if one is dominant, then that could be a problem.

Risk takers are people who are willing to take on virtually any assignment and willing to make decisions for which an element of gambling might be involved. These people may be unsure of exactly what the situation is or what the solution might be, but are willing to take a chance that they can solve it. They are self-starters, in many cases, and rarely back away from difficult assignments. Risk takers are often the type of people encouraged by management to move up in the organization; however, excessive risk-taking destroys organizational protocol and impacts the appearance of responsibility. Risk takers need some self-discipline.

Self-disciplined decision makers may also be self-starters, but they realize that there are limitations, including that of the information on hand. They are going to be more deliberate in making decisions on complex or unanticipated issues; not that they will fail to make the decision in a timely fashion, but they are not going to gamble when little information is known. They are normally less prone to fail than risk takers on certain types of assignments. However, self-disciplined decision makers need some flexibility.

Flexible decision makers are those who are able to perform under unusual conditions or circumstances and are able to adapt to the situation at hand. They perceive what the situation is initially, make decisions to move forward, find out that conditions might have changed, and are able to change course. Risk takers and self-disciplined decision makers may have difficulty in this environment. Flexibility is something that needs to be incorporated in those types.

Self-centered decision makers are ones who focus on how things will reflect on themselves. Rarely is this type of decision maker useful to an organization because the organization is where the focus needs to be, not on the individual making the decision.

Supervisory Styles

Most supervisors have a dominant style, but one style does not work for all people or all situations. No two people are alike, so the supervisor needs to address the particular needs of the staff. Gaining insights about the staff provides the opportunity or incentive for staff to perform optimally. Supervisory styles range from autocratic (X type, i.e., one-way command, military style, no discussion, and no input) to participative (Y type, i.e., allows input from others, decisions are group based, and shares responsibilities). There are many shades in between.

The autocratic style is oriented to command and control. It is efficient and works well in crisis mode. Instructions are usually clear and direct. Punishment is used as a deterrent, which provides the employees with limited freedom. All responsibility falls on the supervisor. The autocratic style should be used with a new crew, during emergencies, in hazardous working conditions, and where there is a low skill level with the crews or the supervisor has specialized knowledge. The autocratic style should not be used where there are highly trained employees (who may have more skills than supervisor) or where field solutions require ingenuity and judgment.

Task leadership is focused on the performance of the task. The supervisor seeks employee input to reinforce decisions, but delegation is rare. The task leadership style is used where there is a clear separation of authority between supervisor and staff. Large groups of people will be directed. The task leadership style should not be used when the supervisor does not possess the required skills, the supervisor's decision making is questionable, or tasks require field ingenuity and judgment. Unlike the autocratic supervisor, the supervisor under task leadership may not be on-site with the working crews.

The relationship-oriented style assumes the welfare of employees is the highest priority and tasks get done as a result of motivation, human relations skills, and incentives. Behavioral punishment is rare. The relationship style is used when the tasks are routine and simple, but the employees are difficult to evaluate. The relationship style should not be used during emergencies or crises, or where there is poor employee performance or education.

The delegating supervisory style requires trust in the abilities of subordinates and assumes that the employees are well trained. It assumes that employees can work independently and with little supervision and that employees understand the organizational goals and have bought into them. This style should not be used during emergencies when coordination of work is required, or the employees have limited skills or training. The style does not work when evaluating employee efforts are difficult.

The teaching style focuses on development of employee skills. Goals are employee based and the overall goal is to make the employees independent. It is effective when defined goals and processes are available to measure employee performance, employees can work independently, and have clearly defined goals and assignments that contribute to organizational goals and performance measures in place. The teaching style should not be used during emergency or crisis conditions, where performance measures are not effective, or where the supervision cannot provide needed direction.

The role model style depends on the supervisor's personal high standards of performance to influence the behavior of employees in their standards of performance. This is a difficult style to implement since supervisors can rarely select employees who all have the same ethics as the supervisor. Complicating factors include changes in organizational structure, workload, and/or work tasks. Employee support, counseling, and interaction with the supervisor are required. The goal is that employees can work independently.

The participatory style assumes that the supervisor and staff will share responsibility, provides for feedback so all issues are known, and relies on employee motivation. The participatory style is relationship based.

Supervision and control

Supervision and control are often intertwined. Control is sometimes misunderstood by supervisors, especially new ones. It is also a major area for potential conflict between the employees and the supervisor and within the organization in general. There are a variety of management styles that involve the control issue. Some of them work against accomplishing organizational goals.

Control from a supervisor's perspective should be a general oversight of the work, not micromanagement and not hands-off to the extent that nobody understands that there is any supervision actually occurring. However, micromanagers and overly controlling, autocratic types will kill any enthusiasm or feedback that might come from the employees. The employees will resent the fact that their opinions are ignored and resent that what they deem are the proper equipment and tools are not provided. Micromanagers do not delegate any responsibility or accountability to their employees (they may believe they do, but they actually do not).

As a result, the micromanager takes complete responsibility for everything that is occurring and is much more likely to fail than a supervisor who delegates responsibility and accountability. This supervisor needs to understand that he or she cannot do everything in the organization; that is why supervisors have staff beneath them. The supervisors are depending on employees to get things done; therefore, they are relying on the judgment and the efforts of those employees to accomplish the work. This means that the employees have to be held accountable but need to be allowed to do their tasks in the best way that they see fit, as long as it meets the goals of the organization.

Supervisors have to have knowledge of human resources, materials, equipment, and budgets in order to accomplish their tasks. Supervisors also need to understand that personalities and attitudes may impact their ability to make decisions. Heredity, personal traits, history, and other factors may impact how they perceive any given situation. Likewise, their environment, upbringing, neighborhood, and so forth, may impact their perception of how situations may be solved. Life experiences such as friends, jobs, education, and cultural values also affect the perception of the appropriate action. This is evident from discussions of politics. Typically, two people from different backgrounds view a political action or a perceived need for action in different ways. In a workplace with a very diverse population, this also should be expected.

Supervisors are not judged by what they do, but by what their charges do. A part of this relates to whether or not employees are happy or whether or not they have a negative attitude (because attitude is important) toward the organization. Positive attitudes will be productive and allow people to work together. Those with negative attitudes often initiate issues with colleagues or interfere with the direction of supervision and management. This creates a conflict that needs to be dealt with so that it does not become a larger problem within the organization.

Positive attitudes are reflected in axioms such as "let's make it work, let's give it a try, we can do it, anything is possible, let's work together, take advantage of opportunities," and "let's exceed our goals." There are a variety of ways to determine how employees will react and what skill sets they have to offer. Myers-Brigg, DASI, and other tests are used for this purpose.

CHOOSING A SUPERVISOR

The organization will not succeed without good supervisors. Thus, if general management sees fit to establish a program, it should certainly see fit to select a qualified person as its leader. To accomplish the program, the supervisor is given the authority to cut across organizational lines and talk to other divisions and departments. He/she is permitted to do this because that is how resources are acquired; whether they be monetary, personnel, or materials.

The supervisor is the person who (it is assumed) has the best understanding of what it takes to accomplish the job. This may not always be the case. Good managers may recognize that they do not have all the answers, which is why hiring a good team of employees benefits them—somebody on the team probably has the answer. It is a management prerogative to try to select supervisors who will buy into and pursue the organization's goals. The expectation of management is that these supervisors will in turn hire and apply the resources in the most efficient and effective manner to accomplish those organizational goals.

A supervisor is far more likely to accomplish the goals of the organization if he or she perceives support from the management. Management appoints supervisors and trusts them to proceed in a judicious manner. New management may want to bring in a new supervisor because they want someone they trust. It is much more difficult to try to work with unfamiliar people who may be loyal to the previous manager. The organization suffers when the working relationships between management and supervisors is difficult or lacks trust and confidence.

The supervisor selection process includes both internal and external sources. The organization needs to determine where potential supervisors exist, how to recruit them, and what selection criteria should be used. Ultimately, it is unlikely that any new person will have exactly the skill set required, so clear development, training, and managing the organization or the portion of the organization is an issue that should be evaluated. Successor development is also an important part of the selection process. An obvious successor should be available on each team. Executive management will then evaluate the manager's performance in light of these processes.

Other skills may be needed. These other skills may include knowing whether or not complex technical expertise is required and, if it is not on the team, if there a means to get it. The opportunity may exist within the department, the organization, or through contract sources. There needs to be sufficient backup and strength in the line organization to accomplish the mission, i.e., enough operators or other personnel to accomplish the task if there is a loss of leadership or technical skills among key individuals. Ultimately issues for selecting the supervisor include determining the priorities for the program and from whom the necessary technical skills may be garnered. Pathways to obtain that expertise must be delineated.

CONFLICT MANAGEMENT

At some point, every organization and every supervisor will encounter some form of conflict. Conflicts arise from differences of opinion between two or more people, or organizational conflict involving groups of employees, departments, or others. Conflicts may arise over people, direction, resource allocation, or any of a number of other issues. How supervisors deal with conflict indicates much about their ability to manage and lead. There are a number of sources of conflicts including:

- Limited resources
- Unclear organizational goals
- Personality differences
- Failure of the supervisor to understand an effective approach toward employees
- Compensation issues

Conflict is caused by accusations, bias, favoritism, harassment, suspicion, unrealistic expectations, workload, and changes in work. Conflicts will arise in a variety of situations and from a variety of sources, including those outside or between departments. Supervisor-caused conflict is caused by the supervisor's failure to make decisions, or make them promptly; limited communication skills; and setting a poor example for the employees. Supervisors who are inflexible, rigid, inaccessible, and/or create an unrealistic expectation for employees are also sources of conflict because employees are frustrated and do not feel they are treated fairly.

Conflict is unavoidable even under the best conditions, but escalation of disagreement to conflict is not good. A good supervisor will defuse conflict with compromise and counseling before this escalation takes place. When difficult conflicts occur within the organization, the supervisor will gather data, review risks, define results, set deadlines, and reach a solution. To successfully resolve conflicts, the supervisor must be able to understand the interaction of the organizational units and the exhibited behaviors within it to build an environment conducive to motivation of his/her charges to accomplish work in a timely fashion.

The ability to recognize and to deal with conflicts is a demonstration of the leader's skill in minimizing counterproductive disruptions. In many cases, it is the supervisor's responsibility to deflect attention and to head off diversions from his or her employees' true mission, and protect the staff from conflicts from other levels. At the same time, the supervisor is responsible for minimizing conflicts within his/her staff and the rest of the organization by communicating effectively at all levels regarding how the mission of those under his or her supervision are meeting the directives and organizational goals that have been set. This can be accomplished through regularly scheduled status review meetings. Such meetings should have agendas and should be specifically oriented to minimize unproductive time. Effective project planning, contingency planning, securing of commitments, and involvement of top management can help to prevent or minimize many conflicts before they impede project performance.

Conflicts usually do not get better with time. The faster they are resolved, the better. Quick resolution means less frustration and ill will. To resolve conflicts efficiently, it is important to:

- Listen with understanding rather than evaluation
- Clarify the nature of the conflict
- Understand the feelings of others
- Suggest procedures for resolving differences

- Maintain relationships with disputing parties
- Facilitate the communications process
- Seek resolution

To solve personality conflicts, supervisors should exhibit fairness, show neutrality, and maintain privacy. To accomplish these goals, the conflict manager will:

- Pause and think before reacting
- Build trust
- Try to understand the motives in the conflict
- Keep the discussions under control
- Listen to all involved parties
- Maintain a give-and-take attitude
- Educate others tactfully on the organization's position

Where personnel matters are concerned, complaints, frustration, and unhappiness are often short term. The employee needs somebody to complain to and rather than complain to other employees, the employeee complains to the supervisor. No action is needed in many of these personnel issues and it is extrordinarily important to determine when these issues are occurring and when a legitimate problem exists.

Activities that can negatively affect employees include poor supervisor accessibility, excessive or insufficient meetings, and confusing orders and directives, all of which serve to frustrate the work effort. To prevent conflicts, the supervisor needs to improve communication, create social activities, provide training, and support self-esteem. Supervisors who see some value in conflict usually pit employees against one another, which clearly can be destructive to the organization. Such supervisors are not helpful in creating a positive working environment and may defeat organizational goals.

AUTHORITY, RESPECT, AND POWER

It is useful to conclude this chapter on supervision by discussing three concepts that are often related to supervision: authority, power, and respect. Authority is not power. This is a common misperception among new supervisors. Authority is a right or privilege given to the supervisor by management to direct activities and it assumes that he or she is reasonable and judicious in accomplishing that. Authority is limited to the position and it is limited to the ability to discipline the employees.

Given that the supervisors will provide employees with indications of what should be done to improve their work efficiency or to correct unsatisfactory results, that authority must be both granted by the employee and delineated by management. The employees will respond when they feel that the organizational goals are clear and comprehensible and make sense and that the supervisor has a direct interest in their ability to achieve satisfactory results. The more the supervisor indicates his/her belief in their ability to be successful and make progress, the more likely it is that the employees will respond in kind.

Power is a completely different issue. Power accumulates to the person, not to the position. Power is usually amassed through respect; if employees respect somebody, then that person is generally powerful and the question is whether they use their power properly or attempt to abuse it. Those who abuse power quickly lose respect and then lose the power.

Power is the ability to achieve expected outcomes. It is a confidence-building tool. If a supervisor comes in, sits down with employees, plans activities, gets the proper equipment, gives the crew the ability to accomplish it all, and they are routinely successful, then they will have significant confidence in that supervisor. That supervisor then will have not only authority, but power with the employees and they will depend on that supervisor to accomplish things. Power then provides the ability to influence results. If the supervisor says that "we can do this because we've done this in the past," the employees will be motivated to accomplish it because that's the track record.

Respect is something the supervisor has to earn. The fact that you are a supervisor does not mean that anybody cares or respects anything that you do or likes you any better than they did before. Sometimes supervisors need to prove themselves to the employees. For example, a college-educated supervisor in North Carolina was working with a water distribution field crew that was not highly educated (and in some cases could not read). The supervisor was trying to teach them how to lay water pipe correctly and efficiently, something they had never done before, but something in keeping with the organizational goals. The employees distrusted the pipe material, lacked education or training in the field, and were unsure the supervisor really knew the installation of this type of pipe, respected their abilities, or understood the field conditions. The supervisor had not earned their respect, and they needed him prove himself to them, i.e., get in the ditch and show them. As soon as the supervisor did that, anything he needed done, got done. If there were problems, that's who the staff called. They respected him and he then had power when he told them that they could get the project done. They believed him and it was accomplished. He earned their respect and with it power and authority. Twenty years later, the story remains a legend of the utility because the employees did not believe he would actually get in the ditch and work with them. Mutual respect was given both ways.

REFERENCES

Collins J. 2001. *Good to Great: Why Some Companies Make the Leap...and Others Don't*. New York, N.Y.: HarperBusiness.

Diamond, L.E. 2007. *Team Building That Gets Results*. Naperville, Ill.: Some Books Inc.

FWPCOA. 1997. *Supervision*. St. Lucie, Fla.: Florida Water and Pollution Control Operator's Association.

Taylor, F. 1911. *Principles of Scientific Management*. New York, N.Y.: Harper & Brothers.

CHAPTER 7

Management of the Utility

There are three areas of the organization that can prevent it from reaching its goals and objectives. These three are staffing problems in the areas of human resources, budgeting, and purchasing. Interestingly, in each case, these are support services, not line services. The support services, if not competently performed, will negatively impact the entire utility operation. In each case, the wrong people in those positions ensure that either the right people are not hired, resources are not obtained, or money is not available when needed.

CHANGING WORKFORCE

A major issue associated with management of a utility is the organization and staffing to accomplish the utility's mission (Olstein et al., 2005). Prior chapters outlined the process for hiring employees. From an organizational and staffing standpoint, matching organizational goals with operations defines how the work will actually proceed and by whom. In order to determine organization and staffing, certain issues need to be identified. First, there needs to be a manager and it must be determined what the manager's assets and skills should be. There may be assistants to the manager and their skills and assets also need to be determined. An office or some sort of facility to work in must be secured and then a team of employees hired to actually accomplish the tasks needed. The manager may be the most important of these positions because it is through the manager that the direction and coordination will occur. The manager may be the supervisor or vice versa. Supervisors will certainly need management skills.

Honesty and integrity are the most important characteristics of any manager. In addition, managers should have the following skills or knowledge:

- An understanding of personnel problems
- An understanding of project technology
- Business competence
- An ability to communicate
- Alertness and quickness
- Versatility
- Energy and toughness
- Decision-making ability

This is a fairly extensive list of issues and every manager will have more or less skill in each of these areas. Where skills are lacking, training should be pursued. Most managers play to their strengths, but it is often the weaknesses that ultimately stymie the organization.

From an executive management perspective, the expectation is that all managers and supervisors will attempt to acquire the necessary skills and improve those skills with time. Good managers will hire people who are perceived to be better qualified than they are, the reason being that these people will work hard, be successful, and make the manager look good. By making the manager look good, the organization also looks good and more work gets accomplished. Having people who work together in a good, productive working environment makes for happy employees.

Workers' skills are an issue that must be delineated as part of the management's responsibility. Workers need to have a goal in mind so that they know what their tasks are and what they're expected to accomplish. Employees must have an understanding of their authority, the decisions they control, and the limitations for both (which may be more important). Employees should be empowered to make decisions about how they accomplish their jobs and, perhaps, the order in which tasks are accomplished. In this way they can take ownership of their work and, therefore, responsibility. Employees should understand their relationships with other employees, who has authority, who does not, and the skill sets that they have. They should also know what the expectations in relationships with others might entail. Feedback with employees through evaluations or other means is important to ensure that employees understand where they are being successful and where they may not be (see chapter 6).

Change is coming to the utility industry. While technology and training have changed with time, so have the demographics of employees and their workplace attitudes (Olstein et al., 2005). There is more education available and more information exchanged. A few facts about changes in utility worker demographics make the necessity of changing management approaches clear. The average utility worker is several years older than the typical worker in general (Olstein et al., 2005). For example, the average water plant license holder in Florida is 57, and 80 percent of them exceed the age of 40. Clearly, a large portion of these people will retire in the next 10 years. A recent AwwaRF study indicated that in the next 10 years, up to 50 percent of the current workforce in water and wastewater utilities will be eligible to retire. The study also found that most people retire within one year of eligibility, so trying to keep these people on board for a longer period of time is not likely to be successful. Engineering, management, and operator positions will be the hardest hit. There is a relatively critical shortage of engineers nationwide, especially civil engineers. As a part of this retirement pattern, the historical knowledge and skill sets are likely to be lost (Olstein et al., 2005).

The pool of desirable workers is not as large as it has been in the past for a variety of reasons. In part, the workforce has changed because the negative perception of field labor outweighs the potential for benefits derived from public utility systems. Because of the salaries paid and the scrutiny that is placed on salaries by local governments, utilities may have trouble competing with other businesses for this skill set. The skill set typically available to utilities will be nondegreed workers with limited expertise in the area, yet the skills required are increasingly more technologically advanced. Few utilities have a succession plan in place to engender a long-term planning solution for replacement of retiring workers.

Utilities that are most at risk for these succession issues are those that have a high average age of the workforce; lots of long-term employees (the average retiree has over 26 years on the job); very limited training budgets; a reliance on operations and maintenance manuals that are often out of date, dusty, and may not be locatable; a lack of drawings; no succession plan; and an inability to secure institutional knowledge transfers. These are usually smaller utilities but it is surprising how many large utilities may be in the same situation. In smaller utilities, historically the concern has been that if you teach somebody else to do your job, you will not be needed to do that job. In fact, just the opposite is happening now. The fact that there are not enough people trained to do the jobs is likely to place the organization in jeopardy. This is not an employee issue, it is a utility management issue.

Loss of institutional knowledge will be significant and it will be difficult to retrieve. In many cases, these are important, but relatively simple issues, such as where valves are located in the water plant and distribution system, where access openings are located, and piping diagrams. It is surprising how few utilities actually have maps showing locations of their valves, fire hydrants, utility lines, and access openings. All of these are easily covered up through paving and other construction activities and may be lost. Connections or repairs that are made in pipe, or field modifications that need to be made to solve a given problem, are also likely to be lost because very few pictures have been taken of these repairs. This is a flaw in the current practice for engineering oversight of construction projects. Record drawings rarely include photographs, which could easily be inserted electronically. The fact that institutional knowledge will be lost indicates that additional training will be required immediately. Training of employees by employees who are getting set to retire is an important knowledge transfer.

The loss-of-knowledge problem is typified by an employee(s) who:

- Knows where all the valves are, but has never put them on a map
- Knows just how to get the pumps to keep running, but never has told anyone exactly how
- Knows where pipes, tanks, and so forth were buried, but it's not on any drawings
- Was there when they reconfigured the piping on the plant site, but the record drawings were never modified
- Knows why the operational changes were made 20 years ago, but there is nothing in the file to back it up
- Knows why specific polyphosphates or other chemicals are used because he/she did the tests, but can't put his/her hands on the tests that were done
- Knows where the left-handed valves are, the odd-diameter pipe, which line is abandoned and why, and why the extra equipment was added to the elevated tank to keep people from falling off, but no one ever recorded it anywhere

Typically, this person assumes he/she has job security by knowing all these things, and is retained. All of this knowledge needs to be documented for future staff.

Training gaps may exist in the following areas:

- SCADA
- Drive systems
- Infrared technology
- Computer tracking of data
- Report development
- Regulations on chemical handling
- Geographical information systems (GIS)

All are high-tech programs with limited training opportunities.

An AwwaRF report (Olstein et al., 2005) notes that the pool of trained workers in the United States is getting smaller, as more and more are filled by immigrants. The generation born between 1900 and 1945 includes 75 million people. The baby boomer generation from 1946 to 1964 includes 80 million people. Generation X (which is 15 years after the baby boomers) only had 46 million people, and the Millennials from 1981 to 1999 had 76 million (Olstein et al., 2005). In the latter two cases, both are less than what would be expected in the population after the baby boomers, especially Generation X, which has dramatically fewer people than either of the prior two generations in spite of spanning as much time as the boomer generation. Recovery of these workers will be difficult to achieve without the utilization of immigrants. That the current vacancies are being filled in large part by immigrants, not US-trained workers, indicates that there is a significant need for additional training and incentives (Olstein et al., 2005).

Many of those likely to become water and wastewater engineers are women, but their participation in these industries is not high on their list of options. Utilities are less gender diverse than most organizations, in part because, historically, they have been viewed as relying significantly on hand labor, and women have not typically pursued labor-intensive positions.

Most utilities are not very racially diverse unless they are in communities with significant minority populations. Training is a problem and relevant courses are being developed in order to train workers who are otherwise difficult to find. AwwaRF noted that Hispanics are expected to increase from 12 to 23 percent of the population by 2050, and Asians from 5 to 11 percent of the population (Olstein et al., 2005). African Americans are expected to remain at a steady 13 percent for many years (Olstein et al., 2005). As a result, there should be a significant increase in Hispanics and Asians in utility industries in the future (Olstein et al., 2005). This indicates that training will be required to deal with different ethnic issues that may arise with these nontraditional, diversity-driven employees.

Recruiting college students is virtually impossible for operations positions. The positions do not pay enough, and college students today expect to obtain white-collar, indoor jobs using computers, as opposed to field labor. The utility industry has not spent a significant amount of time dealing with high school and junior high school student recruitment, especially those who are unlikely to go to college. Mentoring programs would be a potential solution. Mentoring programs with high school students generally do not exist, and the outside work and the requirement for limited math and science knowledge are barriers to attracting certain types of students. The only college students who are participating in utility systems are civil engineers, although chemical and mechanical engineers may be inclined to do so, especially in down economic conditions. Most operations positions require a high school diploma but many high school graduates, especially those who are not going to college, are unaware of the potential opportunities.

The supervisor and management responsibility is to figure out who the workforce is, the skills that are actually needed, how many personnel with those types of skills will be needed, and how that meets projections of what the anticipated workforce will look like. Once those questions are answered, the utility must identify barriers to accomplish the goals. The first is the percentage of employees involved in unions. Currently 60 percent of the employees are unionized, which may hinder the resolution of workforce planning issues, because unions tend to be more supportive of the person who has information but does not write it down (described earlier) than management. Utilities need to become an employer of choice by demonstrating the benefits of employment with the utility system. Over 70 percent of current employees have benefits and perceive utilities as a good place to work (a fact that, unfortunately, is not generally known outside the utility).

Promoting a utility as an employer involves highlighting the benefit plans and leadership training aspects, as well as the recognition that hard work and knowledge are valued. However, the work (working in a water plant or a "sewer" plant) is often not viewed as a positive career choice. However, the amount of this work that needs to be done will increase with time and, since much of the dirty work is now done through technology, the old perceptions are generally not accurate. Utilities should be promoted as providing interesting and challenging work, sometimes routine, but often with unexpected challenges. There is a value to public service as well as pride that can be exhibited in protecting the public health and the environment.

Retention strategies, especially where there is potential for employees to migrate toward retirement, can include flexible schedules, job sharing, and increased recognition and training opportunities. Promoting talented people who can meet the goals of the organization, leadership development, and increased workforce communications are all important to employee retention.

Leadership development includes mentoring, feedback, the assignment of challenging jobs with support to see how well the potential managers perform, the use of projects and task forces to help these people work together, and the opportunity to lead or supervise such projects. Staff development includes the transference of knowledge through training and observation, and the use of skills and abilities learned through other education and experience.

One of the issues that has been identified as a significant barrier to the ongoing efforts of utility systems in general is lack of training. Most operators obtain 20 hours of training a year just to meet the requirements of their license, not to get advanced training. Employers are often reluctant to send their employees off for any significant period of time for training and, as a result, only local training is pursued. If an operator lives in a place that is many miles from a large city, that training may be nearly impossible to get except via computer. Therefore, training opportunities should increase significantly to ensure proper workforce planning.

Among the things a utility can do to improve training is to increase training budgets and to develop internal procedures to capture institutional knowledge. This would include videotaping and photographing the repair procedures; ensuring that all repairs in the field have appropriate photographs taken; using a microteaming concept, where older employees are teamed with younger ones so that the younger ones can learn what has been done; job rotation and cross-training; and the double filling of a position where a retirement is expected. To accomplish this successfully, managers should be thinking about what the organization should look like in 10 years (i.e., how many employees will be needed, will economies of scale or technology help, what is the technology required, and what flexibility within the organization can be used to increase productivity).

To capture the potentially lost institutional knowledge, someone, generally in a leadership or supervisory role, must identify all the processes for which information is lacking. This is a challenging task. The tasks must be ranked in the order of importance and risk (i.e., those that have the highest risk for potential failure if the knowledge is not tracked down and documented should be the first priorities). Those that are less important or for which there may be other opportunities for learning should receive a lower priority. Failure modes and the potential risks of those failures should be a significant alternative measure. Once identified, a protocol needs to be developed on how to capture the knowledge and through which means or methods. This includes getting the employee into the room and having him/her explain what he/she does.

BUDGETS

Budgets are a necessary part of the utility operations. All utilities should be set up as an enterprise fund in order to allow them to pay their own way. Budget preparation is usually done by line managers and reviewed by supervisors, utility management, a budget/finance person, and the management of the utility or city. The budget is a planning tool and an organizational tool. It should not be set in stone. Many people have difficulty understanding that the budget is simply a plan, and, especially with an enterprise fund, the plan can be adjusted upward or downward to accommodate unforeseen conditions.

Table 7-1 is an example of a budget that includes capital within the budget. It is important to note that for many of the items, a simple increase in percentage each year is not appropriate. Past costs may be indicative to some extent of future costs, but at the same time, note that certain line items in this budget have varied significantly, mostly in the maintenance area. If a pump goes down, it needs to be repaired, but repairing multiple pumps in a given year may not be planned for. As a result, having retained earnings is an important part of the utility operation. Retained earnings is the amount of money that is set aside as "surplus" money that is not expected to be spent as a part of the current budget. Retained earnings are important from an audit perspective, from the perception of the financial community, and from an operating perspective. It's clear from a political level that governments that do not have retained earnings have difficulty responding to the needs of their constituency during times when the economy deteriorates.

Many depression-era and modern-day economists note that when the economy takes a downturn, there is a need to continue spending levels to keep people employed. The problem is that most local governments and utilities, when overcollecting revenues, tend to reduce their tax levels and reduce their fee charges to balance the budget instead of adding to retained earnings. The result is that when the economy becomes difficult, they need to raise taxes and raise fees precisely when their constituency has little ability to pay for it or change levels of service. Additionally, when the economy is in a down cycle, there is a tendency for construction projects, the major cost of capital for utilities, to decrease. Having significant retained earnings allows the utility to operate and perhaps even act as a mechanism for increasing jobs in the construction industry through a likely backlog of projects. As a result the utility may find itself saving a significant amount of money by constructing needed infrastructure during poor economic conditions.

Table 7-2 is a project budget. The project budget is used to look at initial costs and life cycle costs. These are necessary in order to evaluate different options. The budget process is what links the operation, management, and finance aspects of the utility. As the interactions of these three management concepts basically parallel one another, capital management similarly is interwoven among these three groups.

While the budget is a plan, it is also a mechanism to plan expenditures and allocate them among different expectations of expense categories, commonly referred to as *line items* (see Table 7-3). An allocated budget for operations and maintenance will permit the

TABLE 7-1 Example budget that includes capital within the budget

		2007	2008	2009	2009	2009	2010	2011	2012	2013	2014
					Y-T-D						
			Last year's	Original	5/30/09	CMPR	Prop	Prop	Prop	Prop	Prop
Account Number	Account Description	Actuals	Actuals	Budget	Actual	Anticipated?	Budget	Budget	Budget	Budget	Budget
	Salaries & Benefits	$ 388,302	$ 425,013	$ 471,512		$ 427,077	$ 458,486	$ 541,410	$ 568,481	$ 596,905	$ 626,750
403-3801-538.31-10	Professional Services	14,876	5,879	26,271	7,499	11,300	11,300	6,000	6000	6000	6000
403-3801-538.31-30	Consultant Engineers	20,218	33,109				50,000	35,000	35000	35000	35000
403-3801-538.32-10	Accounting & Auditing	1,132	867	960	917	0	1,050	1082	1114	1147	1182
403-3801-538.34-10	Contractual Services	3,661	1,190	1,056	740	1,040	20,840	21465	22109	22772	23456
403-3801-538.40-10	Training & Per Diem	1,306	2,090	1,988	1,188	2,000	2,000	2060	2122	2185	2251
403-3801-538.40-30	Expense Account	601	603	600	420	600	600	618	637	656	675
403-3801-538.41-10	Telephone	387	636	780	375	648	648	667	687	708	729
403-3801-538.43-20	Electricity	8,076	7,133	9,966	3,967	9,966	7,409	7631	7860	8096	8339
403-3801-538.44-10	Equipment Rentals	274	255	255	170	255	255	263	271	279	287
403-3801-538.44-20	Trailer Rentals	900	900	900	675	900	900	927	955	983	1013
403-3801-538.44-30	Misc. Rentals	977	992	1,010	1,042	1,094	1,094	1127	1161	1195	1231
403-3801-538.45-10	Insurance Coverage	6,271	6,139	5,142	4,982	5,553	5,855	6031	6212	6398	6590
403-3801-538.46-10	Equipment Maintenance	15,549	8,113	8,000	14,500	10,000	10,000	10300	10609	10927	11255
403-3801-538.46-30	Vehicle Maintenance	13,111	15,747	12,000	8,182	12,000	12,000	12360	12731	13113	13506
403-3801-538.46-50	Grounds Maintenance	21,569	18,573	18,000	10,608	18,284	18,248	18795	19359	19940	20538
403-3801-538.46-60	Utility Maintenance	14,498	10,040	52,000	1,350	60,000	60,000	61800	63654	65564	67531
403-3801-538.49-30	Permits & License Fees	4,149	5,543	4,984	13,527	20,000	20,000	20600	21218	21855	22510
403-3801-538.51-10	Office Supplies	145	150	150	0	150	150	155	159	164	169
403-3801-538.52-10	Gasoline	8,782	11,125	15,848	4,859	11,662	11,663	12013	12373	12744	13127
403-3801-538.52-20	Misc. Supplies	1,037	-24,034	7,000	2,648	7,000	7,000	7210	7426	7649	7879
403-3801-538.52-50	Uniforms	540	559	750	700	750	750	773	796	820	844
403-3801-538.52-60	Building Supplies	275	0	0	0	0	0	0	0	0	0
403-3801-538.62-10	Buildings	0	0	162,488	18,244	0	0	0	0	0	0
403-3801-538.64-00	Machines & Equipment	46,919	$ (1,843.00)	0	0	0	0	0	0	0	0
403-3801-538.64-20	Vehicles	23,219	0	0	0	0	0	0	0	0	0
403-3801-538.64-30	Capital	23,700	1,843	3,500	0	0	3,000,000	600,000	125,000	230,000	145,000
403-3801-538.71-10	Principal	0	0	16,446	12,366	17,183	23,507	$ 227,763	$ 267,876	$ 267,876	$ 267,876
403-3801-538.72-10	Interest	4,029	5,694	2,666	2,812	1,930	1,930	incl above	incl above	incl above	incl above
403-3801-538.91-10	Transfer to General Fund for purchases and services	0	0	30,000	20,000	0	45,000	$ 47,250	$ 49,613	$ 52,093	$ 54,698
403-3801-538.99-10	Contingency	0	26,030	181,195	0	0	50,000	0	0	0	0
	TOTAL OPERATING	$ 624,503	$ 562,346	$ 1,035,467	$ 131,771	$ 619,392	$ 3,820,685	$ 1,643,299	$ 1,243,421	$ 1,385,069	$ 1,338,435

TABLE 7-2 A project budget

Cost Item	2009	2010	2011
Planning	$25,000		
Phase I—Sewer investigation and repair (G7 Program)	$312,000		
Phase I follow-up		$75,000	
Letters to homeowners		$5,000	
Identification of areas to televise		$25,000	
Phase 2 Infiltration repair (televising and lining)			$975,000
Phase 3 Point repairs			$150,000
Total cost of project			$1,567,000
Annual cost savings			$575,000
Present worth of cost savings (20 yrs)	$8,554,548		
Debt cost for program (4.5% for 20 yrs)	$120,465		

utility to evaluate its competitiveness and determine areas where costs might be saved. For example, if energy costs are significantly higher for the utility than one might expect, then the utility can start looking for areas where it may have pumps, motors, drives, aerators, or other equipment that is not very energy efficient and can be replaced with more efficient equipment, thereby saving money. In some cases, power companies and the federal government may offer grant programs to help utilities lower their energy costs.

The outcome of the budget process indicates the necessary revenue needs for the utility and provides some indication about whether or not the maintenance costs of the system are increasing, perhaps because of deteriorating conditions of the assets.

The budget will be adopted by a resolution of the governing board and can only be modified through amendments approved by the governing board. However, the mechanism for approving the budget should not make each line item a not-to-exceed amount that needs to be modified through board action. That way operations managers and personnel can move costs between line items, many of which are small (see Table 7-3), without having to go through the process of gaining approval for moving monies within a given budget.

The annual audit (comprehensive annual financial report, or CAFR) comes after the budget year has been completed. The intent of the audit is to have an external accounting group evaluate whether the revenues and expenditures were appropriately categorized, money was maintained in the appropriate accounts, and revenues and expenses were fully accounted for and appropriately spent. The audit will contain useful information about the revenues and expenditures for the prior years, the amount of debt, the capital assets, and the retained earnings. Auditors will note areas where the utility can improve its accounting on finance methods. It also possible to note areas where bond issues and asset management may be improved. Ultimately the audit provides information about whether or not the financial operations are in compliance with generally accepted accounting principles, they adequately present the financial condition of the utility, and appropriate laws and regulations have been complied with. It also provides some indication of whether or not expenses appear to be appropriate for labor, energy, chemicals, and other expenses of the utility.

TABLE 7-3 Summary of a proposed budget showing prior expenditures and proposed expenditures for the coming year

Account Number	Account Description	2007 Actuals	2008 Last Year's Actuals	2009 Original Budget	2009 Modified Actual	2009 Changes to Line Items
	Salaries and benefits	$388,302	$425,013	$471,512	$427,077	$44,435
403-3801-538.31-10	Professional services	14,876	5,879	26,271	11,300	14,971
403-3801-538.31-30	Consultant engineers	20,218	33,109			0
403-3801-538.32-10	Accounting and auditing	1,132	867	960	0	960
403-3801-538.34-10	Contractual services	3,661	1,190	1,056	1,040	16
403-3801-538.40-10	Training and per diem	1,306	2,090	1,988	2,000	(12)
403-3801-538.40-30	Expense account	601	603	600	600	0
403-3801-538.41-10	Telephone	387	636	780	648	132
403-3801-538.43-20	Electricity	8,076	7,133	9,966	9,966	0
403-3801-538.44-10	Equipment rentals	274	255	255	255	0
403-3801-538.44-20	Trailer rentals	900	900	900	900	0
403-3801-538.44-30	Misc. rentals	977	992	1,010	1,094	(84)
403-3801-538.45-10	Insurance coverage	6,271	6,139	5,142	5,553	(411)
403-3801-538.46-10	Equipment maintenance	15,549	8,113	8,000	10,000	(2,000)
403-3801-538.46-30	Vehicle maintenance	13,111	15,747	12,000	12,000	0
403-3801-538.46-50	Grounds maintenance	21,569	18,573	18,000	18,284	(284)
403-3801-538.46-60	Utility maintenance	14,498	10,040	52,000	60,000	(8,000)
403-3801-538.49-30	Permits and license fees	4,149	5,543	4,984	20,000	(15,016)
403-3801-538.51-10	Office supplies	145	150	150	150	0
403-3801-538.52-10	Gasoline	8,782	11,125	15,848	11,662	4,186
403-3801-538.52-20	Misc. supplies	1,037	–24,034	7,000	7,000	0
403-3801-538.52-50	Uniforms	540	559	750	750	0
403-3801-538.52-60	Building supplies	275	0	0	0	0
403-3801-538.62-10	Buildings	0	0	162,488	0	162,488
403-3801-538.64-00	Machines and equipment	46,919		0	0	0
403-3801-538.64-20	Vehicles	23,219	0	0	0	0
403-3801-538.64-30	Capital	23,700	1,843	3,500	0	3,500
403-3801-538.71-10	Principal	0	0	16,446	17,183	(737)
403-3801-538.72-10	Interest	4,029	5,694	2,666	1,930	736
403-3801-538.91-10	Transfer to General Fund for purchases and services	0	0	30,000	0	30,000
403-3801-538.99-10	Contingency	0	26,030	181,195	0	181,195
	TOTAL OPERATING BUDGET	$624,503	$564,189	$1,035,467	$619,392	$416,075

The utility should have some form of vision statement about where it desires to go over a period of time and the audit will often delve into whether or not the expenditures and programs from the year being audited have complied, or are moving toward compliance, with those goals and objectives. There are alternatives that might be useful to improve compliance or there may be policies that need to be adjusted to meet the goals of the governing board.

PURCHASING

The appropriate tools must be in place in order to get work done. Field crews need the right tools and materials, and operators need the right chemicals. The utility should have a purchasing policy in place to acquire these materials. Implicit in this process is the need to spend public monies judiciously, meaning that bidding and quotes need to be defined. For example, for small purchases, perhaps under $5,000, three or more quotes would need to be gathered. For larger projects, most states require that sealed bids must be secured.

A defined procedure to acquire the quotes, issue purchase orders, and designate how vendors will be paid for services should be in place. This includes the amount of time needed to secure bids and advertise requirements, and the time and place for how bids will be opened. For construction contracts, bid bonds or other securities may be required. The procedure to obtain governing board approval, or a delegation to management or supervisors for approval of small purchases should be defined. Dealing with emergency purchases should also be defined.

One solution for many purchases would be to secure competitively bid, ongoing contracts for routine items such as chemicals, contractor repairs, and repair parts. These could be secured for several years at a time with limitations on price increases to ensure that the utility does not overpay for goods. This eliminates a significant amount of ongoing paperwork for operations staff. Another option is to use state contracts, which often come at reduced costs due to high volumes of sales. Vehicles and furniture are typical examples where state contracts are often available.

A consensus of how items will be acquired both routinely and in emergencies is particularly appropriate for entities expected to produce around the clock. Spending time to obtain a part or a chemical may cause damage to the system or increase the likelihood of public health concerns.

ORGANIZING FOR OPERATIONS

Getting the right people working together in an organization is an important mission. This not only includes personnel, budgeting, and purchasing, but also the interaction between the employees, management, and the governing body. This means that each group must understand its role and perform it appropriately. Technological changes are rapidly making the operation of public water systems more complicated, emphasizing the need for greater expertise to deal with ongoing operations, maintenance, and plans for the future. Staff must have the knowledge and abilities to deal with complicated electronics, machinery, and new treatment methods, and must understand regulations well enough to maintain compliance. Increased documentation means that administrative staffing needs are greater, as is the need to organize and manage the flow of information among staff, management, and other entities. As a result, it is important that the utility management team has professional credentials and a demonstrated ability to deal with complex utility issues. Such individuals will not be successful if they do not have skills to work with and manage people and resources. While the three items below are required of managers, it should be noted that not all people have or can develop these skills. Failure to be able to perform these functions will ultimately lead to failure of the manager:

- Measuring: determining through formal and informal reports the degree to which progress toward objectives is being made.
- Evaluating: determining the cause of and possible ways to act on significant deviations from planned performance.
- Correcting: taking action to correct an unfavorable trend or to take advantage of an unusually favorable trend.

To implement these functions, the manager needs to lead, motivate, and reward subordinates. AWWA has published AWWA G100-05, Standard for Water Treatment Plant Operation and Management, to outline some of the important issues. (AWWA, 2005).

Pulling people together means meeting their basic needs. To understand what this means, Abraham Maslow looked at what motivated people and ranked these motivations (Maslow, 1954). Certain needs, such as shelter and food, must be addressed before all others. These are instinctive, physiological issues. After these needs are met, safety and security arise. Job security would be in this arena for many. Managers need to understand how their actions are perceived in Maslow's triangle. If a manager can motivate using the basic issues, then the manager will be successful. But getting everyone excited over every little item is counterproductive. Monthly birthday lunches get old quickly. Maslow noted that people need to be appreciated and that their efforts need to be recognized. This is easy to do in many cases. Some examples would include a little time spent talking one on one with employees in an unstructured setting, visible management (i.e., can be seen walking around), and comments to upper management or the board that let employees know their work, regardless how small, is recognized. Rewards include the following:

- Security
- Compensation/pride
- Achievement
- Belonging

As a result, the utility organization can help meet these rewards through:

- A feeling of pride or satisfaction of one's ego
- Security of opportunity
- Security of approval
- Security of advancement, if possible
- Security of promotion, if possible
- Security of recognition
- A means for doing a better job, not a means to keep a job

The goal for management is for utility workers to adopt a positive attitude and to not criticize management. As a result, management needs to ensure its vision is implemented. Utility management can do this by:

- Making promises that can be kept
- Circulating customer reports
- Giving each person the attention he/she requires
- Giving assignments that provide challenges
- Defining performance expectations
- Giving proper criticism as well as credit
- Giving honest appraisals
- Providing a good working atmosphere
- Developing a team attitude
- Providing a proper direction

Again, too much of a good thing may be counterproductive even if it costs little. Everyone winning an award makes the award meaningless and therefore it does not motivate anyone. With nonfinancial awards, employees may receive cash-equivalent items, but not cash in hand. Professional needs of employees may be different than for line employees. Professional employee incentives typically include the following:

- Interesting and challenging work
- Professionally stimulating work environment
- Professional growth
- Overall leadership (ability to lead)
- Tangible rewards
- Technical expertise (within the team)
- Management assistance in problem-solving

Clearly defined objectives:

- Proper management control
- Job security
- Senior management support
- Good interpersonal relations
- Proper planning
- Clear role definition
- Open communications
- A minimum of changes

Many people are motivated professionally. Managers appeal to this since it often has limited financial value, but a lot of perceived self-worth to employees. Training and conference attendance are excellent tools for professional employee rewards.

Characteristics of High-Performing Groups

There are many texts on leadership (e.g., Diamond [2007]). Virtually all of them point to the areas identified below for success. Leaders will take risks, take responsibility for failures, acknowledge and reward others, protect their employees, and create a vision that everyone can buy into. Leaders earn the respect of their charges who will want to follow the leader. Leaders hire people who are better than they are, and who will move forward in the organization. They realize that the success of any manager is dependent on the efforts of their charges. If the employees are successful, then the leader will be successful. Good leaders will provide their charges with the following:

- Good planning
- Well-defined roles and responsibilities
- Accountability
- Leadership
- Means to manage/resolve conflict
- Good communication among members

Everyone must be allowed to participate, and innovation must be encouraged.

The same texts on leadership note that employees are team members. The team is only as successful as its weakest link. The manager will attempt to protect the team by trying to arrange work so that team members' weaknesses are not exposed. The obligation of team members is to put their best efforts into the tasks, doing their share and more, and supporting the manager in his/her efforts to get work done. Bad team members are obvious. Everyone knows who they are, and no one wants to work with them.

Good team members take responsibility, deliver on their commitments, and contribute to discussions. They will listen and ask helpful questions, give and receive useful feedback, and do more than their fair share of the work. Management should attempt to address obvious deficiencies on a team through training, counseling, and similar measures.

MANAGERIAL ISSUES

Managerial issues that arise when evaluating the utility involve the credentials of the management team and the operations policies and procedure in place. The following policies should be in place:

- Standard specifications
- Water quality sampling plan
- Cross-connection control
- Connection and billing policies
- Emergency operations plan
- Record management
- Personnel
- Water and sewer extension policies
- Safety plan

- Risk management plan
- Training programs to educate employees about these plans and how they should conform with them

In addition to these policies, operations and maintenance manuals should be available for all pieces of equipment and all treatment facilities. Operations and maintenance manuals provide information on the equipment, parts listings, and frequency for preventative maintenance and replacement programs that should be pursued by field crews and plant operators. Included in the manuals should be shop drawings and flow schematics that will help engineers and other staff in the future when improvements are made.

Records of operation and field activities must be kept. Records will permit utility managers to make decisions and recommendations on improvements and programs that need to be made. Records must be accurate as to work performed, parts used, and findings. Different pipe and fittings may be found, and the record drawings and files should be updated continuously to reflect these differences. Daily logs and monthly summaries of work are useful to track field crew performance. Governing bodies appreciate this information as well.

The most efficient methods of record keeping are a work order tracking system and photographs. Work order tracking programs are available commercially, but the problem is getting field crews to accurately show the information on the work order and staffing a position to enter the data. Field crews working in mud and water would likely be a poor choice to use remote computers for data entry. Also supervisors may wish to review the data prior to entry to decrease confusion. Inventory should be tied to the work order tracking program to provide some control over the quantity of parts and supplies that are available and used in the field. This has the side benefit of reducing losses of inventory.

Those maintenance tasks that are specialized or occur infrequently can be considered for contract work. Examples may include activities like fire hydrant maintenance. Repairs on large pipelines are appropriate for consideration as contract services. However to keep costs down, routine service line repairs, small pipeline repairs, maintenance of pumping systems, and similar activities are more appropriate to retain internally for larger systems since the activities occur continuously.

Management Responsibilities

Management responsibilities include the following:

- Organizing
  - Organizes developed information
  - Uses information to make decisions
  - Allocates resources (human, financial, material, and equipment) to implement decisions
  - Structures the work
  - Manages allocation to ensure desired results

- Directing
  - Delegates responsibility and authority for decisions (this can be a huge problem if too much or too little delegation occurs)

    - Defines accountability/expectations
    - Supervises the assignment

- Control
    - Uses status reports to monitor progress and make decisions on how resources can be more effectively used
    - Cost control
    - Time control
    - Resource reallocation
    - Information to adjust future decisions

A major strategic management goal should be to establish climates that unite staff and encourage a commitment to success and a willingness to provide leadership with stewardship of the assets and with utility–customer relationships. Strategic management goals include the following:

- Operational excellence
- Customer-centered focus
- Leadership
- Enrichment of experience (internal and external with training and outreach)

To ensure these goals are met, there is a need for management to foster the goals in the organizational culture. While every organization has its own culture, culture describes the organization by

- Developing consistent direction
- Establishing an internal way of life
- Determining internal effectiveness
- Setting expectations

Following is a list of types of culture and characteristics within an organization.

- Collaboration—teamwork-driven, customer satisfaction, coordination participation
- Control—predictable, accurate, success oriented, directive driven
- Competence—oriented to unmatched service or products, long-range vision
- Cultivation—focus on personal enrichment, highest customer satisfaction, optimistic, charismatic

The public and governing bodies expect that the management of the utility system will effectively manage the system on a day-to-day basis and ensure regulatory compliance. They expect that managers will resolve problems and implement policy as directed. This is accomplished through interaction with customers and the governing board. The following are questions the rating agencies try to answer to discern the utility's managerial capabilities:

- Are there institutionalized management systems for controlling cash, debt, and budgets?
- What is done to attract and maintain qualified managers and middle managers within the organization?
- How are costs controlled, i.e., to what extent have alternative service-delivery methods been examined that might promote economies and efficiencies in providing services to the customers?

Management has the responsibility to enact appropriate financial controls, i.e., financial policies that tend to reflect financial performance. Management has the responsibility to articulate the negative public health effects of applying inappropriate financial planning models and goals to the utility. It also is responsible for developing appropriate and customized financial management goals and standards for the utility that take into consideration current and future public health obligations, and preparing presentations to managing boards that compare financial indicators in a way that takes into consideration fundamental differences between utilities and recognizes public health needs.

Bond rating firms analyze the financial policies of a utility as a reflection of financial performance. Of concern are not only the financial controls and policies, but also the quality of management. Reliability and continuity of information are important elements of management and exhibit a willingness to make hard choices.

Questions to determine effective management include the following:

- What is the appropriate response to demands for higher productivity and competitiveness?
- What is the best way to accommodate rapid change?
- What events will impact operations?
- How does population diversity affect service delivery expectations?
- How does the community evolution affect recruitment?
- How does volatility of expenditures affect the ability to deliver services?
- How do the economy and community evolution affect revenues?
- How does community evolution affect capital infrastructure and improvements required?
- How does the utility expand the knowledge, skills, and abilities of staff?
- How does the utility improve working conditions?
- How does the utility increase the employee pool?
- How does the utility impart ethics, responsibility, and public service to staff?
- What processes are in place or needed for workforce/succession planning?
- How does the utility address needs for flexible work schedules and home life conflicts, while empowering employees?

Governing Body Responsibilities

One of the major issues in dealing with local governments is defining what are and what are not the responsibilities of various people involved with the utility. The governing board has an important role, but do they understand it? Does management understand it? This is a major point of conflict at times. Failure to clearly define and understand the delineation between management and the governing board often leads to poor direction, inability to move forward, and personality and leadership problems.

The board and stakeholders are two entities that the staff must work with. These two entities are not necessarily always on the same page, although it would seem logical that they should be. Board and stakeholder needs include communicating water issues to the public and balancing between "water is a right" and "water is a commodity."

Administrative measures are policies that should be in place as adopted by the governing board. Every utility should have clear policies in place to monitor money, buy supplies, and train employees. Unfortunately, budget cuts too often affect maintenance expenditures, which leads to potential failures or accelerated capital expenditures in the future. If it is perceived that the utility does not do a good job on these items, then it may raise concerns about the ability of the utility to carry out its plans.

Governing board responsibilities include setting overall goals/vision of the utility and approving the annual budget, financial plans, strategic plans, and capital and special programs. Only the board can approve agreements, the issuance of debt, or the use of reserve funds. The board is responsible for setting policies and procedures that govern operations, including such actions as the billing cycle, payment of bills, late payment, turnoffs, and other customer service policies (see appendix D). The board also provides staff with limits of their authority and standards of performance and accountability for daily operations.

Stewardship and ethics are keys to the board, stakeholders, and staff. Communication, consensus, and ethical behavior build trust between boards and management. All are necessary to show the public the leadership needed to trust the utility, which is especially important during a crisis, such as when rate increases are needed. Responsible stewardship includes

- Clear delineation of expectations of staff with regard to hours, work, and duty station
- Clear fiscal policies, purchasing requirements, and authority for purchases
- Transparent bidding and procurement standards
- Clear policies with regard to use of equipment
- Clear delineation between staff, management, and board authority and liaison

The governing board is responsible for setting the following personnel policies.

- Define hiring practices
- Set compensation plans that are competitive and can attract qualified employees
- Ensure compliance with employment laws
- Adopt personnel policies
- Hold staff accountable for human resource processes

Contract responsibilties include approving contracts to

- Engage an auditor
- Engage engineers and consultants
- Accept contracts to deed property to the utility

- Approve developer agreements or arrangements
- Engage legal counsel
- Engage lobbyists and legislative liaisons
- Approve agreements with other utilities, governing bodies, and private entities

The governing board has the responsibility to hire and fire management. Firing managers is needed at times. However, in many cases, the difficulty between the manager and board stems from miscommunication and misunderstanding of responsibility, not from competence issues. Where this is a problem, firing the manager likely does not resolve the problem, but may indicate that there is a stability problem on the governing board.

Because this process is similar to a city council firing a city manager, let's use that as an example. City managers are typically hired by a city council or board of commissioners. The group hires and can fire the manager, often at will, meaning they do not need a cause for this decision. This makes city managers nervous. Then an election will come and suddenly the manager will become an election issue. One of the first tasks the new board wants to do is fire the manager. The reasons are often unclear (otherwise the prior board would have done it). Some reasons may include "we need a new direction," "we need new blood," "he's too close to the old board," "he doesn't communicate well," "he doesn't tell us what is going on," "the employees don't like him," and so forth. Again, the issues are often miscommunications or politically driven, not skill driven. As noted in chapter 6, the board/supervisors must clearly outline expectations in order for any employee to meet them. This is not to say that some managers shouldn't be dismissed; then, how it is done becomes the challenge.

Ethics

Ethics is an extremely important issue for both staff and governing officials. There are far too many examples of public officials trying to profit from their position. There are hundreds of examples of communities where elected and appointed officials are indicted, sometimes for apparently minor infractions. Most ethics problems are associated with legal issues. Professions such as engineering and city management have specific codes of ethics that must be followed. Failure of professional engineers to follow their ethical codes can cost them their licenses. Ethical standards typically include requirements such as

- Give utmost performance
- Participate in honest enterprises only
- Live and work in accordance with the highest standards of professional conduct
- Place service before profit
- Honor and standing above personal advantage
- Public welfare above all else
- Enhance public trust during your work
- Report violations of ethical practices

Many local jurisdictions have strict ethical standards with regard to things like accepting gifts or lunch from people who could possibly be construed as trying to influence staff. Many professions (like engineers) have similar ethical obligations. Gift giving or acceptance is the most commonly associated ethics problem, but by no means the only one. Accepting gifts such as golf games and expensive dinners from anyone outside the organization should be avoided. Those accepting gifts from vendors often subject themselves to charges of favoritism, as the public does not perceive that they can be independent under such circumstances.

Lobbying elected officials for awards to vendors is also a problem for both staff and the board. It is not likely that a project will be successful when there is intense lobbying behind the scenes. This is a practice that may be contrary to ethical standards in many professions (including engineering).

SECURITY ISSUES

In the wake of the Sept. 11, 2001, attack on the United States, security has become a major focus of local governments. Water systems are no exception, especially since both water and sewer systems have potential entry points for security breaches. Several issues, including those listed below, need to be kept in mind in regard to water security problems.

- Planning is important. Every utility should have a ready plan to react to major incidents on the utility system.
- A great many chemicals are present on treatment plant sites. Gaseous chlorine may be the most significant one. Chemical storage facilities should be secure and protected. There are other options to gas chlorine that should be investigated.
- Having backup systems, redundant treatment trains, excess storage in the system, and excess capacity at any given time may improve the chances of getting the utility back on-line quickly. Excess storage in the system is essential in the event of a plant problem.
- Adequate maintenance is required. If systems are poorly maintained, preventive measures are not taken, and inspections of the system not routinely made, then problems can develop and intruders can access the system easily. Fire hydrants and valves are especially important. Having adequate spare parts or access to them is imperative.
- Security cameras, fences, and telemetry will allow operations personnel to detect people trying to access the system. Telemetry is especially important when well fields and surface water supplies are remote from the operations of the utility.
- Tank covers should be welded shut.
- Staff training from experts (the FBI provides this type of training) should be arranged. Practice drills will help personnel react rather than think about the response needs. Speed is important in emergencies.

Federal rules require utilities to perform a vulnerability assessment (see chapter 4). While there are many consultants willing to do this work, it is nearly impossible for a good assessment to be performed without significant input from the operations personnel. Managers should attend appropriate training, offered regionally on emerging threats, and create relationships with local law enforcement agencies. The need for law enforcement response to the utility may be significant, but the utility may have equipment and personnel that can also be of benefit to law enforcement personnel during emergencies.

PUBLIC COMMUNICATIONS

Communication between management, line personnel, governing boards, and the public is fragile. It depends on experience, confidence, reliability, and other intangible factors of the utility staff on the part of the recipients. In part the medium used depends on the recipients. Different groups rely on different media. In most cases the public relies on verbal communication in 15-second sound bites, which is important for the management staff to understand. Board members likewise rely on verbal communication as opposed to written communication. Internal communication is both verbal and written.

Communication is the most important issue in the workplace, so it cannot be left to chance. There must be a purpose for any communication. It is important to put thought into both the medium chosen and the content. Since communication is a two-way process, if no one is listening, no one receives the message. If there is no feedback, then the utility will not know if anyone received it or if they understood it.

Outreach and credibility are intrinsically linked to customer service. This is where the customer is most likely to contact the utility. If this is a bad experience, then the utility will not be looked upon favorably. At the same time the expectation of customer service is evolving. People expect more, faster, and 24/7. They expect electronic communication and access.

People who cannot communicate cannot be promoted and cannot be exposed to customers. Reading, writing, and speaking are all important tools that supervisors and managers need. This section outlines some tools that students can use to prepare of presentations for the next class. Management must know what kind of message to send, to whom to send the message, and how to translate the message into a language that all can understand. There are a number of variables in communication, including:

- Listening skills
- Culture
- Intelligence
- Knowledge base
- Semantics
- Situational consideration
- Emotional status
- Authority or position
- Common sense

As society broadens, develops, and includes new people, this becomes trickier. Perception is important. For example, of the group *monkey*, *panda*, and *bamboo*, from the perspective of people in the United States, the closest two are monkey and panda—both are animals. In other parts of the world, panda and bamboo may go together because pandas live among and eat bamboo. The perception depends on experiences and societal references.

For writing to be persuasive, it must convince the reader that yours is the best solution. To justify a position, the writer can either present arguments and then the conclusion (recency), or present the conclusion and then the arguments (primacy). People typically remember the most recent thing they perceive; so many persuasive arguments reiterate the message three times. In presentations, visual clues gain attention. This is the reason for attractive people in advertisements.

Writing forms are similarly important. Written communication can be in the form of letters, pamphlets, e-mail, memos, or technical reports. All interim reports should be marked as "DRAFT" or "INTERIM" since the report will otherwise be deemed to be final and to represent the final solution by which the utility will be judged. With written communication, it is important to know the audience. The information cannot be too simple (boring), assume too much knowledge (frustrating), or be the wrong reading level. Writing well takes time and practice.

Presentations should include proper visuals, be professional, and include a speaker who speaks clearly, coherently, and makes the points. Visuals are secondary to what is said. They become a distraction if they cannot be read, have spelling errors, or are not comprehensible. For a good presentation, know the material, know the audience, keep the message simple, provide eye contact, and encourage the audience to listen and learn. The impression given during a presentation is what people will perceive about the utility. If the presentation is not good, then the assumption is that the speaker is unfamiliar with the material. If the presentation is sloppy, it indicates that the presentation is not important.

REFERENCES

AWWA. 2005. AWWA G100-05, Standard for Water Treatment Plant Operation and Management. Denver, Colo.: American Water Works Association.

Diamond, L.E. 2007. *Team Building That Gets Results*. Naperville, Ill.: Some Books Inc.

Maslow, A. 1954. *Motivation and Personality*. New York, N.Y.: Harper. p. 236.

Olstein, M.A, D.L. Marden, J.G. Voeller, J.D. Jennings, P.H. Hannan, and D. Brinkman. 2005. *Succession Planning for a Vital Workforce in the Information Age*. Denver, Colo.: AWWARF.

CHAPTER 8

Utility Planning

Figure 1-1 in chapter 1 showed a typical utility organization chart. Planning was not shown on this chart because planning occurs throughout the organization. Planning is a management function, although much of the technical work involves engineering, and so many planning functions are included in the engineering areas of the utility. But planning is a process that should be used by utilities to reach a vision of the utility as defined by the customers and the governing board, and to meet certain demands for service projected to be required in the future.

Understanding and managing the utility's assets provides important information related to the ongoing future direction of the utility system. However, the only method to develop that future direction is through the planning process. Planning should be undertaken on a regular basis by all enterprises in an effort to anticipate needs, clarify organizational goals, and provide direction for the organization to pursue, and to communicate each of these to the public.

With water and wastewater utility systems, it is imperative to have ongoing planning activities, as many necessary improvements and programs take months or years to implement and/or complete. Without short- and long-term plans to accomplish future needs, the utility will suffer errors in direction, build unnecessary or inadequate infrastructure, and pursue programs that later are found to provide the wrong information, level of service, or type of treatment (Bloetscher, 2009).

Planning can provide for a number of long-term benefits, including improvements in International Organization for Standardization (ISO) ratings to lower fire insurance rates, renewal of improvements as monies become available, rate stability, and most importantly a vision for the utility.

In creating any plan for a utility system, efforts to understand the environment in which the utility operates must be undertaken. Second, the needs of the utility must be defined, typically from growth projections and analyses of current infrastructure condition from repair records or specific investigations. Funneling this information into the planning process will permit a set of clear goals and objectives to be defined (Figure 8-1). However, the types of goals and objectives may vary depending on the type of plan developed. There are four types of plans (listed below) that may result from the planning process.

1. **Strategic plans**, which are action oriented for management-level decision making and direction.
2. **Integrated resource plans.** These are actions for utility management to tie all parts of the system together.
3. **Facilities plans**, including state revolving funds (SRF) loans support.
4. **Master plans** to support capital improvement programs.

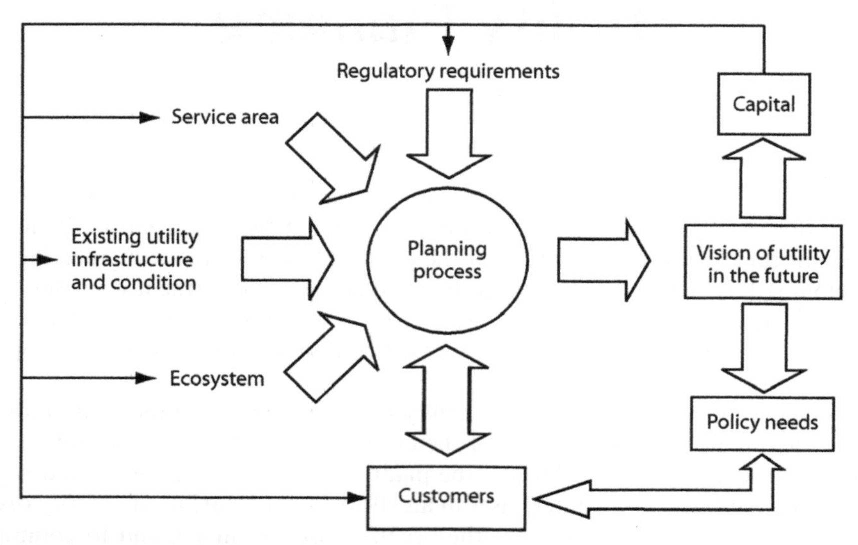

FIGURE 8-1 Planning process diagram

Any utility planning effort should start with a description (and understanding) of the local ecosystem. Understanding the environment from which water is drawn or into which it will be discharged is important. Both water quality and available quantity, whether surface water or groundwater, are profoundly affected by demand. A reduced demand for surface water helps prevent degradation of the quality of the resource in times of low precipitation. Reduction in the pumping of groundwater improves the aquifer's ability to withstand salt-water infiltration, upconing poorer-quality water, contamination by septic tank leachate, underground storage tank leakage, and leaching hazardous wastes and other pollutants from the surface. Overpumping groundwater leads to mining the aquifer or to contamination of large sections of the aquifer. Planning for potential surface contamination is necessary for surface water systems. Therefore, source water protection must be a part of any water planning efforts, including the appropriate application sites and treatment needs for reuse and residuals.

Regulatory issues should be included in the planning process as the projected changes resulting from regulatory requirements may cause major changes in utility operations or require significant capital expenditures. Forecasts need to be made predicting demands based on growth patterns and likely new customer bases and potential utility or service area acquisitions. Water use patterns and projections of variables such as population and land use can be used to predict future demands based on a clear understanding of the service area. Tables 8-1 and 8-2 are examples of average and aggressive growth projections for a utility. The difference over 20 years is over 1,200 units, or 10 percent of the total demands. Table 8-3 shows the population projections for several service areas. One of the issues to note is that the past may not be an appropriate indicator of the future. For all of these examples, population projections are clearly not possible.

Any planning document needs to outline the existing utility system, including evaluation of the current asset conditions. Analysis of the current distribution system pressures and water quality will point out problem areas that need main replacement or reinforcements. An example of a set of proposed improvements stemming from the growth in Tables 8-1 and 8-2 can be found in Figure 8-2. The distribution system analysis will lend itself to modeling of water quality transport in the future. Comparisons of water quality between existing and proposed regulations will indicate areas where treatment should be improved. Concepts to reduce peak demands (thereby lowering estimates of future capacity needs) and to define when major capital investments are necessary are vital to any plan. Infiltration and inflow analyses and surveys of piping might involve the sanitary sewer system. The need to integrate operations data (like work order results) with longer-term asset needs, engineering alternatives, regulatory needs, and design processes leads to better capital improvement plans and policies.

If policy changes like water conservation are implemented, then it is necessary to monitor the effectiveness of the policy/program. There are indications that people make changes in short-term water use behaviors, but eventually return to their former habits. This could be disastrous if facilities are not on-line and demand increases drastically over a short period of time. This is one reason a public relations program is so essential to the continuing water conservation effort. Utilities will experience a general decrease in total operations costs as a result of a properly designed and implemented conservation program (see chapter 15). The result of the plan should be the direction for the future of the utility and an identification of operating, capital, and policy needs to meet that direction. Among the more important steps is the capital improvement program. The following sections outline these plan types in more detail.

TYPES OF PLANS

Strategic Plans

Strategic plans are action-oriented, management-level documents that set a vision for the community and offer a series of strategies and action steps to be taken to reach the desired goal (Bloetscher, 2009). They are typically in 5-, 10-, or 20-year increments and may require substantial changes to current operating and management practices in order to meet them. Corporations often use strategic plans to help guide new products into the market or to develop strategies to carve out their share of the market. Utilities are no different.

Strategies and action steps should include not only utility issues, but how the utility issues may affect, or be affected by, the activities of other sectors of the local economy and local governments. Interagency cooperation is a part of the strategic plan. Often details are limited for a strategic plan, since the focus is typically policy oriented, focusing on the macroissues. The details are left to those implementing the policy directives. Strategic plans are a result of consensus among the participants. Consensus can limit implementation of the plan in situations where issues are contentious, but the goal should be to reach action steps and strategies that can be implemented.

A strategic plan can be brief or extensive, and can be part of an integrated resource plan, which tends to be a more detailed document. Strategic plans assume community and elected-official input. Communication is essential since the elected officials are an important part of selling the plan to the public, despite the fact that many of the facets of the strategic plan will come to fruition after elected officials have left office. Appendix A (Bloetscher and Saltrick, 1998) is a series of strategies and action steps created by the city of Hollywood, Fla., for its strategic plan for the utility. Strategic plans are a governing board/senior management–led document (Bloetscher, 2009).

TABLE 8-1 Average growth demands for water service

Year	Accounts	Units	Usage per Unit	CRA Units	Usage per CRA Unit	Total Units	ADF (mgd)	Annual% Increase	MDF (mgd)	MDF/ ADF	Svc. Area Popul.
2003	4,660	7,903	300	0	250	7,903	2.4	n/a	2.9	1.22	15,806
2004	4,665	7,909	300	0	250	7,909	2.4	0.1	2.8	1.20	15,818
2005	4,669	7,914	300	0	250	7,914	2.4	0.1	2.8	1.20	15,829
2006	4,674	7,920	300	150	250	8,070	2.4	1.7	2.9	1.20	16,110
2007	4,679	7,928	300	250	250	8,178	2.4	1.1	2.9	1.20	16,306
2008	4,683	7,936	300	350	250	8,286	2.5	1.1	3.0	1.20	16,502
2009	4,688	7,944	300	550	250	8,494	2.5	2.1	3.0	1.20	16,878
2010	4,693	7,952	300	700	250	8,652	2.6	1.6	3.1	1.20	17,163
2011	4,697	7,960	300	850	250	8,810	2.6	1.6	3.1	1.20	17,449
2012	4,702	7,968	300	1,000	250	8,968	2.6	1.5	3.2	1.20	17,735
2013	4,707	7,976	300	1,150	250	9,126	2.7	1.5	3.2	1.20	18,021
2014	4,712	7,984	300	1,300	250	9,284	2.7	1.5	3.3	1.20	18,307
2015	4,716	7,992	300	1,450	250	9,442	2.8	1.5	3.3	1.20	18,593
2016	4,721	8,000	300	1,600	250	9,600	2.8	1.4	3.4	1.20	18,879
2017	4,726	8,008	300	1,750	250	9,758	2.8	1.4	3.4	1.20	19,165
2018	4,730	8,016	300	1,900	250	9,916	2.9	1.4	3.5	1.20	19,451
2019	4,735	8,024	300	2,050	250	10,074	2.9	1.4	3.5	1.20	19,737
2020	4,740	8,032	300	2,200	250	10,232	3.0	1.4	3.6	1.20	20,023
2021	4,745	8,040	300	2,350	250	10,390	3.0	1.3	3.6	1.20	20,309
2022	4,749	8,048	300	2,500	250	10,548	3.0	1.3	3.6	1.20	20,595
2023	4,754	8,056	300	2,650	250	10,706	3.1	1.3	3.7	1.20	20,881
2024	4,759	8,064	300	2,800	250	10,864	3.1	1.3	3.7	1.20	21,168
2025	4,764	8,072	300	2,950	250	11,022	3.2	1.3	3.8	1.20	21,454

Note: CRA = community redevelopment area, ADF = average daily flow, MDF = maximum daily flow

TABLE 8-2 Aggressive growth demands for the same utility

Year	Accounts	Units	Usage per Unit	CRA Units	Usage per CRA Unit	Total Units	ADF (mgd)	Annual % Increase	MDF (mgd)	MDF/ ADF	Svc. Area Popul.
2000	4,660	7,903	300	0	250	7,903	2.4		–0.1	–0.04	15,490
2001	4,665	7,909	300	0	250	7,909	2.4	0.1	0.9	0.38	15,818
2002	4,669	7,914	300	0	250	7,914	2.4	0.1	1.9	0.80	16,224
2003	4,674	7,920	300	0	250	7,920	2.4	0.1	2.9	1.22	16,632
2004	4,679	7,928	300	0	250	7,928	2.4	0.1	2.9	1.20	17,045
2005	4,683	7,936	300	0	250	7,936	2.4	0.1	2.9	1.20	17,618
2006	4,688	7,944	300	330	250	8,274	2.5	3.6	3.0	1.20	18,295
2007	4,693	7,952	300	660	250	8,612	2.6	3.4	3.1	1.20	18,973
2008	4,697	7,960	300	990	250	8,950	2.6	3.3	3.2	1.20	19,650
2009	4,702	7,968	300	1,320	250	9,288	2.7	3.2	3.3	1.20	20,328
2010	4,707	7,976	300	1,650	250	9,626	2.8	3.1	3.4	1.20	21,006
2011	4,712	7,984	300	1,980	250	9,964	2.9	3.0	3.5	1.20	21,684
2012	4,716	7,992	300	2,310	250	10,302	3.0	2.9	3.6	1.20	22,361
2013	4,721	8,000	300	2,640	250	10,640	3.1	2.9	3.7	1.20	23,039
2014	4,726	8,008	300	2,970	250	10,978	3.1	2.8	3.8	1.20	23,717
2015	4,730	8,016	300	3,300	250	11,316	3.2	2.7	3.9	1.20	24,395
2016	4,735	8,024	300	3,450	250	11,474	3.3	1.2	3.9	1.20	24,712
2017	4,740	8,032	300	3,600	250	11,632	3.3	1.2	4.0	1.20	25,030
2018	4,745	8,040	300	3,750	250	11,790	3.3	1.2	4.0	1.20	25,348
2019	4,749	8,048	300	3,900	250	11,948	3.4	1.2	4.1	1.20	25,666
2020	4,754	8,056	300	4,050	250	12,106	3.4	1.2	4.1	1.20	25,984
2021	4,759	8,064	300	4,200	250	12,264	3.5	1.2	4.2	1.20	26,302
2022	4,764	8,072	300	4,350	250	12,422	3.5	1.2	4.2	1.20	26,619

Note: CRA = community redevelopment area, ADF = average daily flow, MDF = maximum daily flow

TABLE 8-3 Comparison of three utility areas

	County A	Percent Increase	City B	Percent Increase	County C	Percent Increase
1900	1,075		381,768			
1910	3,321	309%	560,663	147%		
1920	4,459	134%	796,841	142%		
1930	8,532	191%	900,429	113%	2,883	
1940	16,414	192%	878,336	98%	5,102	177%
1950	48,289	294%	914,808	104%	6,488	127%
1960	127,016	263%	876,050	96%	15,753	243%
1970	273,016	215%	750,879	86%	38,040	241%
1980	463,087	170%	573,822	76%	85,971	226%
1990	770,280	166%	505,616	88%	152,099	177%
2000	1,394,440	181%	478,403	95%	251,377	165%
2010	2,440,270	175%	406,643	85%	397,176	158%
2020	4,270,473	175%	345,646	85%	579,876	146%
2030	7,473,327	175%	293,799	85%	777,034	134%
2040	13,078,322	175%	249,729	85%	947,982	122%
2050	22,887,064	175%	212,270	85%	1,156,538	122%
2100	375,647,341	175%	94,185	85%	3,125,785	122%

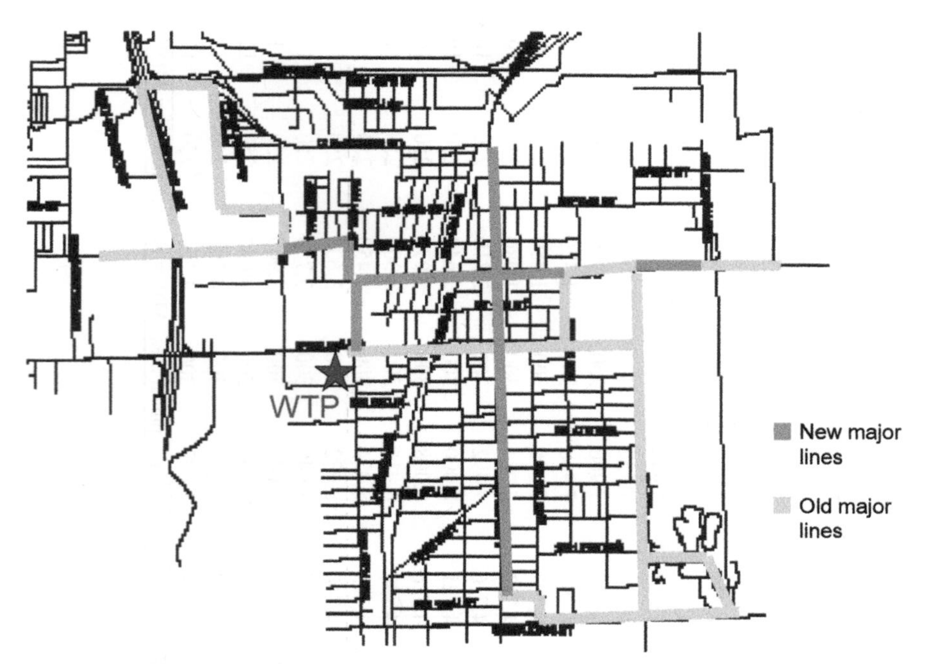

FIGURE 8-2 Water lines planned as a result of Table 8-1 and Table 8-2 projections. The new major lines are in dark gray, while the old major lines are in light gray.

Integrated Resource Plans

Integrated resource plans came into vogue under the auspices of the American Water Works Association (AWWA) in the early 1990s, and are therefore oriented toward the water/wastewater industry, although the concepts can be applied to any infrastructure system. The concept of an integrated resource plan for a utility system requires all aspects of strategic plans to be included, but in more detail, including the full scope of the operating environment, available resources, and ecological system, i.e., total water management. Integrated resource planning involves reviewing alternatives for the provision of service over the long term (generally 20 to 50 years), and providing service to those areas that will be in need of such service. Critical issues like cost–benefit analysis of water allocations to certain users should be considered.

One focus of integrated resource plans that separates them from strategic plans is the development of alternatives for future water sources, as well as improvements needed to provide them. The need for increasing sources of water supply throughout the United States is resulting in a trend within the water industry toward demand-side water management. This is an effort to find a balance between the development of new, varied sources and conservation of current resources (Bloetscher, 2009).

Part of integrated resource planning is the traditional master facilities plan, which requires an evaluation of current and future demands, coupled with a determination of when additional facilities will be needed. These future facilities typically represent major capital investments and additional maintenance costs.

Integrated resource plans are largely influenced by the environmental trends and growth in the area and the viability of conservation techniques. Ideas that should be addressed include steps to reduce water source demands, use of reclaimed effluent water, membrane treatment, and aquifer storage and recovery. All of these are used to increase water supplies during periods of need through better management of the resource, while mitigating demands on current reserves by using other treatment techniques or supplies (or by mitigating peaks, in the case of aquifer storage and recovery [ASR] programs). See chapter 16 for more discussion of the ASR concept.

Integrated resource plans are often complementary, or background, documents developed by staff and/or advisory boards, leading to the strategic planning exercise. Much more detail should be included in the integrated resource plan compared to a strategic plan, and the identification of system needs (such as infiltration and inflow) through development of the appropriate research is a requisite step. The integrated resource plan should identify permitting needs and capital strategies clearly, and define the interrelationships between the water, wastewater, and environmental aspects of the region.

Facilities Plans

Facilities plan is a term coined as a result of the state revolving fund (SRF) loan programs. Facilities plans are the documents commonly required to justify the need for facilities and the recommendations for capital projects where SRF loan programs are proposed to be used. The SRF programs are looking for support documents, their cost efficiency, the alternatives reviewed, and the ability of the utility to repay the loan. As a minimum, a facilities plan should outline the utility and its assets, assess needs, and provide alternatives for meeting those needs together with cost analysis. Capital is a major focus, and the timeline is usually 10 years or less. Regulatory issues must be identified, as must environmental impacts, since these plans go through a state clearinghouse process (Bloetscher, 2009).

There are basic requirements set by the federal government for administration of the SRF program at the state level, including a state match to federal allowances. As a result, facilities plans can be extensive, or not, depending on the state agency administering the program. Financial issues are a major part of facilities plans, which is often not the case with integrated resource plans. An integrated resource plan would meet and exceed the agency goals in most cases, although tweaking might be required. An executive-level strategic plan would not meet the goals due to lack of detail. Many utilities only focus on the SRF requirements, which may ignore important planning issues that might help operations in the long term. Facility plans are typically completed by consultants. Table 8-4 outlines the issues usually looked at by agencies reviewing the facilities plan.

Master Plans

Master planning involves reviewing alternatives for the continual delivery of quality water or wastewater services (or other infrastructure) over a long term and extending service to those areas that will need such service in the future. Incorporated into a *master plan* are alternatives for the sources of this water as well as improvements needed to provide them. A master plan requires evaluation of current and future demands, which indicate when additional facilities are needed. These future facilities usually represent major capital investments and added maintenance costs. They are largely influenced by growth and environmental trends in the area studied. The focus is capital needs over a fixed period, typically 20 years. Master plans are obvious subsets of an integrated resource plan, and the basis for a facilities plan. Master plans are typically compiled by consultants (Bloetscher, 2009).

DEVELOPING THE PLAN

Regardless of the type of plan resulting from the planning process, planning in the water and wastewater fields involves the nine components (Dzurik, 2003) listed below.

1. Development of goals and objectives
2. Data collection and analysis
3. Problem identification
4. Clarification of the problem and ancillary issues
5. Identification of reasonable alternatives
6. Analysis of the alternatives
7. Recommended actions
8. Development of an implementation program
9. Monitoring of results

Development of Goals and Objectives

Before many problems can be identified and solutions sought, a series of relevant goals and objectives should be developed (Bloetscher, 2009). Solutions to be considered to resolve identified problems will be evaluated using these goals and objectives. Therefore if the goals and objectives do not adequately outline the intent of the constituency, the solutions may fail to adequately address the problem needs.

TABLE 8-4 Summary of facility plan checklist

- The boundaries of the planning area, including existing and future project service areas.
- The demographic, geologic, topographic, hydrogeologic, and institutional characteristics of the study area impacting the evaluation of alternatives to the selected facilities.
- The cost-effectiveness of feasible alternatives to the selected facilities, including regionalization of facilities, capital costs, and operation and maintenance costs over the useful life of the facilities.
- The value of potable water resources conserved as a result of reclaimed water alternatives.
- The implementability of the selected facilities from legal, institutional, financial, technical, and management perspectives.
- The environmental effects and nonmonetary considerations associated with the selected facilities.
- The collection, transmission, treatment, reuse, and disposal problems associated with the wastewater system and the local physical conditions associated with those problems.
- The facilities needed to comply with wastewater treatment plant discharge permits and the facilities needed to maintain compliance throughout the project service area.
- Demonstration of the public participation process.
- Capital improvements financing information addressing the following:
  - The source of funds or revenues to be dedicated to repaying the loan and the expenses, charges, and liens against such dedicated funds or revenues.
  - The proposed system of charges, rates, fees, and other collections that will generate the revenues to be dedicated to loan repayment demonstrating that the wastewater management system is to be financially self-sufficient.
  - The proposed rate ordinance or other enforceable schedule for charges, rates, fees, and other collections associated with loan repayment.
  - The applicable actual and projected wastewater management system operating and nonoperating expenses and revenues for the first full year after the project has been constructed and is in operation.
- An affirmation that the selected facilities are consistent with other locally adopted plans.
- The responses generated by a multidisciplined, intergovernmental review, if applicable.
- The executed and fully implementable contractual agreements whenever facilities or services beyond the local government's jurisdiction are involved.
- The following additional information should be provided if the proposed project is shown to be cost-effectives of the recommended facilities as determined by treatment plant flows.
  - The benefits of improving operation and maintenance of existing facilities if the cost-effectiveness of recommended facilities is determined by the level to which wastewater or residuals is treated.
  - A description of the recommended facilities, preliminary design parameters, estimated capital costs, and estimated operation and maintenance costs.
  - The schedule for constructing the recommended facilities.

While defining the goals and objectives sounds easy, this area may be where there is the most conflict. Common areas for conflict include economic issues (as noted on the next page for septic tanks), conflicts between environmental and water use needs (salmon populations in the West are an example), and conflicts between water users. Goals and objectives are general in nature, but clear in application (i.e., whether or not they are met or optimized in a given situation, or used in ranking alternatives). Objectives are more specific subsets of the goals, and several objectives are used to attain the goal.

Data Collection and Analysis

The data collection and problem identification steps may vary in order. Data on water or wastewater quality from a treatment plant may indicate the need to upgrade to meet regulatory standards. There may be a need to increase capacity. Data may indicate where pipelines are in need of replacement or repair, where changes in disposal practices should be considered, or where public health concerns may exist. Data collection will help formulate the scope of the problem. However, once the problem is identified, additional data and analysis will be required to identify the alternatives that should be considered. Data should include a projection of the future. Forecasts should be made of the appropriate variables to consider to meet future needs.

Problem Identification

Problem identification results from defined needs of the system, whether those needs are projects, policies, or programs, and the goals and objectives developed initially. These needs may be defined by the governing board, customers, projected growth demands, regulatory requirements, failures of infrastructure, or some combination of sources. The problem may be defined differently by different constituencies. For example, the replacement of septic tanks with sanitary sewers may be required by regulatory agencies to resolve actual or perceived public health issues, but the residents who are to receive the sewers may view the added expense of sewer service, connections, and assessments as intrusive, unpopular, and unnecessary. Certain projects will require some consensus among the constituents in order to proceed smoothly, while other needs will be obvious. Some consensus about the problem is required in order to properly gather data and consider options.

Clarification of the Problem and Ancillary Issues

Once the initial data and problems have been identified, the goals and objectives should be reviewed and refined where required. When this has been accomplished, additional data have been collected, and public discussion ensues, crystallization of the problem to be solved will occur. Crystallizing the problem to be solved will generate more obvious and pertinent solutions, thereby saving time and dollars. This phase of the planning process will likely indicate a series of solutions that can be evaluated.

Identification of Reasonable Alternatives

Formulation of alternatives is the next step in the process. Poor alternatives should be quickly eliminated from the analysis. Poor alternatives are often of high or very low cost, and only partially meet the goals. They are fodder for political conflict, which is unneeded in solving the complex problems facing utility systems. Alternatives must consider goals and objectives, economic limitations of the utility, environmental effects caused or solved by the alternatives, the impact of operations and maintenance, operational complexity, and regulatory impact. No bad alternatives should ever be presented in a public forum. They are a magnet for elected officials because bad alternatives are often lobbied for by outside entities that see potential opportunities to profit from inadequate planning, policies, or solutions.

Analysis of the Alternatives

Analysis of alternatives involves comparison of economic, social, and other benefits or liabilities of the alternatives. Analysis of alternatives should use the goals and objectives as criteria for evaluation. Economic evaluations are made using cost–benefit analyses or present-worth analysis, depending on the specifics of the alternative being evaluated. With cost–benefit analyses, where the benefits are derived from implementation of the alternative, the option is economically justified. If the alternative passes the cost–benefit test, comparison of alternative economics is pursued.

It should be noted that many times the benefits are unknown or not quantifiable at the time the cost–benefit analysis is conducted. There may be future benefits or the alternative may encourage other action. For example, one reason water/wastewater utilities are typically publicly funded is that the cost–benefit analysis of extending pipelines to undeveloped areas will always fail the cost–benefit test because future development is unknown and there are few current customers. The benefit of extending piping may be to expand development or discourage other utilities that might compete with the utility in the future.

There are several methods to make economic analyses, including present-worth and annual-worth calculations. Present-worth analyses compare the cost of capital construction and the operations and maintenance costs over the life of the project or program among the alternatives if all costs were brought to the present (i.e., a lump sum). The current situation is usually considered the "base situation" (and likely the least expensive option). Appropriate interest rates and annual increases in operations and maintenance costs are required for the analysis to be valid. The same assumptions, especially lifetime of assets or annual costs, should be used when comparing alternatives. Likewise the same life expectancy should be used for present-worth alternatives (Bloetscher, 2009).

Where the lifetime expectations are different, annual-worth analyses can be used, but this may be a harder concept for people to understand. For annual-worth calculations, the capital components are spread over the life of the facility or program, along with annual costs. The annual costs should be adjusted for the appropriate interest and inflation rates. Economic evaluations include four common factors (Blank and Tarquin, 2002):

1. **Interest rates** (i). The rates for borrowed money, inflation, increase in operating costs, and desired rates of return are all included as differing interest rates. Higher interest rates mean higher annual costs to the utility for borrowing. Higher inflation rates will increase operations costs geometrically. Combined interest rates may exist where there is inflation and growth at the same time. Interest rates should be assumed to compound (usually yearly for economic analyses).
2. **Term.** The length of time over which the analysis will be considered (n).
3. **Present value**. The cost of the asset or cash flow series at the current time. Capital costs are present-value amounts (signified by P). Present-worth calculations take cash flow streams and convert them to P.
4. **Uniform payments** (A). The periodic costs of the project converted to a uniform flow stream. Debt service for capital components is typically a uniform cash flow stream. Debt service can be determined from P and i above for a given term n. For instance, many bond issues are issued for 30 years. At a given interest rate, the annual debt service can be found from actuarial tables (A/P, i, n). Operations and maintenance costs are annual amounts, but they are rarely uniform. They tend to inflate at a given percent each year.

Terms for future worth (*F*) and a gradient of annual costs (*G*) can also be used, but rarely are utilities interested in the future value of an asset. A gradient may be useful where a utility is growing by a given number of customers each year. Many rate analysts mistakenly assume that growth of a utility is a given percent each year. This actually means growth accelerates with time, which is rarely case, even in a high-growth areas. A constant addition of customers is more appropriate (*G* in the actuarial tables). A good text on engineering economics defines the workings of these factors.

Table 8-5 shows a comparison of alternatives for a treatment plant. The costs are based on the life cycle of the assets, whereby the initial costs plus the anticipated operating costs over 20 years is developed to yield a present-worth comparison. Table 8-6 shows a comparison of pursuing infiltration and inflow, versus not. This is a life cycle, present-worth cost analysis as well, indicating that the benefit of doing the infiltration and inflow correction outweighs the cost of doing nothing.

Economic evaluations should never be the deciding factor, as purely economic evaluations ignore the social, environmental, and developmental impacts of the alternative. These benefits (or disadvantages) are very difficult to quantify, yet may have the effect of eliminating an alternative due to the damage that may occur in the service area. For example, where a community is dependent on salmon fishing, evaluating a series of alternatives and determining that a dam is the most cost-effective alternative for future water supply may ignore the fact that the dam may eliminate the salmon industry. This would remove most of the jobs from the community and make the additional water supply unnecessary. Table 8-7 shows an example where the salmon industry is clearly the highest priority from an economic perspective, but not from a regulatory perspective—agriculture is. Therefore measuring the positive and negative impacts of the project or program, as compared to the current condition, can be instructive, but not quantifiable.

Evaluation of the alternatives is most easily accomplished in matrix form. Table 8-8 outlines a comparison of three alternatives developed for the continued operation of a wastewater facility. An environmentalist might like the weighting in this model, as the matrix includes the economic, as well as other alternatives. Utility directors would have a very different view and would weigh the criteria differently. This could result in a potentially different answer (see Table 8-9), which is the matrix most utility personnel might agree with.

The criteria used in the matrix should match the goals and objectives defined for the project. Since some consensus on the goals and objectives is usually gained, if they do not match, the potential exists for the consensus to be frustrated. The comparison of alternatives should be clearly delineated using professional judgment. Trade-offs and impacts to social, political, and environmental factors should be delineated as well. Analysis of the alternatives should include the following:

- How well the option satisfies the goals and objectives (acceptability)
- Efficiency of implementation of the program
- Likelihood of success in meeting the goals and objectives
- Economic sufficiency
- Stability and sustainability of the alternative
- Social and political acceptability

TABLE 8-5 Present-worth cost comparisons of four options

Components	Water Treatment Plant No. 3							
	4 mgd	MS Plant	4 mgd	LS Plant	4 mgd	RO Plant	Bulk	Purchase
	Useful Life (yrs)	Amount ($)	Useful Life (yrs)	Amount ($)	Useful Life (yrs)	Amount ($)	Useful Life (yrs)	Amount ($)
New treatment units and disposal (no injection wells)	15	8,000,000	15	6,000,000	15	8,693,000	n/a	17,142,857
Wells		1,500,000		1,500,000		4,500,000		
Injection wells		11,000,000				11,000,000		
Recharge with WW		15,000,000		15,000,000				
Pipe cost	50	50,000	50	50,000	50	50,000	50	3,500,000
Building	30	900,000	30		30	900,000		
Auxiliary generator	10	500,000	10	500,000	10	500,000	10	500,000
Subtotal Item a		36,950,000		23,050,000		25,643,000		21,142,857
Construction (10%) and engineering		3,695,000		2,305,000		2,564,300	$—	400,000
(12%) contingencies		4,434,000		2,766,000		3,077,160	$—	480,000
TOTAL Item a		45,079,000		28,121,000		31,284,460		22,022,857
O&M Values								
O, M & R WTP		480,000		120,000		780,000		2,190,000
(10-yr average) WWTP		5,475,000		5,475,000				
Total Item b		5,955,000		5,595,000		780,000		2,190,000
PW of O, M & R (PWF = 7.403872)		44,090,058		41,424,664		5,775,020		16,214,480
Salvage values								
New treatment units and disposal		50,000		50,000		50,000		—
Pipe cost =$50,000 × (50–10)/50		—		—		—		—
Total salvage value		50,000		50,000		50,000		—
PW of salvage value (SVF=0.5650)		28,000		28,000		$28,000		—
PW = Item a + Item b – Item c		89,141,058		69,517,664		37,031,480		38,237,337

TABLE 8-6 Comparison of infiltration/inflow options

Options Item	Treating Wastewater Excess (est. 660,000 gpd at 2.30/1,000 gal)	Infiltration and Inflow
Annual operations cost	$575,000	$0
Growth rate	1.03	1.03
Capital cost	$—	$1,500,000
Present worth	$8,456,516*	$1,500,000
Assume 6.125%		
Debt	$0	$100,281

*Assumes 6.125% interest with a 3% inflation rate deducted.

TABLE 8-7 Priority to maximize economic value of the Klamath Basin

Industry	2000 value (in million/yr)	Potential Value (in million/yr)	Current Priority Water Rights	Economic Valued Priority
Agriculture	$200	$200	1	4
Tourism	$700	$1,500		2
Commercial fishery	$70	$4,500	n/a	1
Timber	$250	$250	n/a	3
Tribes	n/a	n/a	2	option 1-a

TABLE 8-8 Outline of water use options

	Option 1	Option 2	Option 3
Economic cost (present worth basis)	3	2	1
Environmental consideration (quality and quantity of resource)	1	2	3
Sustainability/renewal of resource	1	3	2
Ease/ability to reuse existing infrastructure	1	3	2
Flexibility/versatility	1	2	3
Ease of implementation and operation	3	2	1
Longevity of installation	3	2	1
Implementation/construction time	3	2	1
Intangible benefits (without economic cost)	13	16	13
All benefits	16	18	14

TABLE 8-9 Weighted evaluation of the same options

	Weight	Option 1	Option 2	Option 3
Economic cost (present worth basis)	10	30	20	10
Environmental consideration (quality and quantity of resource)	2	2	4	6
Sustainability/renewal of resource	3	3	9	6
Ease/ability to reuse existing infrastructure	2	2	6	4
Flexibility/versatility	2	2	4	6
Ease of implementation and operation	2	6	4	2
Longevity of installation	2	6	4	2
Implementation/construction time	2	6	4	2
Intangible benefits (without economic cost)	—	27	35	28
All benefits	—	57	55	38

Recommended Actions

The final solution should be the alternative that best meets the goals and objectives, realizing that no one solution may meet all the objectives better than all others. The recommended solution should be adopted by the governing body as the "plan," regardless of the plan type (as discussed previously). One solution should be recommended to the governing body. Adoption of the plan provides staff with direction to proceed with the implementation phase. In many cases, the approval will permit staff to begin immediate development of policies and procedures and to pursue capital project construction (see chapters 9 and 10).

Development of an Implementation Program

Implementation of the recommended alternative occurs after the plan has been adopted. Too often plans are adopted but never implemented, which wastes a lot of time, money, and the efforts of staff, governing bodies, and the public. Care should be taken not to implement plans that are to be placed on the shelf without further evaluation.

Monitoring Results

Monitoring the implementation and the results of that implementation are important to ensure that the chosen course of action indeed meets the goals and objectives for which the plan was developed and that the problems identified initially are solved. The monitoring phase is where modifications to the course of action or the programs and policies can be implemented to maximize the likelihood of success. The monitoring phase also permits the utility staff to interact with the governing board and public in a positive manner.

EMERGENCY RESPONSE PLAN

AWWA (1984) has developed a manual of practice (AWWA Manual M19, *Emergency Planning for Water Utility Management*) to help utilities develop emergency response plans. The process includes an evaluation as outlined in Figure 8-3. Emergency response plans are utility specific and involve development of

- Notification lists
- Internal employee listing
- External contractor and vendor list
- Protocol for repairing damage to the system and ensuring water service and water quality is maintained for all customers.

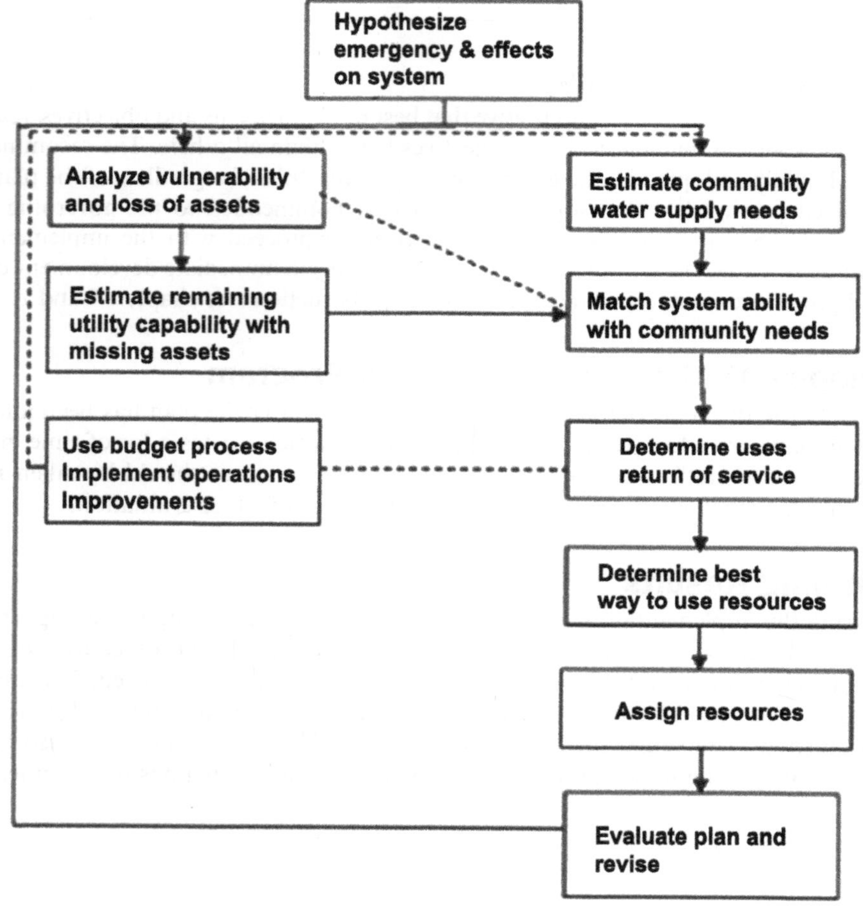

FIGURE 8-3 Emergency response analysis flowchart
Source: AWWA, 1984.

Many utilities have had protocols for service restoration for storms, hurricanes, and other events. How the utility is affected by natural or human-caused incidents drives the steps taken in the emergency response plan. Emergency response plans usually incorporate such measures as

- The use of redundancy treatment equipment
- Backup generators
- Inventory of emergency repair parts kept on hand to expedite service restoration
- Communications systems (e.g., radios, cell phones, e-mail) for staff (note that phone booth telephones often work during natural disasters, so an adequate supply of quarters is recommended)
- Backup computers (and computer backups)
- Written procedures and protocols
- Training of staff

Such plans usually contain contractor information for repairing pipeline and facility damage. Emergency response plans also contain protocols for posting notices to the public about the water quality problems in the piping system, as such contaminants that may have detrimental public health effects. A public relations effort and significant involvement of law enforcement and regulatory personnel would be expected under this scenario. Periodic tabletop drills for both scenarios should be undertaken to ensure that staff (and the changes thereto) are reminded of the scenarios. Vulnerability assessments and emergency response plans are not subject to inspection under open records laws of the federal government or any state.

REFERENCES

AWWA. 1984. AWWA Manual M19, *Emergency Planning for Water Utility Management*. Denver, Colo.: AWWA.

Blank, L., and A. Tarquin. 2002. *Engineering Economy*, 5th edition. New York, N.Y.: McGraw Hill.

Bloetscher, F. 2009. *Water Basics for Decision Makers: What Local Officials Need to Know About Water and Wastewater Systems*. Denver, Colo.: American Water Works Association.

Bloetscher, F., and R.W. Saltrick. 1998. Integrated Resource Planning Helps the City of Hollywood Ensure Its Long Term Water Supplies. In *Proc. Florida Section AWWA Annual Conference.*

Dzurik, A.A. 2003. *Water Resources Planning*. 3rd ed. Blue Ridge Summit, Penn.: Rowman and Littlefield Publishers.

CHAPTER 9

Developing the Capital Improvement Plan

Planning efforts usually provide a series of outputs, some of which include new infrastructure. The need for a utility to construct new pipes, treatment plants, and other facilities is ongoing as a result of regulatory changes, growth, obsolescence, and emergency needs of the utility. Capital expenditures are likewise ongoing expenses by utilities to meet these needs. A capital expenditure is one where there is a significant outlay of funds or other considerations that result in the acquisition or addition of a capital or fixed asset (Vogt, 2004). A fixed asset includes many types of properties that a utility might own, including

- Buildings and land
- Renovations to buildings
- Equipment, vehicles, or other furnishings that exceed a value threshold (often $1,000 or $5,000, depending on the asset and the size of the utility)
- Improvements to property other than buildings that add value to the property, such as parking lots, landscaping, drainage, piping, and so forth

Most utilities include a capital expenditure component in their annual budget for purchasing vehicles and certain equipment, or refurbishing these if they exceed the threshold value. Larger improvements are reserved for the capital improvement plan, which includes much larger facilities directly related to the extension, replacement, or development of the utility's infrastructure (piping, pumping, or treatment). Utility infrastructure is both costly and immobile, with only the local community receiving benefit from it. The intrinsic value of an installed pipeline is basically zero (as an example, piping cannot be dug up and reused).

The planning documents described in chapter 8 typically identify both programs and pro-jects that are required to meet the "vision" of the utility. Unfortunately, most planning documents are oriented to where the utility needs to be in the future, which focuses on expected growth and regulatory changes, and not so much the current condition of the infrastructure. Emergency response plans may identify additional projects, such as those needed to improve security. However, many projects that should be in the capital improvement plan (CIP) often are not, such as those to replace piping and pumps. Determining the needs for repair and replacement of existing buried piping is not a precise science, as discussed in chapter 4.

Water and sewer pipelines are traditionally laid in road rights-of-way. Many utilities were laid in rights-of-way when a road was only two lanes. Expansion to four or more lanes places the utilities under the pavement, creating risks for workers and disrupting traffic. In addition, it has generally been assumed by utilities that these rights-of-way will always be made available for their use. However, road rights-of-way have come under increasing pressure as cable television, electric, telephone, gas, and other conduits are laid under the surface. The lack of as-built mapping information complicates this further. Today, complex systems of piping underlay most rights-of-way.

Transportation departments are also wary of pressurized water and sewer pipes under the surface as they may potentially leak or break as a result of vibrations during construction or from heavy traffic. Failures are a risk that increases as more utilities are installed and as the utilities age.

NATIONAL CONDITION OF INFRASTRUCTURE

There have been a series of studies conducted by the USEPA and others as to the condition of the nation's infrastructure (ASCE, 2003; USEPA, 2002). These reports all have limitations based on extrapolating information provided by solicited respondents. While such respondents may not be representative of a specific utility or locality, the magnitude of the problem is enlightening. The most recent report is the 2002 USEPA "gaps report" (USEPA, 2002), which attempts to compare the condition of the nation's water and sewer infrastructure, using 1980 as a base year. Figure 9-1 shows the condition as assessed by this report. Most of the infrastructure is thought to have been in excellent condition in 1980. Figures 9-2 and 9-3 show the anticipated condition in 2000 and 2020 based on current spending. The picture is disturbing, as by 2020 as much of the infrastructure will be in fair to poor condition. The report presents the shortfall in reinvestment into utility assets: for water systems the shortfall is $151 billion (see Figure 9-4); for wastewater systems the shortfall is $250 billion (see Figure 9-5) (USEPA, 2002)

Yet this picture for a given community may be inaccurate as applied. Figure 9-6 shows a graph of the assumptions USEPA used to measure infrastructure condition, i.e., age of the asset. The assumption was that the assets had a 100-year life, far in excess of what may be reasonable for many areas of the country without additional improvements or repair, and well in excess of mechanical life expectancy. The result is that if the assumed age of the infrastructure was reduced from 100 to 50 years, by 2020 the majority of infrastructure would be poor or failed. Hence before accepting broad assumptions, the utility needs to evaluate its true needs..

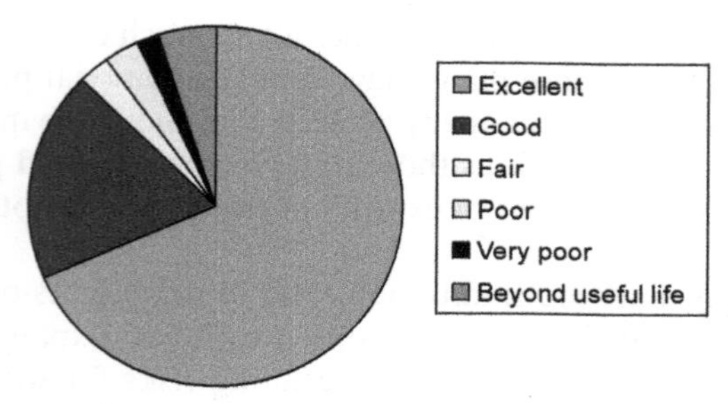

FIGURE 9-1 1980 infrastructure condition
Source: USEPA (2002).

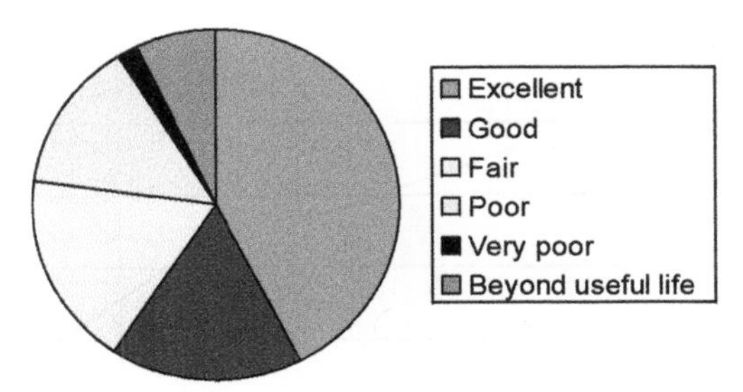

FIGURE 9-2 2000 infrastructure condition
Source: USEPA (2002).

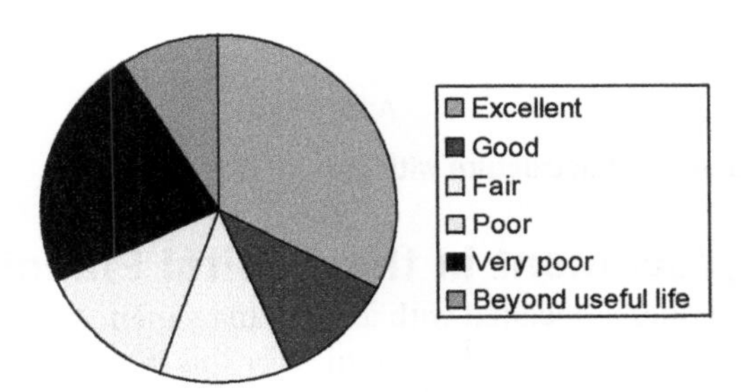

FIGURE 9-3 Anticipated 2020 infrastructure condition
Source: USEPA (2002).

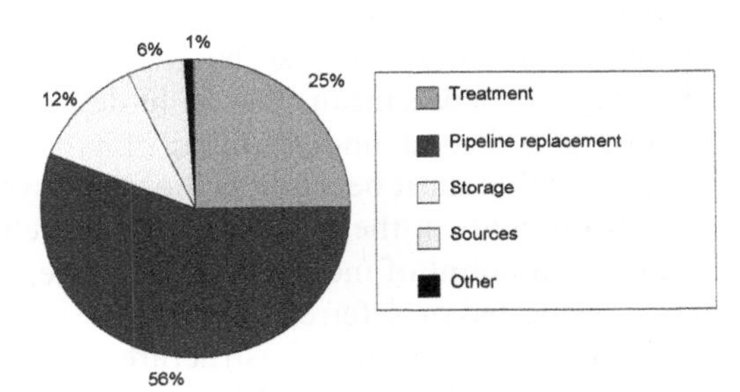

FIGURE 9-4 Summary of national water infrastructure needs ($151 billion)
Source: USEPA (2002).

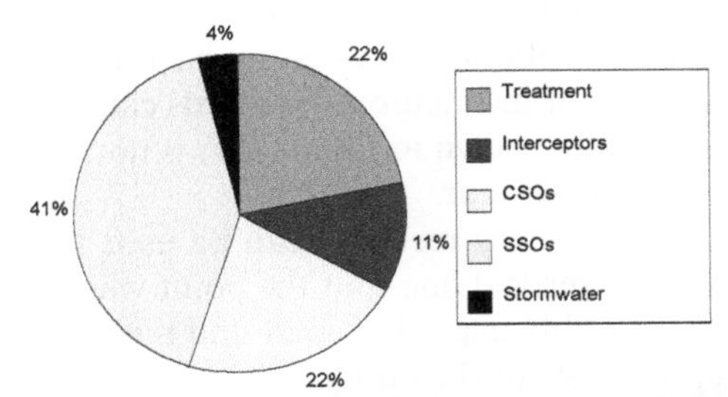

FIGURE 9-5 Summary of national wastewater infrastructure needs ($250 billion)
Source: USEPA (2002).

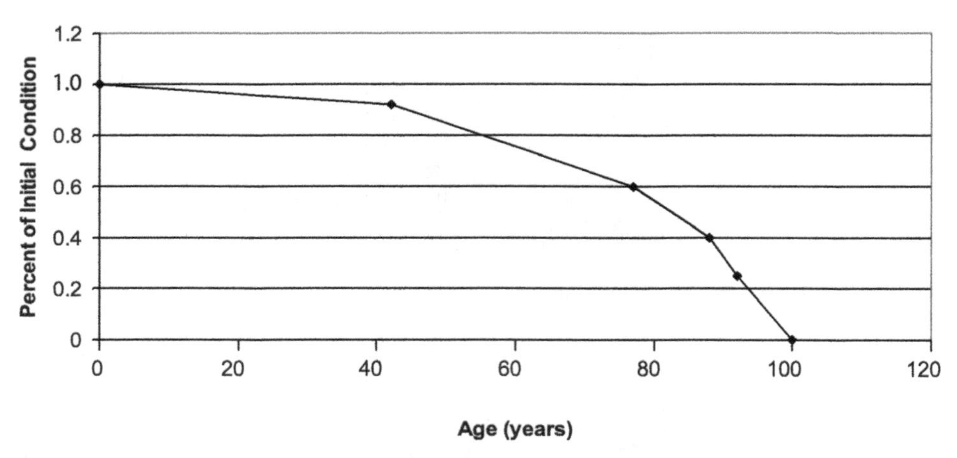

FIGURE 9-6 Deterioration of infrastructure with age

Repair and Replacement in the Capital Planning Process

Chapter 4 outlines the issues associated with asset management and maintenance of the current utility system, and the need to budget monies for this purpose. Repair and replacement funding should become part of the capital improvement process, which should be a result of a long-term planning process (a product of the utility's vision of the future). Unfortunately most planning efforts focus on growth needs, not maintenance needs, which can exacerbate operational problems by deferring replacement of aging infrastructure (Bloetscher et al., 2004). Identifying obvious system deficiencies is easy, but it is the hidden infrastructure that can pose a problem. Estimating these costs means that an in-depth, local knowledge of the history of the system is needed, as well as an understanding of its condition.

Determining the investments that must be made in the utility system can be measured by reviewing the historical investments in the system. Unfortunately, historical data may not represent a realistic value or amount of money to be set aside, particularly if maintenance of the system has been neglected or deferred.

One method of gaining a perspective on infrastructure condition is using the data from the asset management inventory, an estimated replacement value of assets and estimated installation costs for the assets in the future using the inflation indexes. Replacement value of assets of the utility will be substantially more than many expect when compared to the net assets in audit statements because the value of money has changed considerably with time. Annual inflation rates going back in time are required to determine the asset value at the time of installation so records can be developed in the absence of full information. Depreciation, shown in Figure 9-7, is normally calculated by accountants using a constant amount each year.

Figure 9-8 shows how accounting methods can be used to find the optimal time to replace the infrastructure. The vertical line is at the point where the EUAC is minimized. It should be noted that as the utility ages, the repair and replacement portion will increase as a proportion of the capital costs of the utility.

However, these data still provide limited information on the appropriate amounts for repair and replacement funding, especially where historical data are sparse. The number

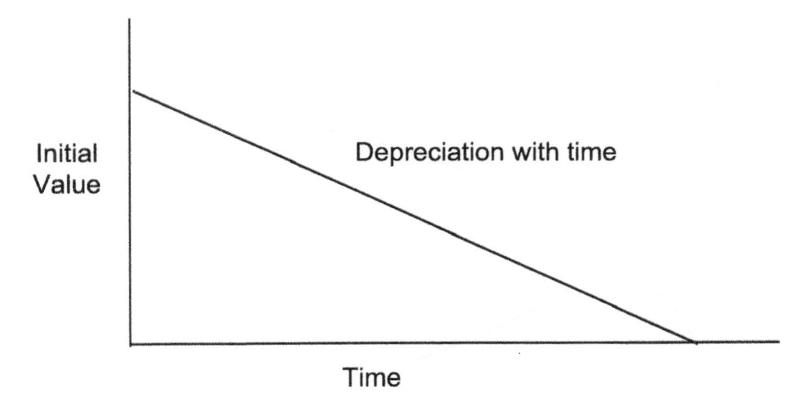

FIGURE 9-7 Straight-line depreciation showing initial value and a steady decline in value until the end of the asset life, when the value = 0

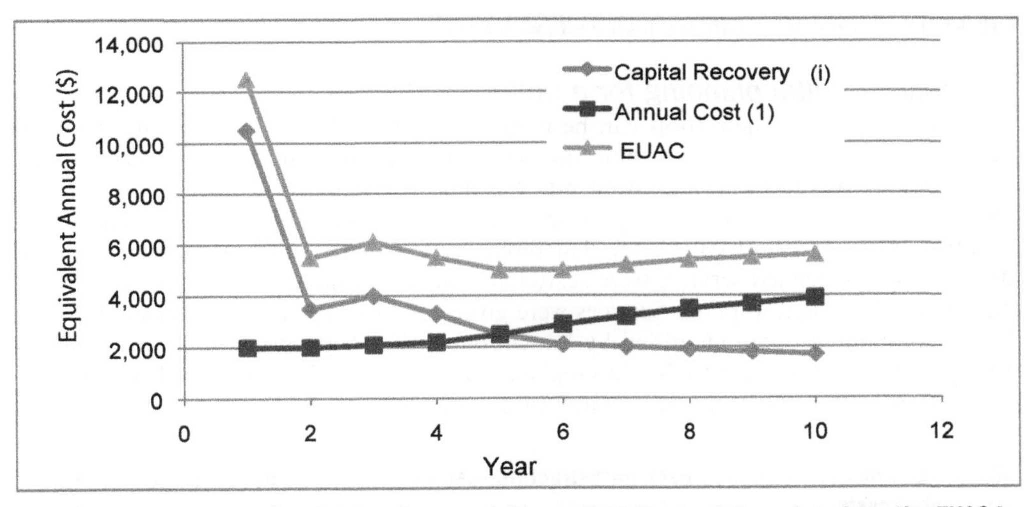

FIGURE 9-8 Evaluation of optimal replacement. The vertical line is at the point where the EUAC is minimized.

of tools to help make these decisions is limited. An example of one tool is CapFinance, developed at the Environmental Finance Center at Boise State University (Jarocki, 2003). This model allows certain scenarios to be inserted for repair and replacement (R&R) and debt funding to help utility managers and engineers determine the appropriate approach to R&R funding for their utility. The information is useful in making estimates of future rate impacts (see Figure 9-9). The model requires tables of data to be input, including age, replacement value, and other information that may need to be estimated. The model is forward-looking, and therefore does not estimate the current deferred maintenance obligation. A deferred maintenance obligation is calculated as initial costs because CapFinance frontloads immediate needs, as would be expected where deferred maintenance backlogs exist. Borrowing is recommended to overcome the initial shortfall.

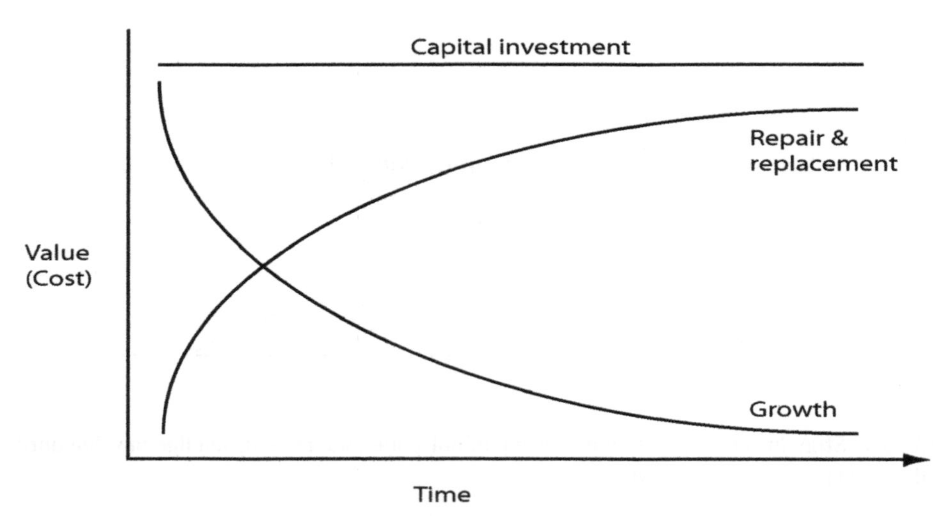

FIGURE 9-9 Change in growth and R&R costs with time

Example—Capital planning for a utility

To demonstrate how these tools can be used, data were collected for a utility in Dania Beach, Fla., that serves 13,600 people via 4,600 water accounts and 3,900 sewer accounts. An inventory of the utility's existing water system was developed from existing maps, plans, and discussions with sewer operations staff. Estimates of utility pipe age were generated from plans and staff knowledge. The piping system, access openings, force mains, lift stations, and associated services were scattered somewhat randomly within three years of the known construction plans. Services were given the same age as the associated piping. Major components were identified by known dates. The appropriate infrastructure and inflation factors were used to develop replacement and depreciated values. The estimated installation costs and current depreciated values are shown in Table 9-1.

TABLE 9-1 Summary of Dania Beach infrastructure—estimated infrastructure installation and depreciated costs

Portion of System	Amount	Installed Cost	2003 Value	Oldest
Sewer Services	3,841	$535,745	$304,614	1963
Piping	229,160 LF			1963
Personnel access openings (PAOs)	767			1963
Total pipe + PAOs		$4,243,572	$2,361,938	1963
Infiltration & Inflow Correction		$1,963,000	$981,500	1993
Lift stations	16	$566,184	$337,348	1963
Total		$7,308,501	$3,985,400	

To help identify the appropriate funding for an R&R account for the water system, CapFinance was used to obtain the annual allotment. Four scenarios were run: repair and replacement funding with debt percentages of 0, 50, 75, and 100 percent. The annual allocations for capital are shown in Figures 9-10 through 9-13. Figures 9-14 and 9-15 show total expenditures and present worth of the expenditure streams, respectively. Some borrowing seems to provide a lower present worth as a result of taking advantage of the reduced cost of money in the future. However, full use of debt was the highest cost in both figures.

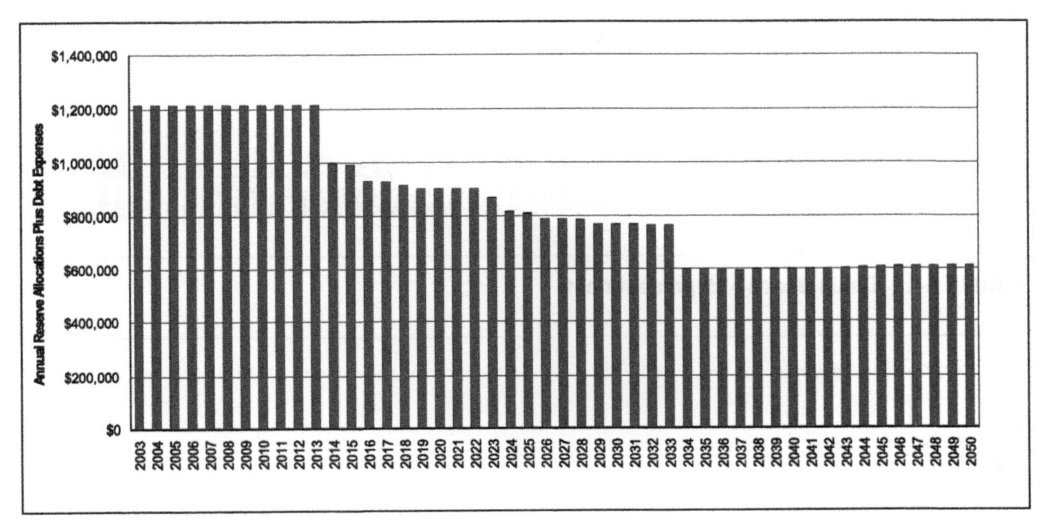

FIGURE 9-10 Dania Beach—no debt

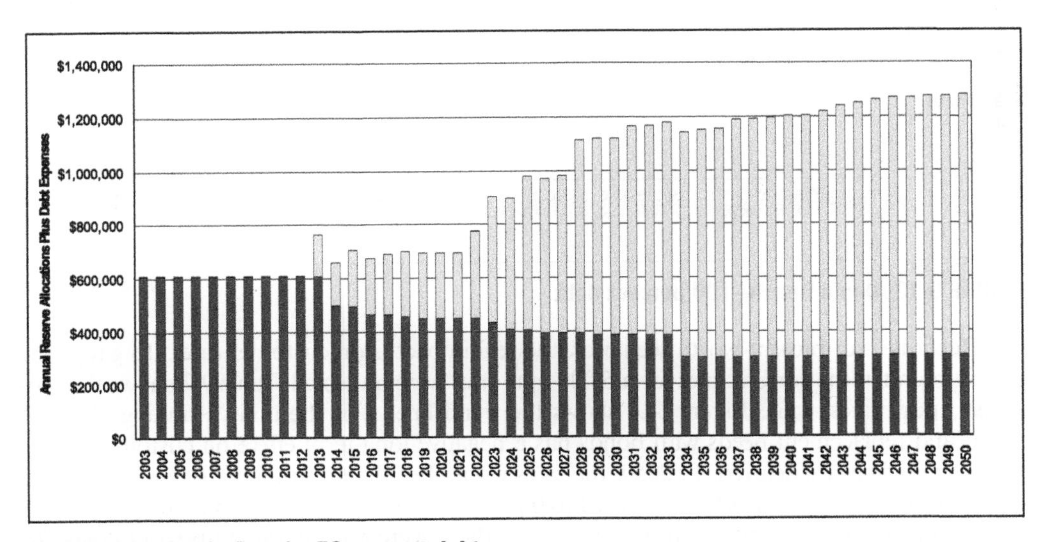

FIGURE 9-11 Dania Beach—50 percent debt

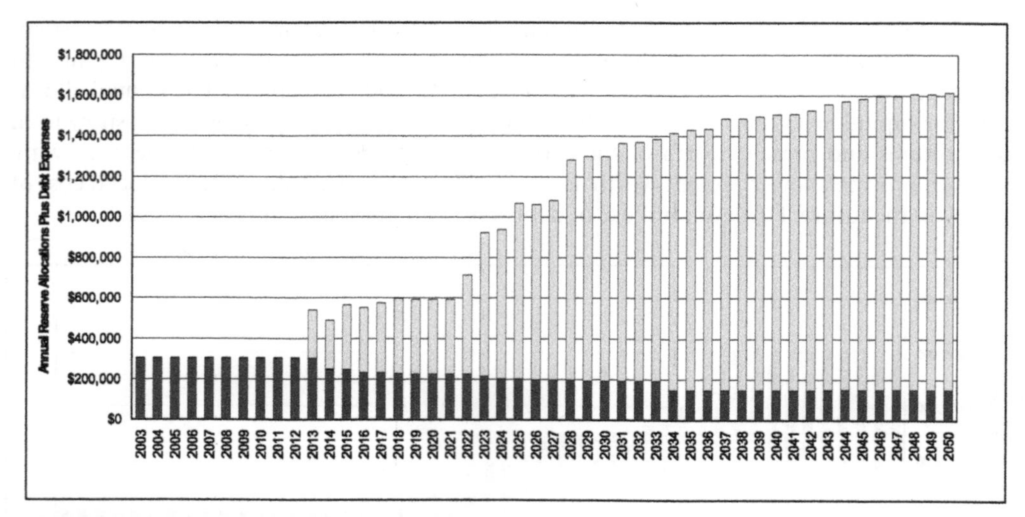

FIGURE 9-12 Dania Beach—75 percent debt

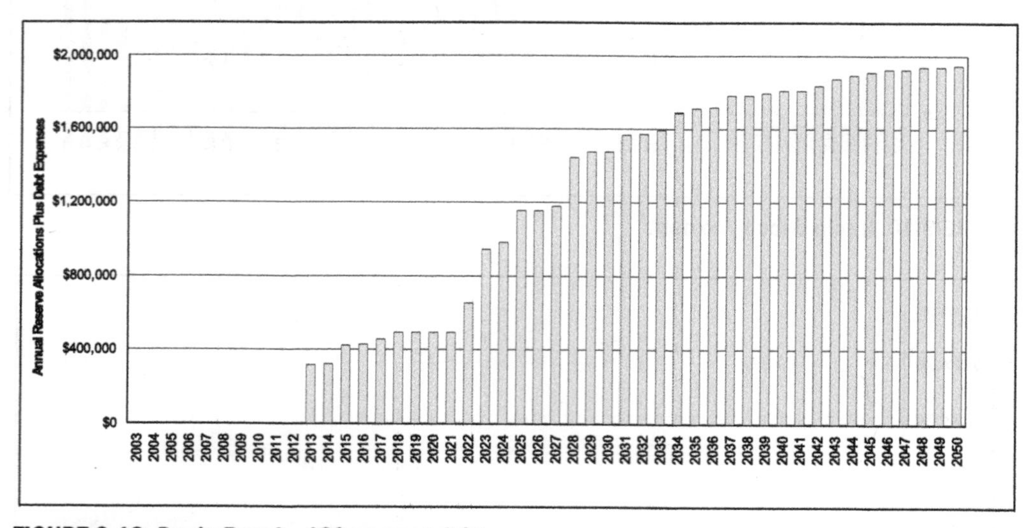

FIGURE 9-13 Dania Beach—100 percent debt

Regardless of present-worth scenarios, the real impact is on customer rates. To gauge the impact on rates, the current rates for a 5,000-gpm user were projected. Table 9-2 shows the projected future rates, while Figure 9-16 shows the trend lines. For the utility, funding all of its repair and replacement needs with bonds has a negative long-term effect on customers, since near-term effects are delayed. In 2050, the rates would be 20 percent higher due to use of 100 percent debt for funding R&R as opposed to 0 percent for 100 percent of R&R funds.

From the no-debt option, it was clear that there were considerable up-front costs. To identify this need, two issues were reviewed—the installation of piping and the deferred maintenance obligations. Based on a replacement value as calculated in Table 9-1, the deferred maintenance costs category was calculated to be only $500,000. However, in

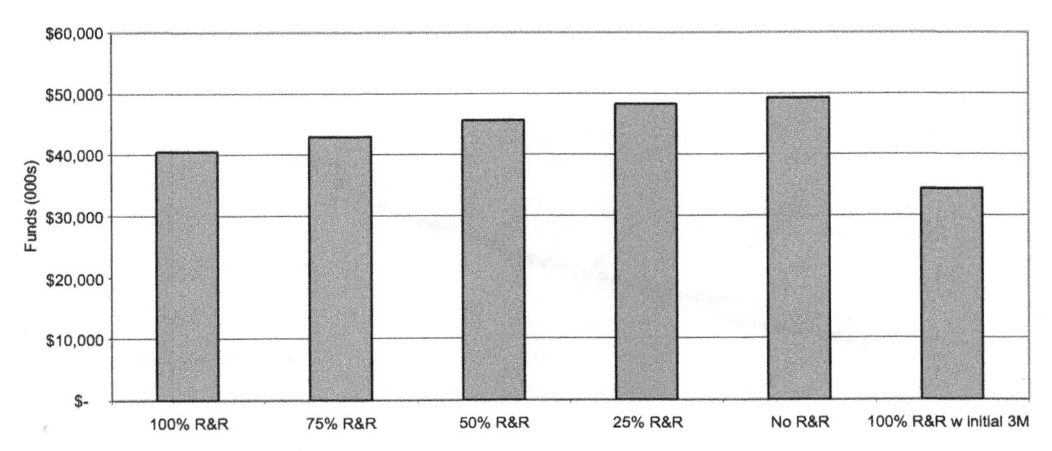

FIGURE 9-14 Total value of expenditure streams

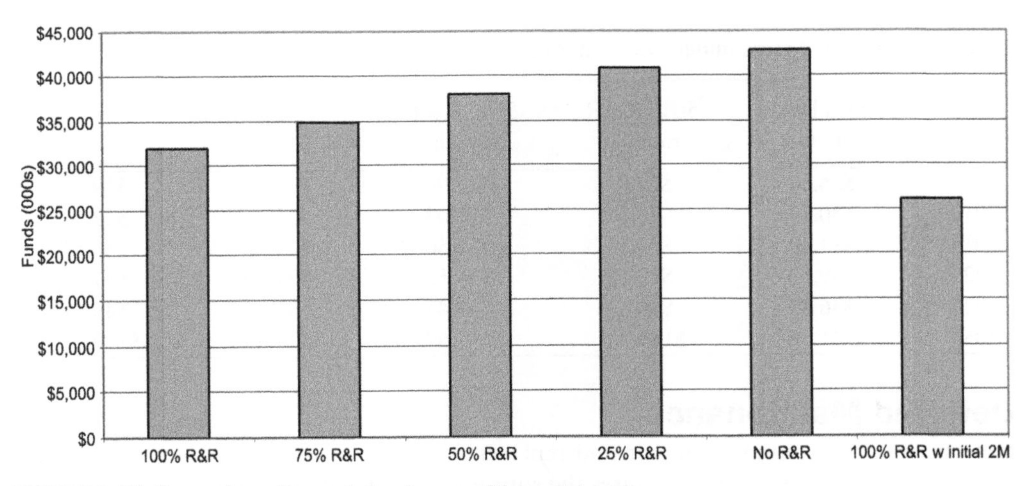

FIGURE 9-15 Present-worth analysis of expenditure streams

looking at the next 10 years, a significant amount of infrastructure would become 50 years old and may need replacement. The total of these obligations is over $2.5 million, which ties in nicely with the front-loading amount. A scenario was designed to employ borrowing to address the backlog and R&R in the future, and the benefit of this method was similar to full R&R funding. Rates were nearly the same as well.

The result for the utility was that it would be prudent to establish a repair and replacement fund as soon as possible (it is in the process of funding a series of projects by borrowing). On a yearly basis, the utility staff should prepare the design and construction budgeting documents for major replacement or upgraded facilities for the coming year and should reprioritize the ensuing years. The planning process should initially identify the specific projects necessary and the expenditures needed to cover their costs.

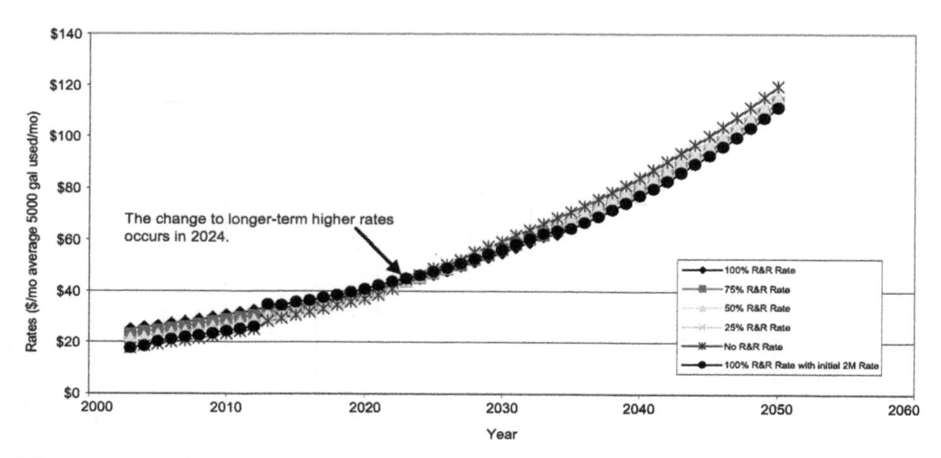

FIGURE 9-16 Dania Beach monthly rate trend line

TABLE 9-2 Projected rates under five scenarios

Year	100% R&R rate/mo	50% R&R rate/mo	25% R&R rate/mo	No R&R rate/mo	100% R&R w/init. $3 M rate/mo
2005	$26.64	$24.80	$22.96	$21.12	$19.28
2010	$30.70	$28.86	$27.02	$25.18	$23.34
2015	$34.40	$33.35	$32.56	$31.76	$30.84
2020	$39.62	$39.13	$38.52	$37.90	$37.04
2025	$46.37	$46.83	$47.45	$47.94	$48.61
2050	$111.55	$113.73	$115.57	$117.53	$119.74

Deferred Maintenance

Determining the amount of repair and replacement money to be set aside is perhaps the hardest decision to make and requires the support of the local government. Growth needs are easier to predict and calculate. In the early years of a utility's life, expansion will be the primary expense. As the utility matures, a larger portion of the capital outlays will tend to be for repair and replacement. Many systems are still growing and adding infrastructure to their inventory. As growth rates slow with time, average infrastructure age increases. To maintain a stable utility with stable operations, fiscal condition, and rates, and low risk, the utility should be investing similar sums of money each year. One way to accomplish this is to try to measure deferred maintenance on the system.

The second model attempts to estimate the deferred maintenance obligation based on the information required for CapFinance. This model is a refinement of one proposed previously to calculate the anticipated deferred maintenance costs, using comparison between replacement value of the total infrastructure and the depreciated cost of the actual investments, adjusted to a present-cost basis (Bloescher, 1999). One theory is that the average age of a properly operated utility system should never exceed half its useful life, i.e., each year investments in capital facilities should be made to replace those facilities whose useful lives have been reached. The formula is as follows:

$$DF = [0.5 \times IV - (0.5 \times RV \times DPW)]/DPW$$

Where

IV = installed value (or estimate based on the deflated value of RV)
RV = replacement value
DPW = deflation from present worth to one half infrastructure life
DF = deferred maintenance obligation

To keep up with maintenance needs on the system and to prevent a significant deferred obligation that can only be funded through bonds or other mechanisms that typically have a significant impact on rates, the value calculated should be positive. Application of this model was done in Hollywood, Fla.

Example—Determining deferred maintenance

The first step undertaken by the city of Hollywood, Fla., to determine deferred maintenance was to assume a lifetime for the improvements. For convenience, a 30-year life for infrastructure was assumed. Based on an inventory of the system components, the replacement value of the infrastructure for the city is over $530 million. The bulk of this amount is treatment facilities, with the wastewater treatment plant being the most significant asset the city has. Applying the theory that a properly operated utility would never allow its infrastructure to depreciate to less than its half-life, the system should be valued at approximately $265 million in current dollars. However, since the value of money has increased over that time, the half-life was reduced by the inflation factor over the past 15 years. This reduction yielded an estimate of $161.2 million in expected value.

According to the audit for that year, the net plant asset value was $143 million. The difference is $18.2 million, which is the unmet deferred maintenance cost. However, the $18.2 million is based on the value of money 15 years ago, so it was inflated to yield a present-worth amount of $29.75 million. To check the veracity of this estimate, the city's 1981 distribution model was consulted. This project indicated that significant work needed to be done on the distribution system; at that time nearly $20 million worth of projects was identified. None of the projects had been completed. Inflating from 1980, these projects were valued at over $30 million, which conformed with the missing component of the existing maintenance deficit given that the membrane softening plant and wastewater treatment plant have received major expenditures in the last 10 years. In looking at the source of the shortfall, the water distribution pipelines are beyond their useful lives.

STEPS IN THE CAPITAL PLANNING PROCESS

The capital improvement planning process is also a product of the future directions, issues, and goals identified during the planning process for the utility system. The capital improvement planning process is the identification of specific project or facility needs to develop, maintain, or more efficiently operate the system and provide better service to the customers, compiled into a prioritized listing of all projects needed to accomplish the master plan goals.

While the master plan defines the overall goals and objectives for the utility and forms that basis for the utility's strategic and operations planning, the utility's goals and objectives must be coordinated with the overall goals and objectives of the larger overlying utility. These goals are then evaluated to identify the impacts on the existing infrastructure and available manpower. New projects are identified and studies are initiated to evaluate the need for new facilities.

Consistent with long-term forecasts of annual customer demands, utilities should determine the appropriate capital projects necessary to address the local needs. This includes both repair and replacement projects and those created by growth. There may be a number of alternative projects that will meet a particular requirement, each of which should be evaluated for meeting the capital funding availability and utility requirements. Engineering and operations personnel should evaluate the expected benefits versus the cost and changes in operational protocols necessitated by the new facilities. Most utilities have not relied on capital budget techniques for investment decisions; however, any utility capital project is an investment in the system.

Instead, many capital improvement programs result from master planning studies that fall prey to other objectives, decisions to defer one project in favor of another that is more popular, or efforts to meet real or artificial budget objectives. Consequently, evaluations of alternative types of projects need to be conducted in-house during the capital budgeting process.

To develop the capital program, the following steps must be accomplished:

1. Establishment of policies through which capital improvements will be identified.
2. Review of facilities to determine their age and condition; providing a ranking methodology to prioritize those projects that have the highest need.
3. Identification of project costs in accordance with the type of construction.
4. Addressing the potential funding sources, such as repair and replacement funds, operations, impact fees, or some form of borrowing.
5. Developing a multiple-year plan.
6. Securing approval of the plan internally and from the governing body.

A single source of funds cannot be used without both legal and economic limitations. Much has been made on the use of impact fees and debt for capital programs. However, impact fees have limitations on use (they must pay for facilities made necessary because of growth). The use of debt should be carefully considered because in many utilities, debt drives the user rates and limits the utility's ability to reallocate its annual resources, which can create operational problems.

Capital assets need to be maintained and repaired to keep them in serviceable condition. Ongoing repair and replacement of projects should occur as part of the ongoing maintenance budget, capital portion, not the CIP (although the values can be shown there). Without this ongoing repair and replacement, the system will fall into disrepair and the risk of service outages will increase.

CAPITAL IMPROVEMENT PROGRAMS

One outcome of asset management and all planning processes is the capital improvement program, which is a plan to build the infrastructure needed to meet the community's vision for the future. *Capital projects* are defined as projects having a long life and a significant expenditure of funds. Typical examples are water lines, sewer collection lines, sewer pump stations, and treatment plants (many other "permanent" installations also qualify). The following are all items that may be included (and recorded for fixed asset purposes) in the capital improvement plan (Vogt, 2004):

- Construction, including labor and materials
- Planning, engineering, and architectural design
- Construction services
- Legal fees
- Property acquisition and associated consultant costs
- Easements
- Equipment
- Interest or other finance charges during construction and planning

Projects should be designed and constructed in accordance with the latest technical and professional standards of the water and wastewater industries. While some improvements may be constructed by developers, many large capital projects must be constructed using utility staff, consultants, and contractors. The utility can control its own projects to a degree, but must establish a set of procedures to ensure that the projects are constructed to meet operational timelines. To meet the goals of a good capital improvement program, a range of issues must be addressed, including:

- Prioritizing projects to allocate available funding among the most pressing needs
- Planning for future projects to prevent conflicts
- Providing elected officials with the technical justification for the proposed projects
- The selection procedure for consultants
- The selection procedure for contractors (prequalification)
- The use of general consulting agreements
- Staff design projects
- Administration of projects in various phases
- Developing standardized legal documents
- Required inspections, testing results, and certifications
- As-built drawings and maintenance literature
- Timing of each of these items

Prioritizing capital projects may be based on a variety of factors (see Table 9-3). The more important priorities should be growth demands, regulatory requirements, and potential for failure. All other criteria should be weighted less significantly.

Table 9-3 outlines the components to be included in a capital improvement program from the Government Finance Officers Association (GFOA) manual on financial planning (Kavanaugh and Williams, 2004). This policy matrix includes policy decisions required to develop the capital program, which will require that the following steps be accomplished:

1. Establishment of policies through which capital improvements will be identified.
2. Review of facilities to determine age and condition; providing a ranking methodology to prioritize those projects that have the highest need.
3. Identification of project costs in accordance with the type of construction.
4. Addressing the potential funding sources, which include repair and replacement funds, operations, impact fees, or some form of borrowing.
5. Development of a multiple-year plan.
6. Development of annual spending plans as part of the annual budget.
7. Securing approval of the plan internally and from the governing body.

Updates of the capital improvement plan and procedures should be performed on an annual basis. The utility staff should prepare the design and construction budgeting documents for major replacement or upgraded facilities for the coming year and should reprioritize the ensuing years. Such an example is shown in Table 9-4. This format shows the type of expenditure, the year, and the operational impact of the improvement. The example is for infiltration and inflow correction that should have an operation and maintenance cost savings. The capital plan should indicate the timing of the projects, as shown in this table.

Many capital plans make the error of budgeting design and construction in the same year. This might be accomplished in small systems, but even then the utility is unlikely to budget all monies in one fiscal year because work simply does not get performed that quickly (see Table 9-5). Thus the financial needs of the plan are distorted. The example that follows summarizes the city of Hollywood, Fla.'s, capital plan, which separates design and construction monies. More discussion of funding and the procedures to construct capital projects is found in subsequent chapters.

Example—Capital Improvement Plan, Hollywood, Fla.

The city of Hollywood has had a capital improvement program in effect in recent years (Bloetscher, 1998). It is estimated that the city has spent $120 million on upgrades and modernization between 1992 and 2000, including $80 million at the wastewater treatment plant, $10 million in sewer force main repairs, $10 million for wells, and over $20 million for a membrane water treatment plant. Despite the focus of Hollywood's expenditures on the wastewater treatment plant, much work still remained to be done.

In reviewing the utility system, the city determined that the older sections have significant needs in the potable water distribution system, and that the large sewer force mains need to be evaluated. During the last infiltration and inflow project, the city spent nearly $9 million addressing potential problems with the gravity sanitary sewer system. However, sewer laterals remain a problem as nearly 30 percent of the flow through the wastewater

TABLE 9-3 Capital improvement policy components

Capital Improvement Policy Components	GFOA Strength of Recommendation		
	High	Medium	Optional
Capital Improvement Plan			
Scope	X		
Project Selection	X		
Designate Responsibility for CIP	X		
Include Operating Impacts in CIP		X	
Adjust Annually for Changes Priorities	X		
Capital Budget			
Moving Projects into Annual Budget	X		
Assign Minimim Capital Expenditure/Allocation (R&R)		X	
Define Capital Budget Year	X		
Project Financing			
Pay-as-you-go		X	
Debt		X	
Fund Balance		X	
Capital Project Evaluation Criteria			X

treatment plant is still due to infiltration and inflow. Major needs at the city's wastewater treatment plant included upgrades to the influent pump station, repairs to the oxygen trains, and an additional clarifier to meet regulatory standards and growth demands in the service area.

Most of the work done to date has been to catch up with system maintenance that should have been earlier. In doing so, the plants were the primary focus, not the pipelines. Future emphasis on the piping system will require additional expenditures. Table 9-6 shows the city's capital improvement program for the water system for the period 2002–2006. It should be noted that the city separated design and construction into sequential years.

Example—Capital Improvement Plan, Dania Beach, Fla.

The city of Dania Beach has a capital improvement program (shown in Table 9-4) with design and construction steps shown graphically in Table 9-7 in Microsoft Project®. Unlike the Hollywood program, the Dania Beach program identifies funding and priorities. With the Dania Beach system, water quality and reinforcement of pipelines are important issues. Fire flows in commercial areas are potentially deficient. Old pipes serve these areas. The Dania Beach water and sewer system is not expected to grow much in the next 10 years.

TABLE 9-4 Example CIP project request and summary form

City of Dania Beach							
Capital Improvement Request Form							
Project:	Sanitary Sewer Infiltration and Inflow Program						
Priority	2			Project Manager:	Fernando Vazquez, P.E.		
Department:	Public Services			Division:	Sewer		
Project Location:	City Wide						
Fiscal Year	FY 06	FY 07	FY 08	FY 09	FY 10	Total	FY 05
Plan and studies	$54,600	$0	$0	$0	$0	$54,600	$150,000
Engineering/ architecture	$0	$0	$0	$0	$0	$0	$0
Land acquisition/ site preparation:	$0	$0	$0	$0	$0	$0	$0
Construction:	$250,000	$350,000	$450,000	$450,000	$400,000	$1,900,000	$0
Equipment/fur-nishings	$0	$0	$0	$0	$0	$0	$0
Other (Specify):	$0	$0	$0	$0	$0	$0	$0
TOTAL COST:	$304,600	$350,000	$450,000	$450,000	$400,000	$1,954,600	$150,000
Revenue source:	SRF	SRF	SRF	Sewer Rev.	Sewer Rev.		SRF
Description (Justification and Explanation)							
Provide annual allocation for repair and investigation of infiltration and inflow into the sanitary sewer system. A 1993 study by Hazen and Sawyer provided a cost delineation for the City of Dania Beach for an infiltration lining program for the northeast, northwest and southwest sections of the city (areas D-3, D-4, D-5 and D-6). The City's approved facilities plan by Public Utility Management and Planning Services indicated that the southeast area of the City needs to be addressed as it remains the most problematic, in part because the land is low and the system remains under water throughout the year. While main lines have been lined in some cases, areas of concentration are laterals and manhole lining. Televising the system and repairing new leaks is suggested. A minimum annual allotment of $50,000 was suggested in the Facilities plan.							

TABLE 9-4 Example CIP project request and summary form (Continued)

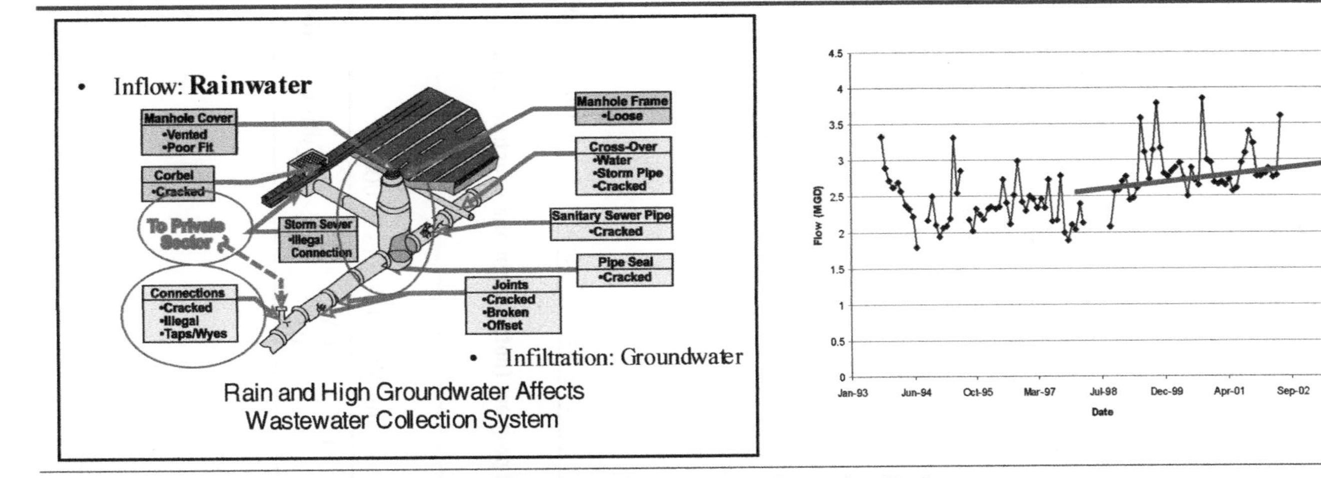

Annual Impact on Operating Budget

Personnel:			Descriptive: Based on the graph above, it is estimated that the City may have as much as 400,000 gallons per day of excess infiltration that is being paid for through City rates. The annual budget is $240,000 for this infiltration, a large portion of which is believed to be correctable. An efficient, continuing I/I program will help reduce our annual revenue losses by reducing the cost of excess infiltration going to SRWWTP.
Operating:			
Replacement Cost:		Year:	
Revenue/other:	–$240,000		
Total	–$240,000		

TABLE 9-5 Typical CIP seen in many smaller communities where full cost of project is in one year

Project Name: Water Supply & Treatment

	Planning Years (Values in 000s)						
	FY 02	FY 03	FY 04	FY 05	FY 06	Future	Total
Water Treatment Objective							
Lime pumps and slakers	740						740
Chemical enclosures		500					500
Filter 7-18 control			330				330
Filter gallery rehab	1,140						1,140
High service pumps		1,500					1,500
Upgrade or replace Reclaim system drier	200						200
New membrane skids				5,700			5,700
Sodium hypochlorite plant	2,000						2,000
Additional storage tanks					5,000	3,300	8,300
Repair R/O capacity		150					150
Filter Gallery mech parts	300						300
MMIS						150	150
VFDs—HSP		344					344
Membrane replacement		1,600					1,600
Painting of water plant						3,000	3,000
Phase II emergency power Generator						1,500	1,500
Portable generator—South well field				150			150
Replacement of fuel tanks			170				170
Upgrade of existing control system @ WTP						580	580
Water treatment total	4,380	4,094	500	5,850	5,000	8,530	28,354

TABLE 9-6 Hollywood, Fla., capital improvement program (note design and construction in separate years)

Project Name: Water Supply & Treatment

	Planning Years (Values in 000s)						
	FY 02	FY 03	FY 04	FY 05	FY 06	Future	Total
Water Treatment Objective							
Lime pumps and slakers	740						740
Chemical enclosures		500					500
Filter 7-18 Control			330				330
Filter gallery rehab	1,140						1,140
High-service pumps		1,500					1,500

TABLE 9-6 Hollywood, Fla., capital improvement program (note design and construction in separate years) (Continued)

Project Name: Water Supply & Treatment

	Planning Years (Values in 000s)						
	FY 02	FY 03	FY 04	FY 05	FY 06	Future	Total
Upgrade or replace reclaim system drier	200						200
New membrane skids				5,700			5,700
Sodium hypochlorite plant	2,000						2,000
Additional storage tanks					5,000	3,300	8,300
Repair R/O capacity		150					150
Filter Gallery mech parts	300						300
MMIS						150	150
VFDs—HSP		344					344
Membrane replacement		1,600					1,600
Painting of water plant						3,000	3,000
Phase II emergency power generator						1,500	1,500
Portable generator—south well field				150			150
Replacement of fuel tanks			170				170
Upgrade of existing control system @ WTP						580	580
Water treatment total	4,380	4,094	500	5,850	5,000	8,530	28,354
Water Distribution Objective							
Water main—West Hollywood/Taft–Davie	600	5,600					5,600
Water system upgrades and assessments	1,000	2,000	3,000	3,000	3,000	17,600	29,600
Port Everglades water main					2,500		2,500
Service Line replacement program	300	300	300	300	300		1,500
Davie water improvements						5,597	5,597
Water main replacement program	300	300	200				800
State Road 7 facility upgrade						1,500	1,500
Water distribution total	2,200	8,200	3,500	3,300	5,800	24,697	47,097
Wastewater Treatment Objective							
Oxygen trains 1 and 2 Rehabilitation	4,975						4,975
Deep well injection system		8,000					8,000
Clarifier 1-4 rehabilitation	7,200						7,200

(continued)

TABLE 9-6 Hollywood, Fla., capital improvement program (note design and construction in separate years) (Continued)

Project Name: Water Supply & Treatment

	Planning Years (Values in 000s)						
	FY 02	FY 03	FY 04	FY 05	FY 06	Future	Total
Effluent pump station upgrade		6,000					6,000
Plant odor/grit	1,500						1,500
Jockey pump—Effl PS	500						500
Replace fuel tank		150					150
Upgrade MMIS					350		350
Upgrade SESS Generator				575			575
Upgrade control sys					600		600
Replace meters						450	450
Rehab Effl PS						5,000	5,000
Clarifier 8			4,450				4,450
Aeration basin 5			5,000				5,000
Ocean outfall Rehabilitation	800						800
Wastewater treatment total	14,975	14,150	9,450	575	950	5,450	45,550
Sewer Collection Objective							
Sewer inspection, Taft St Force main	150						150
Collection system upgrades and assessments	500	1,000	1,000	1,000	1,000		4,500
Infiltration reduction	2,000					13,093	15,093
State Road 7 facility upgrade						7,500	7,500
Sewer collection total	2,650	1,000	1,000	1,000	1,000	20,593	27,243
TOTAL	24,205	27,444	14,450	10,725	12,750	59,270	148,244

TABLE 9-7 City of Dania Beach capital improvement plan

Project	2006	2007	2008	2009	2010	TOTAL
Water						
Water plant upgrade						
Engineering	$200					$200
Construction		$4,230				$4,230
Upgrade Ex. WTP						
Engineering		$36	$96			$132
Construction			$600			$600

TABLE 9-7 City of Dania Beach capital improvement plan (Continued)

Project	2006	2007	2008	2009	2010	TOTAL
Remove elevated tank						
Engineering			$—			$—
Construction			$200			$200
Water main replacement						
Engineering	$18	$42				$60
Construction		$244	$50	$50	$50	$394
Water main looping upgrades						
Engineering	$345	$295	$70			$710
Construction	$918	$1,400	$1,400			$3,718
Construct new well "I"						
Engineering	$325	$75				$400
Construction		$500				$500
Replace Accelerators						
Engineering				$500	$250	$750
Construction					$3,750	$3,750
Chlorine system						
Engineering	$—					$—
Construction	$100					$100
High-service pumps						
Engineering	$75					$75
Construction	$750					$750
Water total	$2,731	$6,822	$2,416	$550	$4,050	$11,969
Sewer						
Infiltration/inflow						
Engineering	$55	$56	$58			$169
Construction	$250	$350	$450	$450	$400	$1,050
Lift station rehab						
Engineering	$61	$61	$42			$164
Construction	$350	$350	$240			$940
Force main upgrades—Dania Beach Blvd						
Engineering	$50					$50
Construction	$350					$350
Sewer total	$1,116	$817	$790	$450	$400	$2,723
Debt						
SRF	$2,941	$7,347	$3,056	$—	$—	$13,343
Other revenues	$906	$292	$150	$1,000	$4,450	$6,799
Total Revenues	$3,847	$7,639	$3,206	$1,000	$4,450	$20,142

CAPITAL AND FINANCIAL PLANNING CONNECTIONS

It should be noted that capital improvement plans and the financial plan for the utility (chapter 14) are intrinsically linked. Figure attempts to define this relationship (Kavanaugh and Williams, 2004). The importance of this figure is that it shows the capital plan is an evolving document, responding to increases in knowledge of the future, such as the actual growth and changes in the economic base of the community, and increases in understanding of infrastructure condition that may reprioritize projects. At the same time, changes in the capital plan will affect the funding mechanisms for the plan. The annual capital planning and budget cycle provides the opportunity to make the modifications necessary to adjust both the financial and capital plans to optimize the capital expenditures.

One important aspect of the capital planning process is the generation of cost data. Cost estimates of construction and engineering are typically developed by engineers. Finance personnel usually accept these numbers and work with them without much real understanding of their genesis. Actual construction costs are often different than the capital planning estimates, which can create friction between engineers and finance staff. Understanding the cost genesis provides a valuable perspective.

There are three types of costs that are estimated in the capital planning process: engineering, construction, and operations. Incidental other costs, such as legal and property acquisition, may also be included. In general, engineering cost estimates are developed from one of four methods listed below.

1. Hourly work estimates
2. Sheet count
3. Percent of construction
4. Comparison to other projects

Hourly cost estimates are developed on spreadsheets where the engineer estimates the hours and personnel to be devoted to the design process based on the scope of services as he/she understands it. For large projects, the hourly cost estimate can be very complicated, which can introduce inaccuracies. If the project will take a long time to complete, then changes in personnel may change the way the project is designed, and therefore the cost. Usually the engineer is paid by deliverable, e.g., report and plan sets. To develop these types of cost estimates, the engineer needs significant experience in the management of design teams. A multiplier for the overhead, profit, insurance, and so forth is usually added to the hourly rate of personnel. This overhead factor is typically between 2.5 and 3.5 times the hourly rate (i.e., a person making $30 per hour would be billed out at $75 to $105 per hour).

Sheet counts are difficult to quantify unless very similar projects have been designed previously. The cost per sheet is $3,000 to $9,000, depending on the complexity of the drawing. Determining the exact number of sheets that will be needed is difficult at the start of the project. This is not a very acceptable way to develop costs for most clients.

Comparing projects from a design perspective to determine the appropriate fee is also difficult. No two projects should be designed exactly the same because soils, foundations, water quality, and other variables are usually different between sites (even those in close proximity). However, this method is often used in conjunction with the percentage of total construction method.

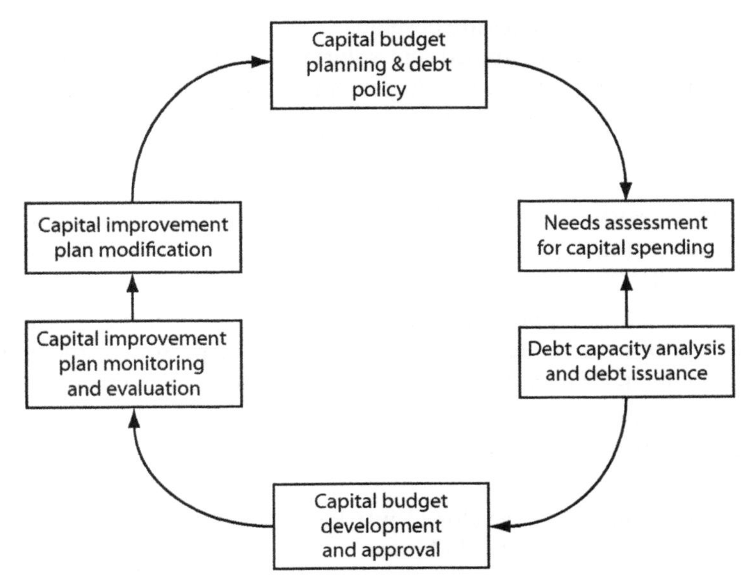

Adapted from: Kavanaugh and Williams (2004).

FIGURE 9-17 Debt capacity and capital budget diagram

The Farmers Home Administration (FmHA) and other agencies will use a percent of construction cost for valuing design. Standard curves have been developed for certain types of projects. Simple projects will find predesign and design fees in the 6 to 10 percent of construction range. More complex jobs will have higher percentages as a result of additional expertise required. Very small jobs will also have higher percentages due to their lack of economies of scale. Large engineering firms often object to the use of the FmHA curves because the overhead numbers are low. The American Society of Civil Engineers (ASCE) and other organizations also publish curves. The major concern with this type of estimating is that if the construction cost is inflated, the design cost may have no relationship to the actual costs for design. Therefore, the curves should be used only for comparison purposes.

The key under any scenario is to negotiate a fair and equitable cost. Total engineering, including construction services, will often exceed 15 percent of the construction cost. If the cost appears to be too high, then another engineer should be used, but negotiating with two engineers for the same job at the same time is prohibited in the public sector under most circumstances. Once negotiation ends, it usually cannot be restarted.

Construction cost estimates can be developed from a variety of sources, e.g., similar projects, knowledge of the market, contractor contacts, and software programs. However, the question that arises is the need for detailed cost estimates in the planning stage. Depending on the use of the estimate and the point in the design process, the cost estimates listed below may be developed.

- **Order of magnitude**. These are overviews of the total job where limited details are known and contingencies must be included to account for potential changes in the design needed as the project progresses. The cost estimates are usually within a range of −30 to +50 percent of the projected actual cost.

- **Definitive.** These cost estimates are developed toward the end of the design phase when most unknowns have been studied and solutions identified. Definitive cost estimates are usually in the range of –15 to +30 percent of the actual projected cost. The exact timing of the bidding of the project may not be known. Timing of bidding is important in cost estimates as a variety of factors may impact the bid price, including those listed below.
  - Amount of work currently under construction or being bid in the market
  - Availability of materials (a major issue in 2004 and 2005 when steel and concrete being used in China drove costs up)
  - Availability of personnel and equipment (after the hurricanes in Florida in 2004, the deluge of work made it difficult to get qualified personnel and equipment in a timely manner, thereby increasing costs for licensed contractors)
  - Experience with the utility (difficult owners or owners that require significant amounts of paperwork and overhead get higher prices)
  - Inflation and cost of goods
- **Detailed cost estimates** are developed just before bidding. Their accuracy is usually within –10 to +15 percent of the bid price.

In any case, experience is a major factor in cost estimating. RS Means, Inc., is a company that provides a book and software showing detailed construction costs. Other such software programs are available, and considerable time can be spent preparing very detailed cost estimates. While software programs may be helpful, construction costs tend to be similar in geographic areas. Engineers with experience in a given area will be able to develop cost estimates based on similar projects constructed in the same area, with appropriate inflation factors. Spreadsheets are a common method, using bid prices from other jobs.

In most cases, the cost to create detailed estimates early in the process provides little tangible benefit, and insufficient information is known about the final project (soil conditions and similar factors requiring major cost adjustments). It is not productive to try to get within 5 percent of the bid costs through highly detailed estimates when there are so many factors that are not under the control of the utility or contractor—this only guarantees the price will not be accurate.

The cost estimate required for the capital improvement plan is created prior to the order-of-magnitude cost estimates, and therefore should be expected to be within ±50 percent of the true cost. Engineers and utility managers will normally include significant contingencies in their capital improvement plan estimates due to uncertainty and the amount of time that will elapse before the project is started (to account for inflation and market uncertainties). Engineers developing cost estimates for capital planning purposes should have experience with construction in the area and should apply their best judgment to the development of these costs.

REFERENCES

ASCE. 2003. *Report Card for America's Infrastructure, 2003 Progress Report.* http://www.asce.org/reportcard/index.cfm. Reston, Va.: American Society of Civil Engineers.

Bloetscher, F. 1998. The City of Hollywood Pursues Integrated Resource Planning to Ensure LongiTerm Water Supplies. In *Proc. American Water Works Association*

Annual Conference and Exposition–Dallas, Texas. Denver, Colo.: American Water Works Association.

———. 1999. Deferred Maintenance Obligations Due to Aging Utility Infrastructure. In *Proc. WEFTEC Annual Conference–Orlando, Fla.* Alexandria, Va.: Water Environment Federation.

Bloetscher, F., W.L. Jarocki, and P. Varney. 2004. Reserve Funds vs Borrowing: The Effects on Customer Rates. In *Proc. EWRI Conference Proceedings–Salt Lake City, Utah.* Reston, Va.: American Society of Civil Engineers.

Jarocki, W. 2003. *CapFinance, Version C2.1 User Manual*. Environmental Finance Center, Boise State University.: Boise, Idaho.

Kavanagh, S., and W.A. Williams. 2004. *Financial Policies: Design and Implementation*. Chicago, Ill.: Government Finance Officers Association.

USEPA. 2002. *The Clean Water and Drinking Water Infrastructure Gaps Analysis*. USEPA Report EPA-816-R-02-020. Washington, D.C.: US Environmental Protection Agency.

Vogt, A.J. 2004. *Capital Budgeting and Finance: A Guideline for Local Governments*. Washington, D.C.: ICMA.

CHAPTER 10

Capital Construction

A utility will always be constructing capital projects to develop and/or continue safe and reliable water or wastewater service. For efficient operation, these new facilities must be developed in accordance with the latest technical and professional standards to protect the health, safety, and welfare of the customers served now and in the future. Some improvements may be constructed by developers to upgrade the system, while other large capital projects must be constructed by the utility using utility staff, consultants, and contractors to meet the utility's long-term goals. The focus of this chapter is capital construction by the utility. A short section on developer extensions is included at the end of the chapter.

BUILDING THE CAPITAL PROJECT

As mentioned, most public utility systems need to construct major water or wastewater facilities on an ongoing basis. With proper planning and anticipation of the needs for construction, as outlined in chapters 8 and 9 the new capital projects can be constructed with minimal impact on the utility operation. Advance planning is required because the current processes used to deliver these facilities, consultant selection laws, and formal bidding procedures often create significant time constraints. Such time constraints limit the utility's ability to have the facilities designed and built quickly enough to meet its needs without significant costs or failure to provide service.

The next part of the capital improvement program is developing a construction procedure (or sequence) for consistency in application, legal sufficiency, and ease of administration by staff. Standardizing forms and procedures that apply to all projects will facilitate the efficient and effective construction of water and sanitary sewer systems to meet the mandates of the utility to provide reliable water and wastewater services to customers. Included in these forms and procedures should be legal documents, contracts, change orders, technical specifications for water and sewer improvements, and various formats for elected-official agenda items. The result is consistency and ease of administration for the capital projects.

The primary issues of importance to public utilities when implementing a capital construction upgrade are cost, construction quality, and completion schedule. All of these issues are separate, but interrelated. For that reason, utilities must choose which objectives are most important because that choice may adversely affect the other two. For example, increasing quality requirements may increase the schedule, and accelerating the schedule and maintaining quality will increase costs. Therefore achieving the ideal scenario for all three is not possible because of the exclusivity among them. A utility can achieve two of the three ideal goals, but not all three—optimization is the key.

While schedule, quality, and cost are frequently the determining issues, they are not the only issues of concern to utilities. For example, control over the final product is also a key concern. Utilities typically believe that in controlling the project, they can achieve schedule, cost, and quality goals and shape the final project. Utility personnel also often have specific ideas on the details of the project, i.e., the more details, the more control. Retaining control means more involvement by the utility in design and construction efforts. However, productive oversight can be lacking because public agencies often build a limited number of new facilities during any given employee's tenure, but outside consulting engineers may have specialists that have built similar facilities many times in the past. By assuming a degree of control, a utility will reduce the responsibility of the contractor because decisions the contractor might have been forced to make will have been made by the utility. Thus, the utility takes the risk, but because the contractor's flexibility is limited, the contractor is constrained as to means and methods. As a result, costs may increase because the contractor assumes risk as a result of the limitations. Utility control may affect

- Response by contractor to change requests
- Ability of utility to make demands on contractor
- Need for utility resources during construction

The time commitments of utility staff can become a more important issue than cost on projects that are not well defined or where the contractor is difficult to deal with. In addition, control by the utility can extend to the transition to operations—either with or without start-up by the contractor or a subcontractor. Other issues to consider in choosing how to provide a capital project, in addition to schedule, quality, control, and cost, are listed below.

- Whether or not operational expectations are met and whether or not the new facility can be easily and quickly integrated within the framework of the existing system.
- Ease of transition to operation by the utility. The transition to operation by the utility is usually improved by having the contractor or one of its subcontractors operate the system for a period of time to "get the bugs out." Start-up of a new system is always difficult since the utility's operations staff must address new and unfamiliar equipment, operating procedures, and protocols.
- Likelihood of change orders from contractor. Change orders are a cost-limiting issue as well as a battleground for control. Change orders can plague all projects, but where utility control is greatest, there is a tendency to have more change orders.
- Responsibility of the parties to one another. Every project will be different, but change orders, the ability to make changes as needed, on-site staffing, and additional effort by the engineer (e.g., shop drawings) result in increased time and cost.
- Operational ability issues must be addressed for long-term facility success. The utility's ability to understand its needs early and clearly will be critical for the project.
- If the new project is an expansion or enhancement of an existing facility, the interface of new and old must achieve continuous service.

The goals of the utility in relation to each of these must be considered when making decisions on pursuing a construction delivery alternative.

When Are Consultants Needed?

Consultants are needed under four conditions. The first is when there is a lack of technical expertise on staff, or the lack of ability to secure the expertise. There are many skill sets that are not likely to be on staff, including specialty engineers, special lab experience, legal skills, certain financial skills, and, in many cases, planning and permit skills. An example would be design engineers for a complex water treatment plant. The utility is unlikely to have someone with such skills on hand because expansions of treatment plants do not occur continuously, and there is the potential for the skill set to diminish if the design engineer is not engaged consistently in design of water plants. As a result, from a skill and cost perspective, having a person on staff to do this work makes no sense.

Temporary workloads are a second area where consultants may be used. Going back to the treatment plant example, a plant expansion might be an intensive effort for a year or two, but not consistently. The utility cannot staff up for short-term workloads.

The third reason for consultants is to obtain an outside opinion on direction of the utility. For example, does a certain water supply planning direction make sense, or is there specialized knowledge or circumstances that the utility is not aware of that might play a role in decision making?

The fourth area relates to costs. In some cases it may be less costly and/or time-consuming to use consultants. Cost is not usually a good consideration for decision making in the absence of the other reasons mentioned.

Utilities need to evaluate their use of consultants. Staff is on hand to accomplish day-to-day activities. A certain number of engineers, managers, operators, and other positions should always be on hand. These are the "go-to" people who respond under all conditions. If consistent capital projects, like small water mains, are constructed yearly, then the utility may want to evaluate having this expertise on staff, as opposed to using consultants. Finances, repetitive legal work, repetitive permitting, and so forth are also tasks that may make sense to have performed by staff. The hourly rate for employees is normally much less than for consultants, because consultants charge for overhead, certain overhead employees, profit, insurance, and the office facilities. However, if the utility investigates the actual costs for delivery of a service, it may find that the costs are not significantly different.

Selecting Consultants

When the utility needs to acquire consulting services, there are often elaborate selection processes. The consultant's staff expertise, ability to complete the work on time and within budget, past experience in similar jobs, and responsiveness are all criteria appropriate for evaluating consultants. Price is one of the issues that cannot be used for architects, engineers, and certain other consultants in many states. This is appropriate since the utility should, when representing the public, want the most qualified people to perform the work, not the cheapest. Hiring the most qualified consultants can result in a lower overall cost because the most qualified consultant is usually one who has both expertise and prior experience with similar jobs. Since these facilities have a long life, a good choice of consultant and resulting project will permit both utility and consultant to have a productive, long-term relationship.

Figure 10-1 shows the relationship that must occur between the utility and consultant.

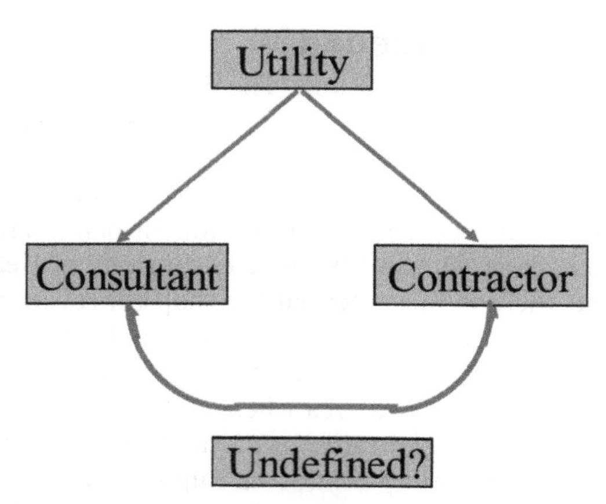

FIGURE 10-1 Relationship between utility, consultant, and contractor

Since most agencies use consultants for construction projects, a third-party contractor is interjected into the equation. Usually a direct relationship exists between the utility and the consultant, and the utility and the contractor; but there is also an informal relationship that occurs between the contractor and consultant defined by the other agreements.

What utilities should look for

Most utilities look for honesty and forthrightness in a consultant, criteria that are subjective (Bloetscher, 1999a and 1999b). Utility officials need consultants who will be truthful, specify what the problems are, and be candid about issues rather than furtive. Because utilities usually want the consultant best qualified to provide the type of work required, trust is of great importance in selecting among the many consultants who submit proposals for the work.

It is often clear that not all consultants are equally qualified, despite the consultant's perceptions. For example, if a utility wants to build a membrane water treatment plant, then it wants a consultant who has already designed a similarly sized membrane treatment plant. Yet, despite this obvious fact, many consultants with no prior experience will apply for the work and some will actually lobby elected officials to obtain that work. This violates the honesty and forthrightness criteria and causes an immediate problem with staff relationships should that consultant be selected. It is far better for consultants who do not have that type of experience to team with consultants who do; thereby, their lobbying efforts are maintained within more appropriate parameters.

Staff looks for a consultant who, in addition to being honest, has the skills to do the job. This includes not only the skills of the firm, but the skills of the personnel who are actually selected by the consultants to perform the work. A common problem is that consultants will submit proposals that include work done by people with the proper experience, but who will never work on the proposed project, in an attempt to show they have the expertise to do a job. This is misleading, dishonest, and an indication that the consultant is not putting a significant amount of work and thought into the proposal process. The likely result is a poor response from the selection committee. Accessibility of the consultant is also important in performing any task; however, this does not mean that consultants need to have an office in the city in which the job is located.

A good reputation among other public-sector bodies and within the consulting field in general is important; therefore, references should include those from other consultants as well as other governing bodies. A poor relationship in working with other consultants is typically a strike against the proposing consultants. Another important criterion is performance, especially in meeting time and budget constraints. This is a measure that can demonstrate that the consultant will apply appropriate staff and resources to complete work in a timely manner and will not unduly burden the projects with what may be termed *gold-plating* to increase fees. It also indicates the consultant's ability to be flexible in meeting project needs, while keeping the project moving.

Finally, the utility should determine whether or not the people assigned to do the job are comfortable in dealing with the consultant. This goes back to the honesty and forthrightness qualities. If the utility staff is not comfortable or does not trust the consultant's personnel, the relationship will be difficult.

What consultants do not need

There are a number of qualities that consultants often think they need, but from a staff perspective, they do not. Consultants do not need to be big national or international consulting firms. Big national firms may be able to access the experts to provide information, but their focus cannot be with one particular agency, especially smaller ones. As a result, a more local consultant who has access or teams with a larger consultant may be a more palatable consulting team. At the same time, the consultant need not always be local, just accessible. If the consultant can be to a job site or meeting within two to four hours, or be available by phone at a moment's notice, that usually meets the accessibility criteria. The utility should not expect that for a water treatment plant, the consultant would be able to have local staff that can do *all* the design work.

A consultant also does not need to have all of the skills to do the job, just the important skills. In this case, the other consultants and persons with whom the consultant teams in preparing the proposal are important, as is their past experience in working together and/or with the utility. If the consultant can bring in experts from other offices or experts from among the team members, this provides a significant help.

There are a variety of ethics issues involved with consulting performance that are beyond the scope of this document. Needless to say, utilities should look for the firm and people best able to do the job and should be wary of using public monies or the public trust to be the "first on the block" to initiate a new process.

Stages in the Design Process

Once the design professional has been selected by the owner and all contract documents have been executed, the design begins. The four stages of the design process are listed below.

1. **Conceptual design** describes what is needed versus what currently exists.
2. **Predesign** is calculation and sizing of the facilities needed.
3. **Preliminary design**, is when plans and specifications of the project are roughed out.
4. **Final design** encompasses the bidding documents.

The first thing required is to refine the needs for the project. The design professional must clarify the needs and define potential options that are acceptable to the owner. This clarification includes getting answers to the following questions:

- What does the client want?
- What is the function of the project?
- What facilities are to be included?
- How many people will be served?
- What codes and regulations must be met?
- What environmental considerations must be met?

Many clients do not have a good grasp on the answers to these questions, which makes the consultant's job more difficult (and introduces risk of changes in direction by the client). The result of these discussions should be a conceptual design or basis of design report. It is important that operations and maintenance staff are involved during the conceptual design stage, because these are the people who will occupy and operate the facility for the next 50 years. Conceptual design reports outline the general principles on which the design of the project will be based. Such reports include

- Introduction to problem
- Background information about the utility, its needs, and its current facilities
- Alternatives considered to resolve the problem
- An evaluation of alternatives
- Recommended alternatives

Conceptual design may also include drawings that outline the following:

- Where the project is located
- Current site usage
- A site survey showing where the improvements are going to be
- Existing site plan showing the current facilities on the site where construction is proposed
- Proposed site plan identifying what the site will look like with the improvements to be designed
- Data on the size of buildings and on location of roadways, parking, landscaping, and similar features
- Identification of utility needs (e.g., water, sewer, and so forth)

The client should provide written acceptance (not approval) of the conceptual design report and drawings. The acceptance provides the design professional with the approval to move forward on the project. The client does not want to "approve" the report or plans as this may cause the owner to assume some degree of responsibility for the design by indicating that some degree of review, calculations, and/or similar analyses have been undertaken, which is rarely the case.

Once this acceptance has been issued, the design professional will develop the preliminary design plans and specifications, including those listed below.

- Determining the utilization rates of facilities (e.g., people per day)
- Finalizing the size of buildings
- Determining the codes applicable to the project
- Calculating the water and wastewater needs, and locating appropriate corridors
- Calculating stormwater and parking requirements and locating appropriate sites
- Updating the proposed site plan to what the site will look like with the designed improvements. This involves adding more detail, including water, sewer, lighting, parking, roads, power, and so forth, as appropriate.
- Architectural layout/models, if desired

One of the results of the preliminary design calculations will be the determination of the following (as applicable to the project):

- Water needs
- Wastewater needs
- Quantity of stormwater to be disposed
- Parking requirements
- Off-site roadway and turn lane improvements
- Mechanical demands

Preliminary design specifications should include

- Building materials to be used
- Construction methods to be used
- Equipment to be installed
- Utility piping; heating, ventilation, and air conditioning (HVAC); and power grids
- Structural components

On receipt of the preliminary plans and specifications, the client should provide comments on the plans and specifications, equipment and materials preferences, and the necessary legal documents for the contract package. Acceptance of the plans should be provided for the reasons noted above. After receipt of this acceptance, the design professional will work toward the final plans and specifications. The final design plan set typically includes the following types of documents (as appropriate for the project, in addition to the specifications):

- Cover sheet
- List of drawings and engineers who did them
- Existing site plan
- Proposed site plan

- Site plan with changes highlighted
- Water and sewer plan for site
- Stormwater plan
- Stormwater disposal details
- Parking plan
- Landscaping plan
- Floor plans
- Elevation plans
- Roof plans
- Structural plans
- Structural details
- Plumbing plans
- Lighting plans
- Electrical plan
- HVAC plans
- Details

Not everything can be described in the drawings, so specifications are required. These are often referred to as the *construction documents.*

Construction Documents

The development of appropriate construction specifications and contract documents is important to ensure that a construction project proceeds as smoothly as possible. The specifications, like the plans, are the responsibility of the design engineer. Too often preparation of the contract documents is also delegated to the engineer. But the contract documents are the basis of a legal arrangement between the contract and the utility. As a result, the appropriate legal help should be enlisted in contract document preparation. Having standardized legal documents and contracts is of significant benefit to the utility when problems occur. The attorneys and staff will be well versed in the content of the documents and the actions they should take. Having to read every contract every time an issue arises is time-consuming and creates the potential for errors and confusion.

The law of contracts defines the rights and duties of the parties to the agreement and expectations for performance. It provides a series of rules for interpretation of the document. The following must be included in a valid contract:

- Mutual agreement on terms
- Exchange of some consideration (money, property, assets, or services)
- Underlying law of contracts must be satisfied
- Purpose of the contract must be legal
- Parties to the contract must be competent

Interpretation of the contract requires that the intent of the contract be defined in the contract documents. In general, the contract is interpreted as a whole to determine the intent. A common problem with construction projects is determining the relationship between the contract itself and the plans and specifications, which may be separate, but attached. Most contracts incorporate the plans and specifications by reference, and therefore the intent of the contract is clear and is controlled by those provisions (Nuechterlein et al., 1991). Interpretations are usually made based on a reasonable interpretation of the meaning or the language. This requires that terms that may be unclear to a judge or jury be defined in the contract.

The contract documents for construction of capital projects include the items listed below.

- Advertisement
- Bid form and instructions
- Addenda
- Contract
- General conditions
- Special conditions
- Specifications
- Bonds (payment and performance bond forms)

Most governmental entities are required to follow state statutes on bidding and purchasing. These may vary from state to state, but they generally require that the utility advertise the project, provide an appropriate amount of time for potential bidders to put together prices, provide a date and time when the bids must be submitted to the utility, and have a public opening (and possibly reading) of the bid results. The advertisement must be placed in a newspaper or other place where potential bidders can see the advertisement and respond to it.

The bid form and instructions outline how the bidder is to respond to the advertisement for the project and the form in which the bid price is to be entered. A bid is an offer by the contractor to perform the service or construct the project at a given price. A bid bond is submitted with the bid to ensure that the contractor awarded the contract will actually enter into the contract to perform the work. If a contractor reneges on his bid, the utility will be able to cash the bid bond to cover the difference between the failed contractor and the next lowest bid. Defective bids or those with obvious errors should be rejected. Addenda are issued to clarify questions that arise during bidding.

The statutes commonly require that the project be awarded to the lowest responsive and responsible bidder. The lowest responsive and responsible bidder may not be the lowest bidder, as governmental entities are not required to enter into contracts they believe are flawed, presented by bidders who cannot perform the project at the bid price, have a poor performance record, or take exception to the bid or contract provisions.

The contract is the most important part of the contract documents. The contract should contain the following provisions:

- Scope of work
- Terms
- Incorporation of other documents

- Notices
- Pay provisions
- Change orders
- Changed conditions
- Time extensions
- Indemnification
- Right of termination
- No damage for delay
- Liquidated damages
- Insurance

The scope simply outlines what the project entails in general, without details. The terms section outlines the meaning of terms used in the contract documents that may be unfamiliar to third parties or might be misinterpreted by a judge or jury in the future if litigation occurs. Note that these are different from the *term of the contract*, which is the amount of time to complete the work. Incorporating the other documents section will define the priority to be given to the specifications, general conditions, special conditions and drawings, and their inclusion as a part of the contract.

The notices section is designed to direct how communication about the contract provisions is to occur between parties and includes addresses and mailing methods. Each party to the contract should designate one person to act on that party's behalf. Construction projects with no clear project representative can deteriorate into confusion when many parties provide input without adequately communicating with others within their party. For example, if a contractor submits requests for change orders to different owner's representatives, and these representatives do not communicate well, then the contractor may get multiple approved change orders from the owner, or may not have their request in the hands of the person who can approve it, thereby delaying the job. Utilities should make it very clear who the owner's representative is and make sure that all official communication goes through that person.

Pay provisions outline how the contractor is to submit pay requests, the number of copies, the required backup information, and how the requests will be processed by the utility. The contractor will submit pay requests, usually monthly, and these will be reviewed by the utility staff. If acceptable, then the utility will normally pay within 30 to 45 days. If the pay request is not acceptable, then it will be returned with provision on how the contractor is to resubmit it.

Change orders are typical of many projects. Change orders can occur for a variety of reasons, including

- Changed conditions, e.g., pipelines or subsurface conditions are not as anticipated by the engineer when performing the design
- Changes requested by the utility staff desiring something different than what was designed or specified
- Changes required because specified equipment is no longer available
- Clarifications that indicate something beyond what was anticipated by the contractor

- More difficulty with construction than anticipated
- Omissions in the plans
- Errors in the plans

No set of plans is perfect, so minor changes should be anticipated. The utility will pay for all change orders except errors in the plans, which should be paid for by the design professional. Changed conditions, constructability problems, and equipment no longer made are issues that often cannot be anticipated. If they were anticipated, then the owner would have paid for them. The same is true for omissions and clarifications in the plans that result in change orders. Changes requested by the owner are fully controllable and should be minimized. There are numerous examples of deterioration of contract relationships and extensive, costly change orders resulting from owner-requested changes. Change orders typically must be approved by a governing board.

Contracts will have a term (e.g., completion within 180 days from execution of the contract). Time extensions are permitted in cases where the contractor is delayed due to acts beyond anyone's control, such as weather (force majeure clauses), by utility staff, or by other entities, such as the power company, building department, or permit agencies. The contract will typically include a "no damage for delay" clause whereby the utility would not pay extra to the contractor for such delays. No damage for delay clauses can be difficult to enforce, especially when liquidated damage provisions are included in the contract.

Liquidated damages are imposed by the owner on the contractor when the contractor fails to meet the contract completion dates defined in the contract. Liquidated damage amounts are supposed to have some relationship to the damage that will be incurred by the utility if the project is not completed on time. Hence, most liquidated damages are based on the cost of the design professionals and other staff to supervise the project. If a separate contract for water or wastewater delivery existed that brought the utility new revenue, those lost costs could be used for the liquidated damages. Punitive liquidated damages are not permitted by the courts.

Indemnification clauses are hold-harmless paragraphs. Most require that the contractor hold the utility, its officers, and staff harmless from damages that may occur during the project. A right of termination clause permits the utility to terminate the contract for any reason, without cause. The contract must have a provision to terminate a contractor for nonperformance, but since nonperformance may be difficult to prove, the right of termination clause removes any question. Finally, contracts will include the insurance requirements that the contractor must satisfy.

The general conditions may include all or some of the above provisions. General conditions are usually consistent among all contracts. There are many versions of general conditions available for use. The National Society of Professional Engineers (NSPE), American Institute of Architects (AIA), and contractor organizations each have their own standard versions. However, utilities should be careful when using any of these documents as they were not written by owners, thus will tend to favor the entities that wrote them. Utilities are encouraged to develop their own standard general conditions with the help of competent construction counsel.

Special conditions are developed by the design professionals to highlight issues or concerns outside the scope of the general conditions, or that require a change to the general conditions for the project. Likewise, the specifications are developed by the design professional to define what is to be built, what equipment and materials are to be used, and how various pieces are to be put together.

Bonds (payment and performance bonds) are specified in order to provide a guarantee of performance of the work by the contractor. Performance bonds are issued at the beginning of the contract, for the full value of the contract, to provide the owner with a guarantee that money will be available to complete the project in the event of a default by the contractor. Payment bonds are used to ensure that the subcontractors and suppliers will be fully paid by the contractor. Unlike private projects, most public-entity projects cannot be liened, so without the payment bond, a subcontractor or supplier has no guarantee they will be paid. The utility has no responsibility toward suppliers or subcontractors. Both bonds are released on successful completion of the work.

Bidding

Once the plans and contract documents have been finalized and accepted by the owner, the next step is to secure bids. Most public entities must competitively bid their construction and operations contracts. The intent is to secure the lowest price for the public benefit.

The complexity of bid documents has increased exponentially with time. There is quite a contrast between filling in a simple bid 20 years ago and bid documents today that may ask for extensive submittals, multiple copies, electronic copies, lists of subcontractors, subcontractor references, identification of disadvantaged business partners, financial data, and supplier costs. These have not traditionally been a part of the final contract and increase the cost to the owner and, ultimately, the customer (Hewitt, 2010). The complexity also causes many bids to be deemed unresponsive, meaning that the lowest or second lowest bids may be disqualified, which is a direct cost to the public. This is a problem in part because most subcontractor and suppliers' bids are delivered within hours of when the bid is due.

The bids are usually opened and read at a public place at a given time. Once a bid is opened, the design professional reviews it for errors, gathers information on the contractor's capability to perform the project, and makes a recommendation to the owner. Selecting the wrong contractor as a result of competitive bidding could lead to major conflicts. As a result, many construction managers and some construction lawyers support greater use of prequalification by all contracting entities. Where used, prequalification of the potential contractors is the most important part of the selection. Cost is not a component of this prequalification. Experience with the equipment/processes to be installed, experience with design and coordination of construction, and the performance record of previously installed equipment carry the weight in the implementation and long-term operation of the system.

Prequalification takes on many of the same issues as a review of bidders for responsive and responsible status, including the following:

- Technical competence
- Experience with similar projects
- Financial capability and status
- Bonding capacity
- Current workload
- History of litigation
- Prior performance

Where there are problems, the contractor may be disqualified from bidding on a project. Prequalification is used on complicated projects, where schedules may be tight, or on specialty work where the number of truly qualified bidders may be limited. Prequalification is part of many alternative project delivery processes, as will be discussed later in this chapter.

Bid methods

Bidding the contract may be done by lump sum, unit price, or negotiated cost-plus methods. Lump sum contracts will contain only one figure on the bid sheet, the price to perform all the work of the contract (see Table 10-1). Buildings and treatment facilities may be bid this way; pipelines rarely are. Once the bid is awarded, the contractor is usually required to break the work down into its components in order for monthly progress payments to be made. The goal is to prevent the utility from overpaying early in the contract, which may encourage the contractor to default at the end of the project, thereby costing the utility much more to complete the work. A typical situation applicable here is where the punch list value exceeds the value of retainage. Change orders for lump sum contracts have no basis in the bid, and must be negotiated on an individual basis.

Unit price contracts allow some flexibility in the project. The bid sheet includes specific dollar costs for all facilities (see Table 10-2). Pipe costs are on a per-foot basis, while valves and fittings are on a per-unit or weight basis, and excavation is on a cubic yard basis. If the quantities change, or the prices are good and the utility desires to add to the contract (with the contractor's concurrence), the prices for such changes are based on the unit costs. No negotiation about price is required. Project expenditure curves are often easy to project. Many contractors will ask for a mobilization line item for unit price contracts. This amount covers the costs of getting set up for the work; no work actually occurs. However, the contractor will often prepay for materials and these funds may be an attempt to recoup their interest costs.

Negotiated or cost-plus contracts are based on the actual cost incurred by the contractor to perform the work, plus some agreed-to overhead and profit amount (typically 10 or 15 percent). There are an infinite number of options, but few meet the competitive bidding statutes. Cost-plus contracts are more common between private parties. For cost-plus contracts to work, the contractor must document all the costs. This is often arduous, and the result tedious for the contractor, utility staff, and engineers to review. The contractor must predict its cash flow, which may be different from what the utility expects.

Fixed-cost contracts may also be negotiated. These contracts require that the contractor guarantee its price against all contingencies. Owners should be aware that the contractor will add enough profit to the bid on a fixed price to minimize the likelihood that it loses money. The result to the owner is normally a higher bid than one received through competitive processes.

Awarding the contract

Once the bids have been received and evaluated, public owners then post a notice of award and send a copy to the lowest responsive and responsible bidder. The next step is an award of the contract by owner (or governing body), assuming there are no protests. The owner and contractor then execute the contract documents. A preconstruction conference is held and a notice to proceed is provided by the owner to the contractor. The contractor then moves forward with construction through submittal of shop drawings and progress schedules.

TABLE 10-1 Example of a unit price bid sheet with a series of allowances for work that is required, but for which the scope is unclear at the time of bidding (this comes from a design/build contract)

	BID SHEET	
Item No.	Item	Cost
1	Design and construction of nanofiltration water treatment plant	
2	Contingency	$250,000
3	Base bid total (note 1)	
4	Deducts to base bid (specify reasons for same) (Note 2)	
4.1	Deduct 1	
4.2	Deduct 2	
4.3	Deduct 3	
4.4	Deduct 4	
4.5	Deduct 5	
4.6	Deduct 6	
4.7	Deduct 7	
4.10	Add 1	
4.11	Add 2	
4.12	Base bid minus approved deducts	
5	Allowance 1—decommission and remove fuel station	$250,000
6	Allowance 2—rehabilitate electrical system and control upgrades from existing lime softening system	$500,000
7	Allowance 3—lead/asbestos/contam. soil	$100,000
8	Allowance 4—LEED certification	$100,000
9	Allowance 5—FPL transformer	$100,000
10	Allowance 6—permit fees/specialty/threshold inspection	$50,000
11	Allowance 7—architectural treatments	$125,000
12	Allowance 8—horizontal well investigation	$150,000
13	Base + allowances	
14	Base + allowances + all deducts + all additions	

Note 1: All bidders must submit a responsive base bid that conforms to the preliminary design report and addendums.

Note 2: The CITY reserves the right to reject any deducts before bid analysis.

TABLE 10-2 Example of unit price bid

Bid Item	Quantity	Units Paid	Unit Cost	Item Cost
12-in. PVC C900 water main	1,365	LF	$35	$47,775
12-in. gate valve—MJ	3	Ea	$1,250	$3,750
Fire hydrant	3	Ea	$2,500	$7,500
12 × 6 tee MJ RJ	3	Ea	$750	$2,250
6-in. gate valves	3	Ea	$600	$1,800
12 × 8 Tees *RJ—MJ (for interconnects)	2	Ea	$1,100	$2,200
8-in. gate valves MJ	2	Ea	$850	$1,700
Jack and bore (w /casing)	200	LF	$375	$75,000
Remove unsutiable materials	65	CY	$18	$1,170
Clean fill	65	CY	$25	$1,625
Bedding stone ($57)	100	CY	$45	$4,500
Pavement restoration	500	SY	$18	$8,750
Unforeseen condition allotment LS ($50,000)	1	Lump sum	$50,000	$50,000
TOTAL BID				$208,020

Note: CY = cubic yards, MJ = mechanical joint, RJ = restrained joint, SY = square yards, and LF = linear feet.

Construction

Constructing the project

While there are a number of other delivery methods for providing capital projects, the five most common processes are

1. Traditional design-bid-build construct (DBB) program
2. Design-build (D/B)
3. Performance-based specifications
4. Design-build-operate (DBO)
5. Design-build-own-operate (DBOO)
6. Design-build-own-operate-finance (DBOOF)

The primary goal of each capital project delivery process is to provide the utility with the right product, while meeting objectives involving time, cost, control, and project quality. Each of these delivery mechanisms involves both a consultant and a contractor, and may include operations personnel or multiple contractors, depending on the delivery mechanism.

Design-Bid-Build (DBB—the traditional approach). Traditionally, the construction of water/wastewater treatment facilities by public agencies has been a process in which a design consultant is selected, plans and specifications are completed, permits are obtained, and then the plans and specifications are used to obtain formal, sealed bids for construction (Figure 10-2). Two separate steps are required: selection of design consultants and selection of the construction contractor through the bidding process (both discussed previously).

When new water supplies or disposal alternatives are required, the preliminary phase can be extensive. Construction occurs after the design has been completed. In the best of scenarios, it is difficult to complete significant projects in less than two to three years. The benefit of employing the traditional DBB method is that one contractor is responsible for all phases of the project's construction, although its responsibility may be limited if specific materials, pieces of equipment, and construction methods are specified. Incompatible equipment can be specified without realizing this is a problem, which can result in change orders and delays. Also, highly equipment-laden contracts can be "jobbed" by the prime contractor (bid out among a series of subcontractors to negotiate a low price to the prime contractor without a similar cost savings to the owner), which may result in insufficient attention to the details.

The utility's control is limited to the design process. Depending on the contractor and contract conditions, the utility may be able to influence some issues, but control after the design is complete is usually limited. Once construction is under way, the more control over construction progress or the more changes the utility makes, the more change orders the utility will see. The DBB method requires the least time commitment from the utility, but still requires staff competent in dealing with construction projects.

A subset of the traditional DBB approach is the use of parallel prime contractors, where different parts of the project are assigned to different contractors depending on their expertise. This method can be used to overcome the construction and equipment provision issues on projects with disparate equipment needs or experience levels, but has the disadvantage of requiring the owner to administer multiple parallel contracts. This may lead to conflicts on the site that could delay the project and increase costs, as a result of change orders and disruptions to the contractor's schedule. However, the benefit of parallel contracts is better assurance that the equipment portions of the project will be provided by a knowledgeable contractor who can make sure that the equipment will work together. Some expediting of work can occur when the equipment can be constructed off-site.

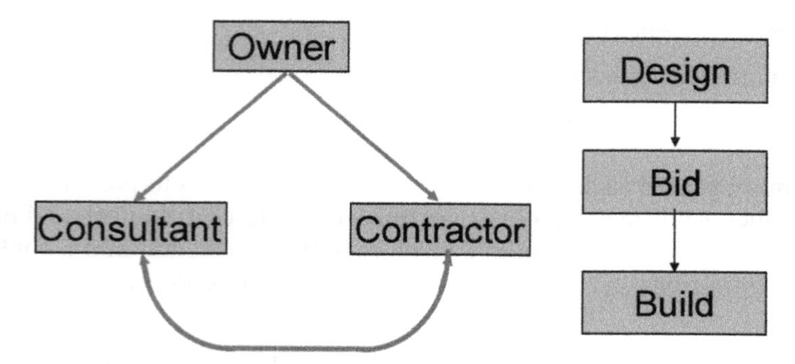

FIGURE 10-2 The traditional design-bid-build (DBB) approach to capital project construction with the process shown at the right

Design-build (D/B). Design-build projects (see Figure 10-3) are permitted under most state and local bidding laws. The bidding process is normally qualification-based as opposed to price-based. A design-criteria professional is required to create the bid documents. If the statutes are followed under the DBB approach, then the bid price becomes a part of the selection process after the unqualified firms are eliminated. The D/B process involves prequalification of a team that includes an engineer to eliminate those not competent to bid on and complete the project. Once a group of contractor–engineer teams is selected, those entities turn in a proposal based on a specific criteria package developed by the utility. A D/B project may range from very detailed requirements (10 to 60 percent design effort) to brief outlines that provide limited guidance for bidding. The extent of the scope of services relates to the amount of control of the project the utility wishes to retain. Once the bid is awarded, the utility's control will diminish significantly, as the designer and the builder will use the established criteria to complete the project. A construction manager (CM) or project management (PM) team is required to oversee the work on behalf of the utility. This is one reason for the utility to exercise continued control of the project.

Some professionals have proposed D/B as a means to reduce time and require the contracting entities to take more responsibility for the project since they control all aspects of the design and construction process. Time savings occur once the full scope of the project is known. D/B methods should reduce the problems periodically encountered with equipment compatibility and changes in the equipment itself. Most D/B projects list the contractor as the lead due to the liability issues with construction, but the engineer can lead the effort if the engineer has the right insurance and is willing to put its capital at risk. Note that the engineer on a design-build project cannot be the design criteria professional.

Within the D/B approach are a number of suboptions that require various levels of input or detail up front from the utility. These include

- **Level of construction criteria package.** This varies from performance-based specifications to 60 percent plans and specifications.
- **Selection criteria framework.** Selection based on qualifications only, then is a contract and price are negotiated. Price is the only criterion.
- **Select team versus select individuals.** The utility selects the engineer and, at the same time, a construction manager. It then requires the two entities to work together.
- **Various "partnering" arrangements.**

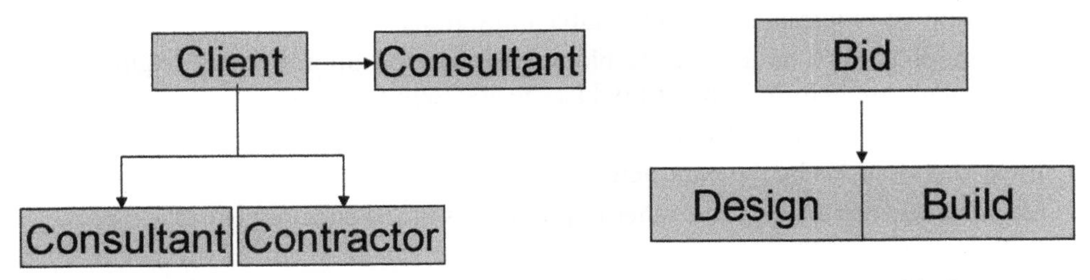

FIGURE 10-3 The design-build (D/B) approach to capital project construction with the process shown at the right

However, where utilities have limited time to construct projects that are equipment-intensive, the use of a limited set of requirements, referred to as *performance-based specifications*, has merit.

Performance-based specifications. Where time may be critical, the project mechanically intense, and control over the project partially relinquished, the concept of performance-based specifications (PBS) is an option worth considering. Performance-based specifications are aptly named, but are a relatively new concept to the D/B water industry. The major difference is that the utility retains more control over the engineer and the contractor as separate but coordinating entities. PBS requires the contract team to deliver a product that meets certain requirements. Knowing the base conditions (i.e., water quality requirements coming out and water quality going in) is necessary, and this knowledge must be gathered prior to creating a performance specification.

The utility and its design criteria consultant set parameters for general quality and delivery time, plus minimum requirements for performance, i.e., roughly summarizing what the project should look like. However, the utility surrenders control over specifics of equipment and lets local building codes define structural and electrical issues. PBS documents may or may not be very detailed, depending on the amount of process control the utility desires. The key is that the contractor must meet basic requirements within a given timeline. For example, a membrane plant could be specified by the performance of the resulting permeate characteristics, concentrate waste product, limits of pressure, material guidance, and control parameters (as follows), without getting into details of the process itself. Such performance-based criteria could be similar to one used in the following membrane water treatment plant example:

- Number and size of skids
- Postchemical stabilization requirements
- Minimum recovery percentages
- Maximum operating pressure
- Materials (e.g., HPDE, Zeron 100 stainless-steel pipe, and CPVC products)
- Fully integrated control system
- Standards of quality in the constructed project
- Minimum/maximum pressure into the distribution system
- Water quality limitations
- Pretreatment requirements or acceptable limitations
- The building, such as a concrete block enclosure on appropriately engineered slab (floor plans are helpful if the utility has specific requirements)
- New equipment manufacturers (assuming limits to achieve maintenance goals)
- Interim operations by the contractor
- Engineering, permitting, and other requirements of the contract entity
- Delivery date

The specifications force the contracting party to sort out these decisions in order to deliver a quality product without start-up or delivery problems, and to reduce time. A period of operation by the contracting entity before transition to utility staff is recommended, as this will ease the operations transition and support any necessary budget modifications (the contractor will minimize those as well during the period when it operates the equipment). The design, typically using the criteria consultant, occurs concurrently with the ordering and assembly of equipment (which can be accomplished with equipment-intensive projects) by using the shop drawings to support the design package presented to the regulatory agencies. This is the reason equipment-intensive projects like membrane facilities or sequence batch reactors are prime candidates for the PBS process.

A limitation with PBS is the unfamiliarity of the concept to municipal purchasing departments and regulatory permit writers. Permitting is one of the biggest time issues in any capital project. Critical time can be lost during the permitting phase of the project; however, matching the regulatory agency requirements with the use of performance specifications in the design report allows flexibility in selection of equipment, equipment manufacture, and options for specialized expertise that would not otherwise be involved in a standard design and construction program. In such cases, up-front discussions with those involved in the decisions should ease the concerns or bring the issue to resolution. Most agencies are initially less concerned with the details than the concept, so follow-up with detailed shop drawings may be helpful.

Performance specifications cannot be used for every treatment facility a utility might contemplate. Most piping projects could not be completed using a performance specification, as permitting agencies want to see completed designs before roadways can be disrupted. However, for equipment-intensive projects, like membrane treatment facilities and sequence batch reactor wastewater treatment plants, the major cost and time constraints involve the equipment that treats the water. The tankage or buildings required for treatment are often ancillary to the equipment, as evidenced by the cost of the equipment in comparison to cost for the rest of the work. In addition, equipment-intensive treatment systems use advanced computer controllers (e.g., programmable logic controllers [PLCs]) and other telemetric equipment for operation and control, which are often difficult to coordinate. The interface between this equipment and the computer controls should remain the responsibility of the equipment supplier and not jobbed, as happens in many traditional construction contracts, often with disappointing results.

Design-build-operate (DBO). Design-build-operate projects have also been used to reduce delivery time and to eliminate the need to staff up a new facility (or as a part of a privatization effort). Three parties are typically involved in a DBO project: a contractor, an engineer, and an operations firm (see Figure 10-4). There are two reasons to pursue DBO: lack of capital and lack of operational experience.

The bidding process is typically qualification-based as opposed to price-based, although the price element can be included in the initial step of selecting the best team and should include prequalification teams. This prequalification process eliminates those judged incompetent to bid on and complete the project. Once a series of potential bidders is qualified, those entities prepare a proposal based on a specific scope of services developed by the utility (and its design criteria professional, who cannot be in any proposal team).

DBO projects may have extensive specifications, including detailed construction and operating requirements for the proposed projects. The criteria for selection include price, determined from a life cycle or annual not-to-exceed cost. These guaranteed costs can be described on a customer basis, where it is up to the success team to provide the long-term services at a specific

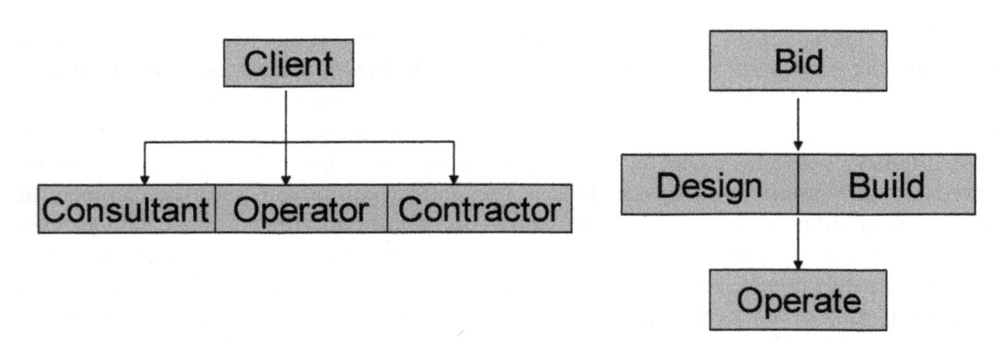

FIGURE 10-4 The design-build-operate (DBO) approach to capital project construction with the process shown at the rig

monthly customer rate. Typically these contracts are for 5 to 20 years or more. The contracting entities are required to be more responsible for the project since they control all aspects of it.

However, guaranteeing costs projected over the contract life is nearly impossible. Since the DBO project assumes the responsibility for all life cycle costs, transference of cost between maintenance and capital can be a source of future conflict in poorly written or poorly understood contracts. Maintenance is traditionally a cost for the operator, but capital is the responsibility of the owner, as in private-sector utilities regulated by public service commissions. As result, comparing costs to provide service in public operations and the administration of operations contracts is somewhat difficult in the DBO scenario.

Design-build-own-operate (DBOO). DBOO projects (see Figure 10-5) are similar to DBO projects, except that the contracting party becomes the owner of the project, instead of the utility. As a result, the only control the utility has when pursuing this option is adherence to a set of performance parameters and, potentially, negotiated costs. This option has been suggested to reduce delivery time and to eliminate the need to staff as a part of a privatization effort. Design-build-own-operate-finance projects (DBOOF) add the financing component to DBOO projects.

Engineering During Construction

An infinite variety of issues can arise during construction, but usually construction projects involve inspections of the contractor's work, progress reports, meetings, and the processing of pay requests. Other issues involve field changes and time extensions, which may lead to change orders. Defective plans and specifications may involve additional changes, but these are not the norm.

Progress reports are developed to track progress on the project. Progress reports have the following parts:

- A review status of the project overall (from a macro scale)
- A review of what has happened since the last progress report (in detail)
- A review what will happen in the coming progress report period (in detail)
- An update of the schedule

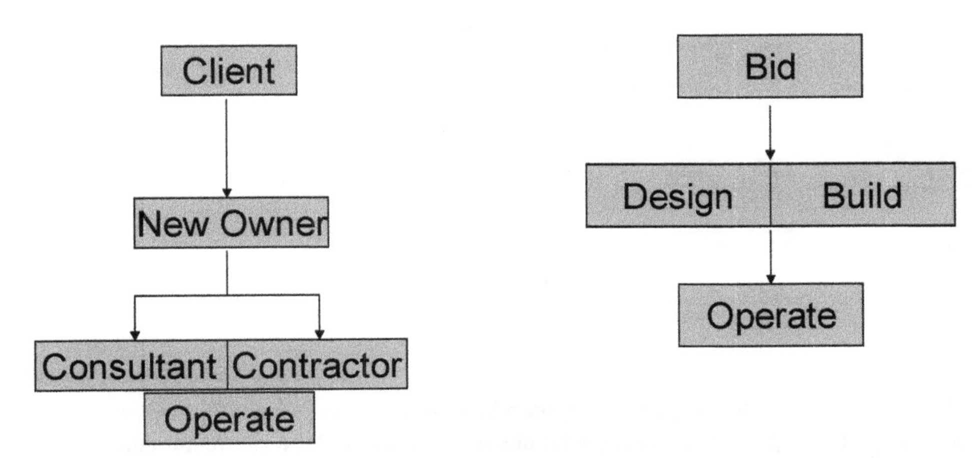

FIGURE 10-5 The design-build-own-operate (DBOO) approach to capital project construction with the process shown at the right

In addition, the owner needs to know about foreseen problems with the plan/project. The progress report needs to identify areas requiring input from the owner and include recommendations on how to resolve problems.

A number of recommendations are available to help the utility during construction. The utility should require record drawings that show how the contractor built a project. The engineer-of-record should verify that the drawings accurately depict the construction. Photographs of the work in progress are particularly helpful.

Analysis of the project will indicate the targets for the project schedule and costing. The initial engineer's project schedule will generally be a bar chart (see Figure 10-6). Bar charts may be useful, but do not identify dependence appropriately and should not be used by contractors as they cannot identify the critical path clearly. Instead, the contractor should use a scheduling system that will create a series of subcomponents of the project schedule. Included for all scheduling activities are the following requirements:

- Understanding the tasks required to complete the project
- Projected timelines for each task in the project
- Projected labor requirements for each task
- Understanding the dependence of each task on preceding or succeeding tasks (order)

Once these items are understood, a series of networks are created. The networks outline the order of the tasks and dependence between tasks. Such programs, called *critical path methods* (CPM) charts, are appropriate to understand the network. This type of scheduling is used to determine the critical path, which is the shortest time in the contract. The remaining tasks all have "float" within them. The schedule is the best tool that the utility can have to determine whether or not the contractor is making adequate progress.

Figure 10-7 is an example of a simple schedule for a simple project. The schedule is used to sequence work and determine the amount of time for each task. As the contractor wants to be paid, progress against the schedule can be monitored. The critical path can also be determined to be nodes as follows: 1-2-4-5-8-10-11, which means the contractor will

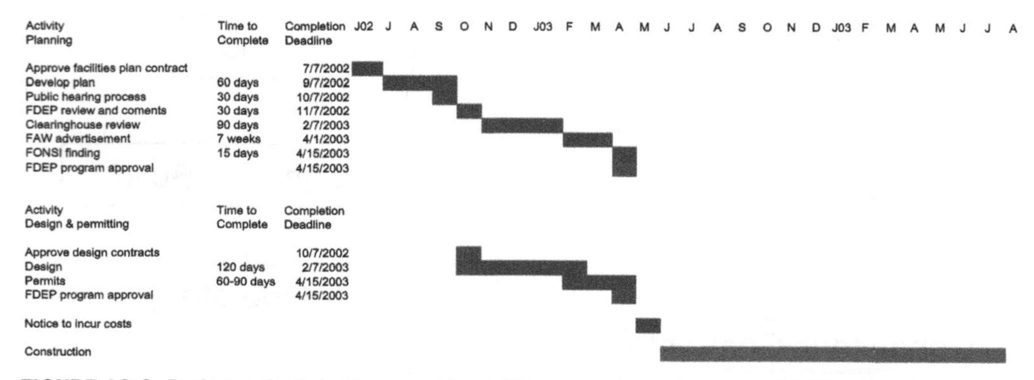

FIGURE 10-6 Project schedule (bar chart) used by an engineer to secure SRF funding for a project, showing full project timeline. (Note: this is not sufficient for a contractor during construction.)

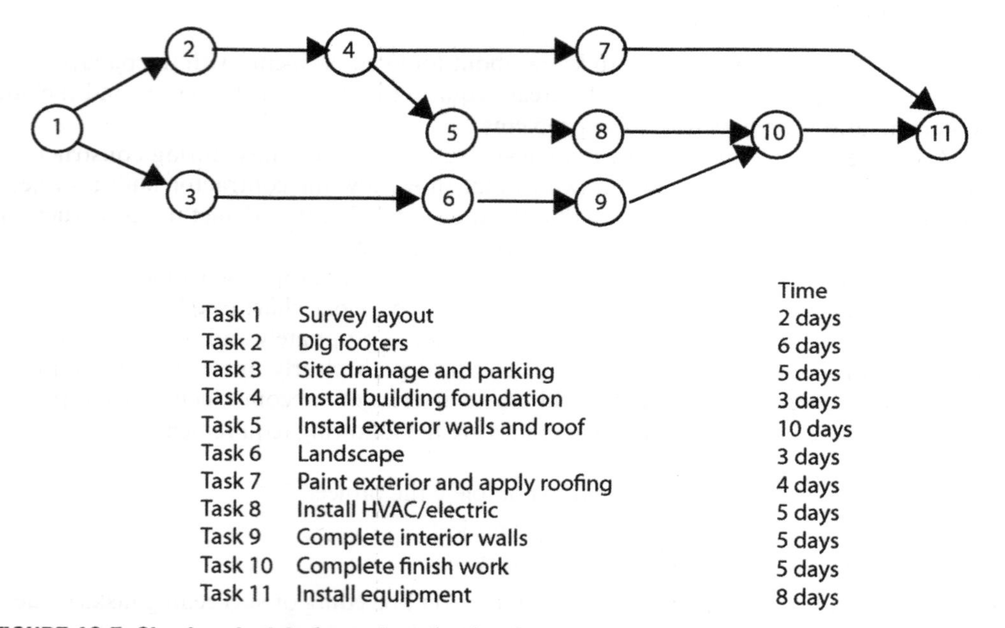

		Time
Task 1	Survey layout	2 days
Task 2	Dig footers	6 days
Task 3	Site drainage and parking	5 days
Task 4	Install building foundation	3 days
Task 5	Install exterior walls and roof	10 days
Task 6	Landscape	3 days
Task 7	Paint exterior and apply roofing	4 days
Task 8	Install HVAC/electric	5 days
Task 9	Complete interior walls	5 days
Task 10	Complete finish work	5 days
Task 11	Install equipment	8 days

FIGURE 10-7 Simple schedule for project that involves a rudimentary building

take 37 days to complete the work. There are a variety of methods and software programs to help contractors, engineers, and owners with scheduling their project (see Figure 10-8 using Microsoft Project®). Every progress report should update the schedule. Any revisions to the schedule should be made in comparison to the original schedule and the changes made in the previous progress report. Explanations should be given for changes.

Estimating the cost of the project relies on the engineer and utility staff's ability to define fixed capital assets. Direct fixed asset costs include the raw materials and equipment, power, and labor for installation. Labor costs include salaries, fringe benefits, overtime, and so forth. Indirect costs include the costs for supervision, security, and similar

Task Name	Duration	Start	Finish	Predecessors	Cost	Fund Source	Resource Names
New Well Construction Schedule	**885 days**	**Fri 7/1/06**	**Thu 11/20/08**		**$660,125.0(**		
Install Monitoring Wells	45 days	Fri 7/1/0!	Thu 9/1/0!		$25,000.0(	Ops/IF	GMWA
Identify New Well site	150 days	Fri 7/1/0!	Thu 1/26/0(		$0.00	Ops/IF	GMWA
Acquire new well site	250 days	Fri 9/2/0!	Thu 8/17/0(	2	$100,000.0(	Ops/IF	City
Modify Water Use Permit	180 days	Fri 9/2/0!	Thu 5/11/0(	2	$0.00	Ops/IF	GWMA
Prepare RFP	20 days	Fri 3/3/0(	Thu 3/30/0(	2FS+130 days	$1,500.0(	Ops/IF	PSD/PUMPS
Issue Notice to RFP	5 days	Fri 3/31/0(	Thu 4/6/0(	6	$0.00		City Clerk
Receive RFP's	30 days	Fri 4/7/0(	Thu 5/18/0(	7	$0.00		City Clerk
Evaluate Proposals	20 days	Fri 5/19/0(	Thu 6/15/0(	8	$1,000.0(	Ops/IF	City + others
Shortlist Interview	0 days	Thu 6/15/0(	Thu 6/15/0(	9	$0.00		City
Recommendation to CC for Award	5 days	Fri 6/16/0(	Thu 6/22/0(	10	$1,000.0(	Ops/IF	PUMPS
City Commission Award	45 days	Fri 6/23/0(	Thu 8/24/0(	11	$0.00		PSD/PUMPS
Negotiation with Selected Consultant	10 days	Fri 8/25/0(	Thu 9/7/0(	12	$0.00		PUMPS
City Commission Award	45 days	Fri 9/8/0(	Thu 11/9/0(	13	$0.00		PSD/PUMPS
Issue NTP To Consultant	10 days	Fri 11/10/0(	Thu 11/23/0(	14	$125.0(	Ops/IF	PUMPS
Design & Geotech Investigation	90 days	Fri 11/24/0(	Thu 3/29/0;	15	$35,000.0(	Ops/IF	TBD
BCDOH Permit	30 days	Fri 3/30/0;	Thu 5/10/0;	16	$2,500.0(	Ops/IF	TBD/BC DOH
SRF Funding Resolution	30 days	Fri 5/11/0;	Thu 6/21/0;	17	$1,000.0(	Ops/IF	PUMPS
Bid Project	45 days	Fri 6/22/0;	Thu 8/23/0;	18	$2,500.0(	Ops/IF	City
Review Bids	10 days	Fri 8/24/0;	Thu 9/6/07	19	$2,500.0(	Ops/IF	PUMPS
Secure SRF Approval on Award	15 days	Fri 9/7/0;	Thu 9/27/0;	20	$1,000.0(	Ops/IF	PUMPS
City Commission Award	45 days	Fri 9/28/0;	Thu 11/29/0;	20,21	$1,000.0(	Ops/IF	PUMPS
Notice to Proceed	45 days	Fri 11/30/0;	Thu 1/31/0!	22	$0.00		PUMPS
Construction	180 days	Fri 2/1/0!	Thu 10/9/0!	23	$450,000.0(	SRF	Undetermined
Construction Services	210 days	Fri 2/1/0!	Thu 11/20/0!	23	$36,000.0(	SRF	TBD
Project Complete	0 days	Thu 11/20/0!	Thu 11/20/0!	25	$0.00	n/a	n/a

FIGURE 10-8 Detailed example of a schedule prepared to monitor a project's progress showing dates for start and completion, responsible party, funds required, and funding mechanism

functions. Home office overhead costs include insurance, management of the contractor, expenses, office space, and others, which are not present or spent at the project site, but which are necessary to manage the project. These are points of contention when projects are delayed by the owner and the contractor desires additional home office overhead costs. All of these costs should be identified. Bids for similar projects are the best source of information for pricing projects. Cost estimates should take into account the economic situation of the area; e.g., if it is growing fast, costs will be high (see chapter 9).

Project cash flow is important for the utility to consider. Contractors want to be paid monthly. However, the amounts the contractor wants will vary from month to month. To develop the cash flow for the project, the contractor needs to define and schedule the workload for each task. Typically cash flow for a project starts slowly, depending on whether or not the contractor can be paid for stored materials under the contract. The middle part of the project includes the majority of the installation, so money is drawn quickly. The final stages of the project include finishing work. This is usually slow work that has little return for the contractor. To ensure that the finish work is actually completed and that all record drawings and other deliverables are delivered by the contractor, most contracts permit an amount to be withheld each month as retainage (typically 10 percent). The final contract payment is the retainage for the project, paid on final completion of the punch list and receipt of all deliverables.

The cash flow diagram uses the schedule. Requiring updated schedules forces the contractor to focus on the workload and what work needs to be advanced to maintain the schedule. It also requires the contractor to identify when there are problems that need the utility's attention. Table 10-3 is a set of cash flows. Figure 10-9 is an example of a project cash flow. The project is typical of most pipeline projects, where progress is greatest in the beginning. The utility's finance personnel can plan their investments of reserve or borrowed funds to match the proposed cash flow plan. The last payment is 10 percent.

Construction and risk

Risk is defined as a certainty that an event or condition will happen. The construction manager needs to determine whether or not he/she can turn a potential risk issue into an

TABLE 10-3 Set of cash flows

Project Month	Amount Billed ($)	Amount Retained ($)	Amount Paid ($)	Cum Amt Billed ($)	Cum. Amt Paid ($)
1	56,904	(5,690)	51,213	56,904	51,213
2	85,355	(8,536)	76,820	142,259	128,033
3	113,807	(11,381)	102,426	256,066	230,459
4	113,807	(11,381)	102,426	369,873	332,885
5	170,711	(17,071)	153,639	540,583	486,525
6	341,421	(34,142)	307,279	882,004	793,804
7	455,228	(45,523)	409,705	1,337,232	1,203,509
8	569,035	(56,904)	512,132	1,906,267	1,715,641
9	591,796	(59,180)	532,617	2,498,064	2,248,257
1	603,177	(60,318)	542,859	3,101,241	2,791,117
0	569,035	(56,904)	512,132	3,670,276	3,303,248
11	512,132	(51,213)	460,918	4,182,407	3,764,167
12	512,132	(51,213)	460,918	4,694,539	4,225,085
13	455,228	(45,523)	409,705	5,149,767	4,634,790
14	398,325	(39,832)	358,492	5,548,091	4,993,282
15	56,904	(5,690)	51,213	5,604,995	5,044,495
16	56,904	(5,690)	51,213	5,661,898	5,095,708
17	28,452	(2,845)	25,607	5,690,350	5,121,315
Retainer pad		(569,035)	569,035	5,690,350	5,690,350
Total	5,690,350				5,690,350

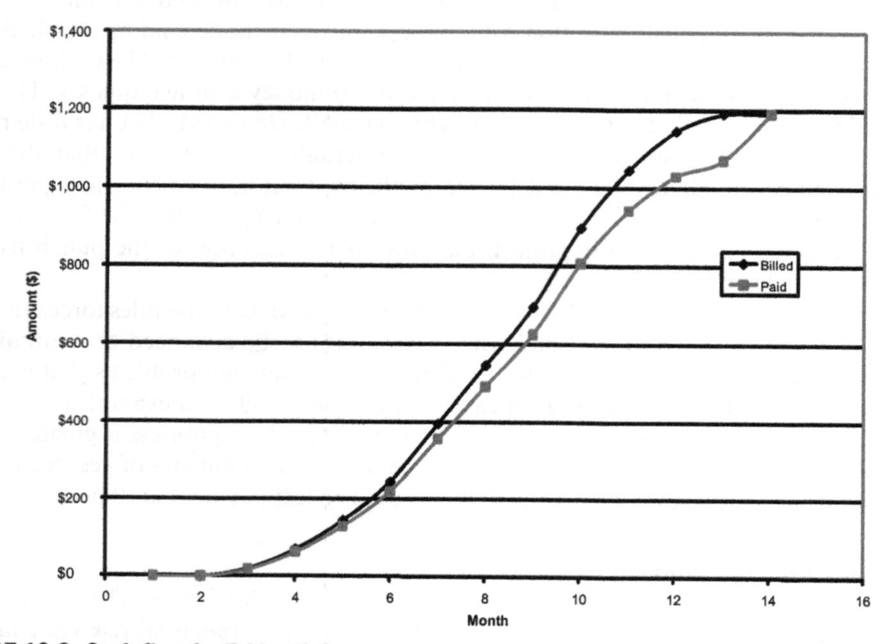

FIGURE 10-9 Cash flow for Table 10-3

opportunity to improve the project. Components of risk include the event that will occur, the probability that it will occur, and the impact in terms of cost. This is very much the same issue as any risk assessment program. From a construction management perspective, the risks are usually either business-related, meaning that the project doesn't get done on time, on budget, or with the quality anticipated; or insurable or pure risk, where the project may not get done on time or on budget, but the utility may make itself whole through insurance or bonds against the contractor. Likewise, if there are damages that occur, such as injuries to people on the site, typically the contract will hold the contractor liable, as opposed to the utility.

The utility needs to determine how much risk is acceptable. Depending on the individuals in the organization, risk tolerances range from low to high. There are people who naturally try to avoid risk at all cost, including regulatory agency personnel and most elected officials and utility managers. In general, utilities are risk-averse and may be extremely risk-averse, whereas subcontractors and contractors are more willing to take on some level of risk. As a result, their prices include costs that are quantified for the risk of the project. The typical sources of risk on a construction project start with the scope and cost, which are usually at the low end of the scale, and migrate through schedule, performance, and quality, to the contract documents and budgets. In most cases, the risk on a construction contract has everything to do with contract matters as opposed to scope of work. Problems may be internal or external, predictable or unpredictable, and there may be some degree of knowledge about problems beforehand. A good example is underground conditions in well drilling. One has very limited ideas of what exactly will happen, even though there may be general ideas of what may be expected in the area. Human error or unexpected problems may create failures in the construction project that may be associated with design or construction.

Typical construction project problems have to do with labor issues, too many concurrent projects, or leadership positions on the site. These are uncertainties associated with unexpected problems as opposed to problems arising from cash flow and other issues that are not anticipated as a part of the project.

There is a balance between risk and reward. Cost, time, and customer satisfaction need to be optimized, but all three cannot be maximized at the same time. This is very similar to the owner attempting to maximize quality, price, and schedule all at the same time; one can optimize them, but one cannot maximize all three. From a risk management perspective, it is important to identify the potential events that may influence performance or the result of the project. Risk can be managed, but the emphasis must be on the objectives that must be made and on addressing them as they come up. During the capital improvement program, most of the risk and opportunity occurs during the initial design phase.

In general, engineers have minimal risk during construction unless they have a defective design, whereas the contractor accepts the highest amount of risk at the beginning of the project when the bid is submitted. The contractor's risk declines with time since more is known about the site, more work is complete, and there is less work to be accomplished.

Managing risk involves identification of likely events that will affect the project outcome, whether these are time-related, personnel-related, or related to acquisition of materials. It is imperative that during the construction process, attempts are made to minimize the potential risk, which can only be accomplished through effective communication. The output of the risk identification process is to identify the risks, their sources, and potential responses and solutions that should be implemented when certain triggers are met.

CASE STUDIES

Perhaps some of the issues encountered in dealing with the client/consultant relationships are best illustrated by two relationships that did not go as anticipated by either the utility or the consultant. One of these problems was created by the consultant; the other was caused by a third-party contractor. In both cases there are actions that could have been taken by the utility and the consultant to rectify the situation and better serve public interests.

Example 1—Wastewater Treatment Plant Construction, Consultant Issues

This example concerns the design and construction of a large wastewater treatment plant for a utility authority. The consultant had five years of experience in doing prior work with the client, including water treatment and wastewater treatment plant expansions. The firm had received over $5 million in fees during that five-year period. The wastewater treatment plant project in question was a new facility with a 24-month construction time and an estimated $10.5 million construction cost. Consulting fees for the project were just over $2 million. The consultant's past experience and working relationship with the utility staff were major factors in their selection. The consultant was not a local firm; it was an out-of-state firm that provided a local presence of no more than two people (which does not constitute a local office). They did, however, get support from local consultants to address some of their limitations.

The consulting services required for the project included design of the facility, contract administration, daily inspection services, services during bidding, negotiation of change orders, and schedule review. The consultant essentially had complete responsibility for the entire project from beginning to end. At the beginning of the project, the utility did not have anyone on staff with significant experience in dealing with projects of this magnitude, and it did not have the staff to do inspection and other services. In fact, the expertise of the staff was so limited that it did not even have the appropriate people to review the engineer's drawings, although after construction was well under way, some expertise was added to the authority staff.

The issue

The construction contract awarded included construction of a large oxidation ditch and two clarifiers. The construction was poured-in-place reinforced concrete, as opposed to precast or prestressed concrete. The walls of the aeration basin were constructed in a predetermined sequence in an attempt to reduce shrinkage problems as much as possible. When the aeration basin was completed, it was filled with water for hydrostatic testing. During a visit to the plant site, the utility staff and its attorney noticed that there was consistent, diagonal cracking every 10 ft (6 m) around the perimeter wall. The direction of the cracking changed at the midpoint of the wall. Water was visibly leaking from the tanks. The consultant was asked to investigate the problem and provide a response. The response was "the cracks will heal" and "it is the contractor's responsibility." The consulting engineer declined any further responsibility for this work.

The utility staff had begun to evolve during construction. The new personnel had more knowledge of construction practices and facilities of this size. Because of the increase in skill of the staff over the construction period, the utility sought the opinion of a second consultant because the cracking appeared to be a design flaw. As is normal practice when a design or construction defect is noted, a second consultant was asked to (a) identify the problem, (b) identify the cause of the problem, (c) identify what needed to be done to repair the problem, (d) identify the cost of repair, and finally (e) assign responsibility for the defect. The second consultant determined that the cracking was caused by insufficient longitudinal steel being placed in the wall—a design problem. Two additional consultants were asked to review the project. Both reached similar conclusions.

In addition, a review was made of the prior two facilities that the design consultant had worked on. Similar cracking problems were noted in each of those, although to a lesser extent. As a result, the utility withheld payment from the consultant pending the correction. The consultant ultimately denied that the remediation was necessary, and sued when the corrections recommended by the second consultant were made under supervision of the second consultant as a change order to the construction contract.

The result

The utility believed that this consultant should have recognized the defect since it was on the site the entire time. In the utility's view, the consultant could have seen the problem and should have suggested corrections and had the work completed. The tanks had been filled for several weeks, so the problem was not new. Despite indications that this was a design problem, it was not uncorrectable, and the utility was looking to the primary consultant to notify it of the problem and take corrective action at its cost, i.e., to take responsibility for the design problem.

The utility should not have relied so completely on this consultant. The utility divorced itself from the design and construction process up to the point that the cracks were noted, which was too late. Ultimately, the result was that the consultant was not awarded any additional work, litigation occurred (which was later settled), and the utility ended up with a tank that had to be repaired prior to being placed in service, the potential reduced life of the tank, and ongoing maintenance costs in the future.

Example 2—Water Treatment Plant Construction, Consultant Issues

The second case is the design of a membrane water treatment facility. This was a project where there were a limited number of consultants in the area with the necessary design experience. The consultant chosen for the project had designed prior plans for similar construction, including one in a neighboring municipality. The project cost was expected to be over $20 million and had a short construction time. The consulting fees were approximately $3 million. The services to be provided by the consultant included design of the facilities, contract administration, daily inspection, services during bidding, negotiation of change orders, and review of the schedule. The utility's staff, while limited, did have some experience with construction projects and was providing periodic inspection and project coordination on less complex projects. As a result, the consultant was receiving some direction from the utility staff of the project on a routine, consistent basis, including reviews of the drawings and discussions with the operational staff.

The issue

The contractor that was awarded the project was very claims and litigation oriented, which created a problem that surfaced immediately. As a result, the contract construction time became unachievable and significant animosity developed on the construction site between the consultant and the contractor's staff. There were several minor things that were not clear in the construction documents, but they were not design defects. The contractor, however, was quick to point these issues out and file claims for time and money due to the lack of clarity in the construction documents. The schedule dragged and delays resulted. The utility management ignored the recommendations of its project management staff to terminate the contractor and decided more personnel should get involved. Finally, the project came in over a year late and litigation with the contractor ensued.

The consultant had on-site staff, but the experience and number of on-site individuals was not sufficient to deal with the significant antagonism that ensued from the start with the contractor. As a result, the consultant got behind in dealing with the claims submitted by the contractor (the contractor submitted as many as 30 requests for information in a given day) and was unable to recover. Additional support personnel were brought in, but because the project was proceeding and there was a start-up time for the additional staff to become familiar with the project, the situation continued to drag further and further behind. Minor items that came up during construction became major items. Discussions between the contractor, the consultant, and the utility's project management and operations staff became strained. Different perceptions of the ongoing problem emerged between the two utility sections. As a result, the utility's staff became split over how well the consultant was managing the staff. The engineering and management staff felt it was being managed, but undermanned; the operations staff was particularly critical. The operations staff also asked for a significant number of changes during construction, which complicated the effort. As a result, the working relationship between the engineers and the operations staff disintegrated.

The result

The project was late and, by the end of the project, fractures existed among all the parties. The utility's operations staff maintained that a portion of the settlement with the contractor was the responsibility of the consultant despite the indication that no design defect existed. As a result, in reviewing proposals for expansion of the facility, the utility's operations staff successfully thwarted the award of the expansion to the consultant, although it was the best qualified to perform. Interestingly, the consultant awarded the expansion project was later fired for failing to follow the design parameters of the first consultant.

The consultant should have addressed the contractor problem more quickly. The utility can be criticized for not speaking with one voice, through its project manager, thereby confusing the situation further.

Lessons Learned from Examples 1 and 2

These two cases provide nearly all the lessons that need to be learned in utility–consultant relationships, including those listed below.

- The utility has the right to expect that the consultant will always deal honestly with the utility.

- The utility needs to be involved in the process, not leave the project completely to outside persons.
- If lacking expertise, the utility has an obligation to the public to secure appropriate expertise.
- The utility must recognize that *no design is defect-free*, and no design should ever be expected not to have field changes.
- The utility has the right to expect that the consultant will take the initiative to correct defects, errors, and omissions.
- In reviewing the case law, any errors that exist within the contract documents should be corrected by the consultant at the consultant's expense, but omissions are items that would have been paid for as a part of the bid. The consultant should keep the client informed of these issues and provide the appropriate documentation that it was omitted and should have been included in the bid. A fixed contingency line item account within the bid will provide an opportunity to pay for these items.
- The utility must appoint one project manager who will speak for the utility. Multiple spokespersons only confuse the direction.

Example 3—Golden Gate Water Plant Performance-based Specifications Installation

Golden Gate City is located approximately 8 mi east of Naples in western Collier County, Fla. Golden Gate City occupies just over 4 mi^2 of developed urban residential and light commercial property. The majority of properties are single-family residential. Commercial development consists of support facilities for the surrounding residents.

In 1999, the Golden Gate water treatment plant (WTP) had a permitted capacity of 1.22 mgd. The monthly average daily flow for the 31-month period ending July 1999 was 1.08 mgd, with a maximum average daily flow of 1.32 mgd. The maximum daily flow was approximately 108 percent of the design capacity. Based on the flow records from January 1997 through July 1999, there was a limited increase in the monthly average daily flow. However, in reviewing the parameters of the water usage, it was clear that the maximum daily flows in any given month exceeded plant capacity, and over the prior three years there was a general upward trend in maximum daily water usage. In comparing the trends of the past three years and projecting over the next seven years, it was clear that both the peak and average daily flows would exceed plant capacity sometime about the year 2000.

In addition, there were several new developments under construction within the existing service area that had requested water service. Numerous people in the western portion of the Golden Gate community desired service, but had no facilities from which to obtain it. These additions would increase the total of single-family customers to nearly 6,000. Thus, it was estimated that the total build-out treatment capacity would have to be approximately 2.5 mgd.

Time was a critical consideration in trying to solve treated water supply problems in Golden Gate City. A plant expansion using ozone was nearing final design, but the cost for the expansion was over $1 million, and the time frame for construction was 12 months. The utility staff and its consultant were directed to evaluate options for providing additional treatment plant capacity quickly, and to make a recommendation for a solution. The use of nanofiltration came to the top of the list. It was determined that a 0.25-mgd nanofiltration skid could be purchased and installed for $210,000, and could be assembled and ready for placement into service within 60 days from authorization. The utility board directed staff and the consultant to create an agreement for the purchase of the skid using the performance-based specification approach.

To this end, a number of assumptions were made about the use of the site. It was thought that an existing building could be used to house the skid and that an existing pump station could pump the concentrate to the sewer system. This would require minimal site modifications and permitting, which were determined to be the limiting factors in providing the capacity. Piping modifications and connections were readily available, minimizing pipe costs. The raw water was assumed to be relatively free of microbial activity, an issue that would have to be evaluated over time.

While the design of site improvements was being completed and construction of the 0.25-mgd nanofiltration skid was going on, it was determined that a new building would have to be constructed to house the skid since it could not fit into any existing structure; thus, modifications to the pumping station would need to be made. This created additional costs, but more importantly created a permitting delay.

The contractor agreed to provide a prefabricated building and replace the existing pumping station. The additional cost was $146,000, but this change would allow the later installation of another 0.25-mgd nanofiltration skid, with only the cost of the skid and minor piping required. Figure 10-10 shows the project during construction. Time for completion, despite a permitting delay on the part of the county building department of 70 days, was less than 90 days. Clearly this could not have been accomplished by the previously proposed expansion of any of the construction method alternatives except the PBS system.

Example 4—Design-Build Water Plant

A city advertised for a design-build project team for the design and construction of a 2.0-mgd nanofiltration water treatment plant on the city's existing water plant site. The building program would involve on-site piping, pretreatment, nanofiltration skids, building, chemical storage, cleaning system, electrical connections, parking improvements, connections to the clearwell, and other ancillary items as contained in two documents produced by a prior consultant. These documents outlined the results of pilot testing of the water from the raw water sources, and a preliminary design of the proposed facilities, including the skids, connections, building floor plan, and piping.

The design-build project team was permitted to propose alternative arrangements for the facilities and equipment shown on the preliminary site plan. The improvements would integrate the new nanofiltration water treatment plant facilities with the current facilities and coordinate construction to minimize disruption to current activities. Wiring for computers and other communication needs, and connection of all instrumentation to one central control center in the new operations building, were included. Parking, landscaping, and other required project elements are also included. The inclusion of US Green Building Council LEED certification parameters was a bonus to the city.

Any construction needed to be undertaken in conjunction with continued operation of the existing water plant, which needed to remain in service. The cost estimate for the project was between $6 million and $8.5 million. The funding source was SRF. Construction was required to be in full conformance with the high-velocity hurricane zone regulations in the most recent version of the Florida Building Code.

FIGURE 10-10 Golden Gate City PBS project (skid built while design was ongoing)

This first phase response was to evaluate the proposals before asking for prices. A city selection committee scored each submittal in accordance with the rating guidelines and could schedule interviews, presentations, or both by the top-ranked firms. The following were the guidelines used for evaluating responses (with associated weighting based on City of Dania Beach Nanofiltration RFP, developed by the author):

1. **Company's expertise (5 points).** Rating is to be based on information provided on experience related to the type of work described within this request for proposals (RFP). Directly related expertise in membrane water treatment plant projects earns a higher rating (5 points) than incidental or limited experience in this type of project. Level of difficulty and the successful overcoming of strategic challenges in similar projects may receive higher ratings. Firms with previous direct work in south Florida may receive higher ratings. Firms with properly defined and functional team and quality control policies may receive higher ratings.

2. **Previous staff experience (5 points).** Rating is to be based on the experience of the staff that will be involved in the day-to-day design and construction of the membrane water treatment plant project. Significant experience in performing substantially the same type of projects may result in more points. Limited staff experience in membrane water treatment projects may result in fewer points. **No staff experience** with membrane water treatment plant construction shall result in **zero points**.

3. **Current and projected workload (2 points).** Rating is to reflect the workload (both current and projected) of the firm, staff assigned, and the percentage availability of the staff member assigned. **Respondents that fail to note both existing and projected workload conditions and percentage of availability of staff assigned shall receive zero (0) points.** Please identify current projects that are ongoing in the same offices as those that will be building the city's project.

4. **Office location (2 points).** Office location is based on ease of proposer to execute any level of the contract work and provide subsequent responsiveness. However, **to receive any points in this rating, the bulk of the work must be performed in the closest local office**. Firms located outside a one-hundred-fifty- (150-) mile radius of City Hall shall receive zero points.

5. **Demonstrated prior ability to complete project on time (3 points).** Respondents shall provide a tentative project schedule for the project described in this RFP in a CPM/ GANTT style schedule. The schedule should include both design and construction portions of the project, and comparisons with projects previously constructed. Respondents shall be evaluated on the logic applied to each timeline, interrelationships between project timelines, and predicted impacts to scheduled projects as well as subsequent know-how in establishing a streamlined and successful delivery process. Respondents will be evaluated based on previous experience in the successful completion of and steadfast conformance to similar project time frames. Specific attention will be given to successful strategic and managerial approaches used to exercise timely project completion, as well as the ability of the respondent to provide full dedicated attention to each workload priority. Respondents that have demonstrated an inability to complete projects on time shall receive fewer points.

6. **Demonstrated prior ability to complete project on budget (3 points).** Proposers will be evaluated on their capacity to establish competitive and technically responsive projects, as well as their capacity to adhere to initial budgets. Comparisons shall be made between initial negotiated task costs and final completion costs. A table that shows initial budget/ award and final cost is the best means to present this. Respondents shall be given the opportunity to explain budgetary overruns and consideration shall be given to scope modifications as a result of expansions or reductions in original scope. Unjustified budgetary overruns shall receive fewer points.

7. **Price (maximum—5 points).** This rating is to be used for valuation of the base price condition minus any deductions in the base price, plus allowances. The price provided is a not-to-exceed base bid. The city recognizes that the proposers may not have a full grasp of the project costs and that a factor will be included to cover unknowns. Partial scores will be given (to two decimal places). Price will be ranked in order of reasonableness (hence bids that are well below expectation will be penalized). Explanations and qualifications for prices should be outlined. Options to limit or reduce costs will be used as deducts if accepted by the city. The project price submitted will be allocated proportionately, and the lowest price submitted receives the total amount of points for the category. For example, since the weight for this category is 6 points, illustrated below is the manner in which the points will be calculated:

   Proposer A: $350,000 = 6 points for lowest reasonable price

   Proposer B: $400,000

   $\$350{,}000/\$400{,}000 \times 6 = 5.25$ points

   Proposer C : $450,000

   $\$350{,}000/450{,}000 \times 6 = 4.67$ points

The committee ranked qualified firms in order of preference, and created a shortlist from which prices were to be solicited (see Table 10-4). In this instance, two firms were ranked well ahead of the others, so the shortlist consisted only of firms A and B.

TABLE 10-4 Design-build results described in example 4

Evaluator/ Firm	1	2	3	4	5	Total	Rank
Firm A	19.00	17	15	19.5	15		
Budget	5	5	5	5	5	110.50	1
Firm B	19.00	17	16	18	16		
Budget	4.344	4.344	4.344	4.344	4.344	107.72	2
Firm C	17.00	17	13	15	11	73.00	5
Firm D	15.00	17	15	16	11	74.00	3
Firm E	17.00	14	14	13	11	69.00	6
Firm F	16.00	15	14	15	14	74.00	4
Firm G	13.00	14	11	11	10	59.00	7

To develop a better base bid, a meeting was held with each of the proposers. The intent was to reach agreement with the city on a plan that would be used as the base bid. The base bids will conform with what was agreed to by the city at the meeting, with explanations as to what is included and what is not. The following were minimum requirements of the base bid.

- Capacity for 140-mph wind loads.
- The existing garage building will be torn down from the south end of the building to the wall depicted on the attached diagram.
- Include either a bridge crane *or* an A-frame gantry and small forklift.
- Plan the building to hold a third skid. The space for the third skid is not required as a part of the building at this time, but the ability to expand the building for the third skid is required.
- Remove hydropneumatic tank.
- Vertical turbines were assumed to be the base bid but split case horizontal pumps may be substituted.
- Lime softening rehabilitation needs to be coordinated.
- Conduits on existing building need to be relocated.
- A 90 percent recovery, three-stage nanofiltration system. A fourth stage is assumed as a future addition. The power capability for this stage is assumed as part of the base bid.
- A minimum building size of 5,000 ft^2 (45 m^2).
- A chief operator's office with a minimum size of 10 × 12 ft (6 × 8 m).
- A lab that may be connected with the operations area, instead of being separate.
- Workshop area under air of no less than 300 ft^2 (28 m^2).
- Access to the lime softening plant site.

- Glass to the process area from the operations area.
- External walls made of reinforced concrete block. The building is assumed to be a prefabricated metal structure.
- Two ADA-compliant bathrooms (male and female) plus showers and lockers.
- Two smaller Lakos laval inertial sand strainers.
- Minimum landscaping requirements of the city.
- Any changes in the site plan that require changes to the management and storage of surface waters (MSSW) permit.
- Generator provided must be able to supply the nanofiltration (NF) treatment process plus high-service pumps plus expansion.
- Painting the existing building to match the new building if the two are used.

The bid sheet, schedule of values, and equipment information were submitted along with a schedule, site plan, floor plan, and elevations. The bid sheet included additions and deducts to the base bid (see Table 10-1). The final contract amount was developed based on acceptance of these items. The apparent low bidder was firm A, as shown in the work breakdown in Table 10-5. The process permitted the city to have concurrence on most of the design issues, site plan, and floor plan prior to the contract being signed, which will accelerate construction.

TABLE 10-5 Summary of comparison of work breakdown from shortlisted bidders

Division	Purpose	Firm B	Firm A	Delta
1	General requirements	$2,035,000	$1,426,745	$(608,255)
2	Site work	$429,500	$487,179	$57,679
3	Concrete	$342,000	$184,231	$(157,769)
4	Masonry	$50,000	$92,757	$42,757
5	Steel	$42,500	$ 74,708	$32,208
6	Wood/plastic	$—	$—	$—
7	Thermal/moisture protection	$9,500	$—	$(9,500)
8	Doors/windows	$45,500	$68,852	$23,352
9	Finishes	$152,500	$138,445	$(14,055)
10	Specialties	$17,000	$16,690	$(310)
11	Equipment	$1,895,000	$2,378,034	$483,034
12	Furnishings	$11,000	$10,675	$(325)
13	Special construction	$800,000	$1,093,467	$293,467
14	Hoists/miscellaneous	$19,500	$37,670	$18,170
15	Mechanical	$1,020,000	$224,297	$(795,703)
16	Electrical	$1,271,000	$913,325	$(357,675)
17	Lighting	$—	$48,848	$48,848
TOTAL		$8,140,000	$7,195,923	$(944,077)

DEVELOPER EXTENSION POLICIES

Not all facilities will be constructed by the utility. Developers will want extensions of the system to serve their developments. These are normally accomplished at the developer's cost, but within certain parameters and with utility approval to ensure that existing customers do not bear the cost of construction associated with new customer growth. To address the issue, a comprehensive water and sewer extension policy must be formulated.

The extension policy is used to establish procedures and policies regarding the extension of water distribution and sewer collection lines into previously unserved areas of the service area, or to areas currently served but that are not served adequately for existing or proposed development.

The main thrust of the policy should be that developers requesting service must pay impact fees up front and construct the infrastructure needed to serve their project. However, provisions can be made to allow for cooperative arrangements between the utility and developers for larger, regional components, such as upsizing agreements, impact fee credits arrangements, reprioritization of capital elements, and development of assessment programs to improve the financing requirements allocated to the future customers.

A set of policies regarding financing of these improvements, and the accompanying responsibility for financing and maintenance, must be addressed in a manner that is neither arbitrary nor capricious. The water and sewer extension policy establishes and presents the minimum utility requirements and standards for development of water transmission and distribution and wastewater collection and transmission facilities within the area served by the utility. The minimum requirements define certain construction criteria that will not aggravate current maintenance demands and will ensure that reliable and economical utility services will be provided to the users of water and sewer systems for all utility construction.

The standards also facilitate the development of water and sanitary sewer systems to meet regulatory and legal mandates of the utility and to provide for water and sewer services to the residents of the service area that has been designed and constructed in accordance with the latest technical and professional standards of the water and sewer industry. Requirements of a water and sewer extension policy should be used in conjunction with, and as a supplement to, local subdivision regulations, land development codes, and growth management plans as they apply to the development and/or subdivision of lands within the utility's service area. An example extension policy is included in appendix F.

REFERENCES

Bloetscher, F. 1999a. What You Should Expect from Your Consulting Professionals (and How to Evaluate Them to Get It). *Water Engineering and Management*, October, pp. 24–27.

———. 1999b. Looking for Quality in an Engineering Consultant. *American City and County*. December, p. 28.

Hewitt, R. 2010. Complex Bid Documents Undermine the Process. *Florida Water Resource Journal*, 62(5):38–39.

Nuechterlein, M.F., L.M. Watson Jr., W.F. McGowan Jr., T.F. Icard Jr., D.E. Hemke, C.J. Cacciabeve, S.D. Marlowe, and R.B. Campbell. 1991. *Florida Construction Law: What Do You Mean....? A Practical Guide for the Construction Industry*. Eau Claire, Wis.: National Business Institute.

CHAPTER 11

Capital Funding Mechanisms

Capital projects occur on a regular basis. Utilities should develop policies for this ongoing funding need. Typically a base level of funding is built into the rate structure through a capital fund, repair and replacement fund, or some similar instrument in the annual budget. Policies for these expenses should be delineated because, in difficult financial periods, a cutback in construction for capital programs is too often politically expedient.

Figure 11-1 shows an example of the variations in capital from year to year. The average is denoted. For the variations in capital expenditures, the utility can do one of three things: (a) accumulate funds as part of a repair and replacement program, (b) borrow the funds, or (c) seek grants (which are usually not available). The horizontal dashed line was added to show where the utility expects to pay for the work without borrowing. The funds above the average would be borrowed. There are a number of borrowing options for utilities.

BORROWING

For large capital projects, most utilities will have insufficient cash on hand to construct facilities as shown in Figure 11-1. Therefore borrowing becomes the appropriate option. There are a number of instruments used for borrowing, from short-term commercial paper and notes to long-term revenue bond and state revolving fund (SRF) loan monies.

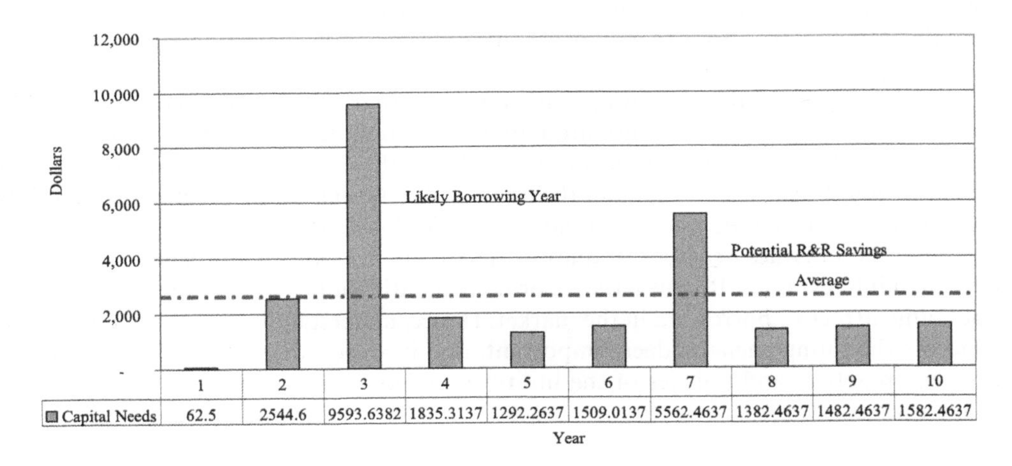

FIGURE 11-1 Example of annual capital funding for a utility

Bonds

Bonds are the most commonly used instruments for borrowing by large local governments. Municipal bonds are tax exempt from federal income and capital gains, although perhaps are not locally or state tax exempt. Between 1989 and 1999 over $2 trillion in long-term debt was issued, with 40 percent of that amount going to local governments, 30 percent to states, and nearly 19 percent to local authorities. Between 7 and 8 percent of the local government and authority debt is issued for water and sewer purposes.

Various types of bonds exist for utility applications, although revenue bonds are normally preferred. General obligation bonds are used by general fund enterprises and typically have numerous caveats associated with their approval (e.g., requiring voter approval). Industrial revenue bonds, housing bonds, and similar types are not appropriate for utilities. Revenue bonds are attractive to utilities because they carry low interest rates, but they rely solely on the revenues of the utility as pledged revenues. The interest rate for revenue bonds will be slightly higher since they are not deemed to be as credit worthy as general obligation bonds, in which the full faith and credit of the local government is pledged. Insuring the bonds can increase their credit worthiness and decrease interest rates.

Revenue bonds can be issued for terms up to, and sometimes beyond, 30 years. However, the service life of the infrastructure may indicate that such a long-term obligation may not be appropriate. Revenue bonds can be used for virtually any purpose for which they are needed, although the official statement must identify the uses of the bond proceeds and detail the repayment schedule. Since a revenue bond issue is for a specific amount of money, one disadvantage to issuing bonds before the project is started or complete is that change orders can modify the amount and leave the utility short of funds. To solve this problem, a short-term note or bond anticipation note (BAN) might be used to fund the infrastructure prior to issuing the full revenue bonds, or additional money can be borrowed.

Where bonds are the chosen option, getting a good interest rate and providing the best possible analysis of the utility are important. Small details can be major concerns. However, flexible terms often allow utilities to lower their debt payments if interest rates drop, or refund the issue and obtain the savings. A financial advisor can keep the utility abreast of the appropriate time to refund or combine bond issues.

Small bond issues or issues of questionable financial backing are hard to place and have high interest rates. Recent lawsuits or financial difficulties can prevent utilities from obtaining attractive borrowing in the market. Hence, understanding and practicing those concepts the rating agencies deem important, and incorporating them into the ongoing financial practices and policies of the utility, are of long-term benefit to the utility (see chapter 4).

Issuing Bonds

A variety of people are involved in the issuance of bonds (Nickerson, 1995), including

- Financial advisor
- Underwriter
- Bond counsel
- Disclosure counsel

- Engineers
- Rate consultants
- Utility staff

Financial advisor

The financial advisor is normally one of the first people hired when a utility wants to issue bonds to borrow money. The financial advisor should be a person or firm with specialized experience in dealing with public-sector debt issuance, and will have contacts within the bond and underwriting community. The financial advisor will help the local utility by providing insight into how the borrowing should be structured, the timing of the bond sale, and coordinating the sale of the bonds to investment banking firms. The financial advisor will make a recommendation on the award of the bonds to one or more investment bankers based on competitive bids or negotiated sales. Usually the focus is the repayment term and interest rates the investment banker will command.

Regardless of the sale methodology, a significant amount of work must be coordinated. The paperwork portion of this work is developed into an official statement. The official statement identifies the issuing entity, what the entity plans to do with the proceeds, and the entity's financial condition, ability to repay the debt, growth patterns, and other items.

While it is not mandatory to hire a financial advisor, the benefits of doing so include the technical assistance that can be provided in the development of the official statement, the evaluation of proposals from investment banking firms, and having an expert who can negotiate with the underwriters while having a fiduciary responsibility to the utility. Since few local governments have experts of this type on staff, the financial advisor provides the expertise and takes the responsibility to make sure the best efforts are made on behalf of the utility customers. There is a price for the financial advisor, although on large issues, the cost is small.

Underwriter

The underwriter is the firm that will purchase the utility bonds and then redistribute them to investors for a fee. The underwriter ensures that the bonds will be purchased, although it usually purchases them from the utility at a discount so resale is easier. For small issues where a financial advisor is not retained, the underwriter can perform the financial advisor function (for no fee beyond the normal underwriter's fees), but the underwriter has no fiduciary responsibility or obligation to the utility. Long-term relationships are important, so most underwriters try to be fair and helpful to issuers.

Bond counsel

Bond counsel is a lawyer or more commonly a firm of lawyers with specialized experience with the issuance of debt. Many bond counsel firms specialize only in tax-exempt public offerings. Bond counsel is retained by the utility and therefore has a responsibility to accomplish the transaction in the terms most favorable to the utility. Bond counsel provides guidance and oversight for the transaction and drafts the resolutions and other legal documents included in the official statement. Bond counsel also has the responsibility to render an opinion about the tax-exempt status of the bonds for governmental utility systems.

If the utility has counsel, the underwriter will have its own lawyers to protect its interests. Like bond counsel, the underwriter's counsel is also an expert in the issuance of debt obligations. Underwriter's counsel, like the underwriters, have no fiduciary responsibility to the utility. Underwriter's counsel usually is responsible for drafting the official statement.

Disclosure counsel

A third set of lawyers that may be involved in debt transactions is the disclosure counsel. Because the official statement is signed by the governing officials of the utility, many local governments are using their own disclosure counsel to draft the official statement and debt purchase agreement. The disclosure counsel provides an opinion to the utility and a letter of reliance to the underwriter. The opinion is referred to as a *10B-5 opinion* and protects the utility and its officials from section 10B-5 of the Securities Act. The letter of reliance provides comfort to the underwriters that the official statement contains no statements known to be untrue or misleading, nor does it omit any factor or data that might materially affect the actions of the underwriters. Most underwriters will accept the 10B-5 opinion, although it may not mean that the underwriter does not retain its own counsel.

Engineers

Engineering consultants are employed to prepare a report that outlines the condition of the utility system. The condition of the utility system includes evaluation and inspection of the system assets, a review of the system operations, permit compliance, and deficiencies that may need to be addressed. The engineer should opine as to the appropriateness and reasonableness of the proposed debt expenditures, as well. Because many assets are hidden from view (i.e., buried), the engineer will normally have a series of caveats to his opinion, unless he is familiar with the system. Engineering reports prepared by engineers internal to the utility are not viewed favorably because the bond market is looking for an independent viewpoint, not the view of an insider.

Rate consultants

A rate analyst will normally be retained to evaluate the sufficiency of the revenues to generate enough funds to permit the utility to repay the debt. The rate analyst will normally perform a multiyear rate forecast based on prior performance of the utility, and make recommendations as to rates that should be imposed to meet the debt repayment needs. Coordination between the engineer and rate analyst are important as the system conditions and deficiencies might impact the financial solvency of the system (such as major improvements needed to meet upcoming permit changes).

Utility staff

Aside from all of the consultants, the staff of the utility must be heavily involved in the transaction. The finance staff is best positioned to answer questions about utility billing policies and accounting practices. The operations personnel can best explain how the utility addresses the repair, maintenance, and facility operations programs. The staff engineers and operations personnel are best positioned to render an opinion relating to compliance with permit conditions and the status of ongoing capital construction. Utility administration personnel and the "utility" attorney can address compliance by the utility with local laws, ordinances, and policies, and state and federal regulations. The utility's attorney can also add information on policies and procedures of the governing board, and submit resolutions and documents to the governing board to secure the loan.

Finally, the governing board must approve the contracts for all consultants, approve the resolutions required to issue the bonds, and issue the final resolution approving the award or sale of the bonds. Once all the documents and resolutions are completed, a date is set for the formal closing of the transaction. The person who chairs the governing board, the clerk to the governing board, and the utility's attorney must execute the final borrowing documents at the bond closing. Representatives of the consultants also usually attend the bond closing.

Commercial Paper

Commercial paper is used for short-term financing (one to five years) or anticipation of some other borrowing or revenue source. Commercial paper is rarely rated, and the requirements may be less onerous to the utility. Short-term paper options might be tax anticipation notes (TANs), revenue anticipation notes (RANs), and bond anticipation notes (BANs). These notes are usually very short term and will be paid off fully in a matter of months. BANs are used to fund capital projects when the utility does not wish to risk change orders. The bonds can be sold for precisely the debt required after the project is complete, which solves one disadvantage to using bonds early in the construction process. The requirements to secure commercial paper and notes is significantly less than for selling revenue bonds, albeit they are not rated and the interest rates are higher. Access to large sums of money may be limited.

Letters of credit are one- to five-year notes usually negotiated with a local bank. The paperwork to complete the transaction is limited in comparison to a bond issue. The local bank will normally hold the paper until it is paid off. A later bond issue may be used to pay off the letter of credit. Many banks will negotiate lines of credit with local governments.

SRF Loans

State revolving fund loan programs grew out of the 1970s' 201 grant program administered by USEPA via Section 201 of the Clean Water Act. During the Reagan administration, grants were transitioned to low-interest loans, and the responsibility for program implementation was delegated to the states. Most Americans expect the federal government to address defense, public works, and regulations associated with health and safety (Kotlin, 2010). However, the federal government provides significant monies toward state revolving programs (80 percent federal with 20 percent state match).

Each state has rules for its SRF loan program focused on funding water and wastewater improvements. In most states, much flexibility is built into the program. Those activities that may be funded by SRF loans include the following:

- Land use, including easements and rights-of-way that will be used for the ultimate disposal of water or wastewater treatment plant sludge (or other residuals) or will be an integral part of the water or wastewater treatment process
- Construction and related procurement, the contracts for which are executed after a loan is made or if an authorization to incur construction costs has been given
- Demolition and removal of existing structures

- Contingency for project cost overruns
- Legal and technical services after bid opening
- Project start-up services
- Loan repayment reserve
- Interest accruing during construction for payments under binding agreements between a local government and contractors, manufacturers, or suppliers
- Multiple-purpose facilities, such as those for sludge and solid waste handling resulting in cogeneration of power

Project costs usually not allowed include the following:

- Acquisition of all or part of existing wastewater treatment works
- Facilities not in conformance with an approved facilities plan
- Facilities not included within the approved project scope
- Construction using local government personnel (except in emergencies)
- Facilities or services for which the requirements of the facilities plan are not met

To receive consideration for a loan, the local government should submit a complete loan application to the appropriate state agency. Loans for project contingencies, at the time of loan approval, generally do not exceed 10 percent of the estimated sum of the costs for allowable land (when the actual costs are unknown), equipment contracts, materials contracts, and construction contracts. The contingency remaining after accounting for contract change orders is eliminated when project closeout occurs and is returned to the state. Contingency funds cannot be used to purchase equipment or pay for construction work not described in the loan agreement. Increases in the amount of the planning allowance initially included in a preconstruction loan (where allowed) are also not usually permitted.

The loan repayment period for construction loans is limited to 20 years under the Clean Water Act. Preconstruction loan repayment periods are limited to 10 years. However, when a construction loan is executed to finance projects that have been included on the fundable portion of the priority list for a construction loan, after facilities have been planned and designed under a preconstruction loan, the local government usually has the option of converting the repayment period to that negotiated for the construction loan. Requested increases to the design and administrative allowances initially included in a preconstruction loan are made only in conjunction with a construction loan that provides funding for facilities planned and designed under the preconstruction loan. A project or any portion thereof that is financed with an SRF loan cannot be refinanced at a lower interest rate via a subsequent SRF loan or bonds.

While the nature of a project will determine specific requirements, the general requirements for a construction loan are listed below.

- A cost-effective facilities plan with environmentally and financially sound facilities consistent with local comprehensive plans
- Provision for public participation in the planning process
- Establishment of how the loan will be repaid

- Demonstration that the necessary project sites have been acquired
- Design facilities consistent with the planning recommendations
- The necessary permit(s) to enable construction

The following outline the requirements for the determination for readiness to proceed:

- Complete facilities plan adopted by the local government
- Evidence that the 30-day public comment period has expired (following publication of the results of the state's environmental review process), and environmental concerns, if any, identified during the 30-day comment period have been resolved
- Biddable plans and specifications that include evidence that the plans and specifications are in conformance with the facilities plan
- Value engineering, if required
- Notice of intent to issue the construction permit
- Certification of availability for all project sites necessary for the purpose of construction, operation, and maintenance over the useful life of the facilities

Once the utility is ready to proceed, revised construction and preconstruction loan priority lists are adopted at a public hearing each year. The lists become effective after adoption. The priority for funding is based on a priority score that considers public health effects and the potential for surface water and groundwater pollution control. The quantity of existing flow that will be collected and treated, or that otherwise generates the need for the project, is used as a further priority determinant. An example of the priority score information for a state revolving fund program is included in Table 11-1. Special consideration shall be given to projects that will assist in the restoration or protection of a sensitive water body.

The funds available to a local government for all its projects listed on the fundable portion of the construction loan list is limited to no more than 25 percent of the funds allocated each year in most states. Projects are listed on the fundable portion of the construction loan priority list only if sufficient funds are expected to be available. When a project is segmented and included on the fundable portion, the unfunded remainder is placed in the contingency portion to be financed with funds allocated in the next and, if appropriate, successive fiscal year(s). A project may consist of various facilities, all of which qualify for the fundable or contingency portion. Projects included on the preceding year's contingency portion advance, maintaining the same relative order, to the fundable portion when the annual updating occurs. However, only projects qualifying for fundable portion listing may advance.

There is no limitation imposed on the amount of funding assigned to contingency portion projects. A project shall be segmented for deferred funding of the unavailable funds when a local government qualifies for funding in excess of that available to it in any one fiscal year. The extent to which segmented projects qualify for deferred construction loan funding is evaluated each year. Projects with equal priority scores are subranked and may be listed on the fundable or contingency portions. Other projects may be those consisting of facilities that qualify for the planning portion as a result of differences between the projects in readiness to proceed. The scope of a fundable or contingency portion project described on the construction loan priority list is not increased to encompass additional work except where such increases have been subject to the prioritization procedures.

TABLE 11-1 Example priority score calculation for an $820,000 infiltration/inflow project for SRF loan priority ranking purposes

7. Project Category (Base Score)	
Reduce documented public health hazard (500 points).	
• Must be annotated and signed by the Director of the County Health Department.	
• Examples: elimination of failing septic tanks or failing package plants; elimination of SSO	
Protect groundwater or surface water (400 points).	
• Examples: upgrade in level of treatment, elimination of SSO (in the absence of a documented public health hazard), laws requiring the elimination of a discharge from specific water bodies, I/I correction.	
Promote reclaimed water or residuals reuse (300 points).	
• Examples: constructing new or expanded reuse facilities for beneficial reuse of treated wastewater/stormwater or residuals.	
Compliance with enforceable standards or requirements (200 points).	
• Examples: state or federal Consent Orders or Administrative Agreements, temporary operating permits with enforceable schedules and requirements, or new state or federal regulations.	
All other (100 points)	
• Example: treatment plant rehab/expansion, major sewer rehab/replacement, new collection sewers, or new transmission mains.	
8. Cost-to-Benefit Index	
Project priority scores (base category scores adjusted for water body restoration and protection) are multiplied by an index as detailed below. Note that Bureau of Water Facilities Funding personnel are available to assist in the computation of the index.	
• Total project cost (item 6 above) in units of $1,000 (e.g., $1,000,000 is 1,000) expressed to the nearest whole number is	$820
• Highest base category score (from item 7 above) multiplied by the water body restoration/protection factor (from item 8 above) is	400
• Cost-to-benefit ratio (item a divided by item b) expressed to nearest 0.01 is	2.05
• Natural logarithm of cost-to-benefit ratio (item c) expressed to the nearest 0.001 is	2.33
• Index computed as [1.20 – 0.021 × Natural logarithm] (from item d) and expressed to nearest 0.01 is	1.19
9. Priority Score	
• Base Score from 7 above.	400
• Restoration and Protection of <u>Special</u> Water Bodies multiplier from 8 above.	1.00
• Cost-to-benefit Index from 9 above.	1.190
• Small Community Economic Hardship points from 10 above.	0
• Priority Score is (a × b × c + d) or if default minimum is selected (100 + d).	4,760

Source: Form 62-503.900(1) (FDEP, 2007).

The revenue source to be used to repay the loan must be identified. In general, pledged revenues are those resulting from the operation of water systems or water and wastewater systems. Pledged revenue should not be less than 1.15 times the amount required to make semiannual loan payments. In addition, for utilities that have not demonstrated the ability to service long-term debt, special loan security provisions can be negotiated to provide assurance toward fulfillment of debt service requirements. Such assurances include the following:

- Additional escrowed reserve funds (equivalent to, but not less than, five semiannual loan repayments) and a lien on the assets of the entity
- A letter of credit from a bank or trust company and a lien on the assets of the entity
- A personal or corporate, as applicable, obligation ensuring that all semiannual repayments can be made
- Other security features equivalent to those previously described

Standard loan agreements are set up for a pledge of water and sewer utility revenues.

MUNICIPAL LEASING

Municipal leasing is an option for securing funds for the use and possible acquisition of capital equipment and facilities (Vogt et al., 1983). Interest in municipal leases increased after the Deficit Reduction Act approved by Congress in 1984. Vehicles and construction equipment, both significant assets of utilities, were among the first assets to be leased or leased-purchased because the leasing arrangements allowed the cost of acquisition to be spread over a number of years without the need to go through extensive negotiations for bonds or other borrowing.

Lease agreements can be relatively simple, and can be entered into in a relatively short period compared to the bond market. Since leases are typically year-to-year, the agreement can be terminated and the assets returned if significant financial stress affects the utility, a benefit when the economy may have significant uncertainty. The financial condition of the utility is rarely a significant issue when entering into a lease—the allocation of monies in the budget is what is required.

Leases require two parties: (a) the lessee, which is the entity acquiring use of the asset, and (b) the lessor, which owns the asset and is allowing the lessee to use it in exchange for some financial benefit. Other participants may be involved in the transaction, depending on the complexity of the program. For instance, a lender might also be involved if the lessor does not fully own the asset or needs to acquire it in order to lease it. A broker might be used to package one or more "deals." Lawyers will be present, regardless of the complexity of the transaction. It might also be useful to include rate analysts, financial advisors, and engineers in the transaction, depending on what is being leased. An underwriter might also be involved when significant borrowing is required to pull a complex transaction together. This is usually the same group that is involved in bond issues.

Leases can be set for virtually any term or interest rate, and in the case of certain assets, the asset may never be acquired by the utility, but instead exchanged for new assets at set periods. Vehicle leases of this nature are common, as are such arrangements for computers. Hence the utility could obtain new vehicles or computers every three years, while using a multiyear leasing arrangement obtained through the bidding process. This type of arrangement has the potential to reduce personnel costs for maintenance of assets that never approach their life expectancies.

Leases have a number of other benefits. In politically charged environments, leases can be an option if bonds are unfavorably perceived by the public, or where special equipment is needed for a limited period of time to make changes or upgrades to treatment or transmission facilities. Leases are rarely a mechanism to save on interest rates because such rates are often higher than those on the general obligation or revenue bond market.

This is because leasing arrangements are not tax-free like the interest on bonds is for governmental entities, although there are certain tax benefits that Congress has created for municipal leases. But the term can be much shorter.

Temporary situations, like office space, are common lease arrangements. Leases may also be attractive where technology is changing rapidly and the utility does not want to "get stuck" with antiquated equipment within a few years of acquisition (computers might be an example).

There are several types of leases. A true lease is one where the utility would acquire the use but not the ownership of an asset. Office space in a commercial building is an example. The federal tax laws may provide some benefit with regard to accelerated depreciation and investment credits to the lessor if a utility were to lease office space in the lessor's building. However, the lease payments would be taxable income. The tax arrangements may affect the cost of the lease.

Lease-purchase arrangements are agreements whereby at the end of the lease, the lessee would own the asset. In the interim, the lessee obtains use of the asset. An example might be a backhoe that the utility acquires via a lease-purchase agreement over three or four years. The utility would pay monthly lease payments to the equipment company. It would not own the backhoe until the last payment was made. If the utility failed to appropriate funds to pay for the backhoe, it would simply return the asset to the lessor and terminate the agreement. Where this is permitted, the interest rate is generally higher to protect the lessor. Lease-purchase agreements also normally include a principal and interest portion of the lease payment so that if the lessee wants to pay off the lease and take ownership of the asset at any given time, it can do so at a defined amount in the contract (as opposed to negotiating an amount).

A leveraged lease is one that includes a third party that is a lender to the lessor. Under this arrangement the lender loans the lessor money to acquire an asset that is intended to be leased to the lessee. The lease payments are used to repay the lender. An example might be an arrangement where a small treatment facility is built under a design-build contract (the design-build contractor would then be the lessor), and the facility is leased by the utility. This type of arrangement is used for larger projects that may have multiple users.

A sale-leaseback is one where the utility might sell an underutilized facility (like an unused or only partially used warehouse) to a private party from whom the utility then leases only the space it needs. This option may be attractive to a utility if it needs money for pressing projects and/or the facility it is only partially using is expensive to maintain or is shortly to be phased out. The sale-leaseback allows the operations costs of the asset to be moved to another party. Often buildings may need substantial changes to meet new technological guidelines. It may be easier to construct a new building that includes these amenities than try to retrofit an existing building and keep it operational at the same time. A sale-leaseback of the existing structure would be appropriate in such a situation.

Certificates of participation (COPs) are tax-exempt lease-purchase arrangements where there are fractional interests in an asset. A regional well field, such as the one that Broward County, Fla., installed in the early 1990s, that benefits a number of municipal water systems, is an example where COPs might be appropriate. COPs are placed privately by lenders in fixed denominations. COPs often look very much like revenue bonds, but rarely would the lender want the asset back or be able to derive any real benefit from it.

Lease agreements are further subdivided between capital and operating leases. These divisions are distinguished for accounting purposes by the term of the lease in comparison to the life of the asset. Operating leases typically have terms significantly shorter than the life of the asset, while capital leases do not. Leases may be defined as (a) net leases, where the lessee is responsible for all operations, maintenance, tax, and other costs, excepting the repayment of any loans to acquire or build the asset, and (b) master leases that allow the lessee to add to the property under certain conditions.

REPAIR AND REPLACEMENT (R&R) FUNDS

Aside from borrowing money, it is suggested that the utility create revenues sufficient for the operation and maintenance requirements of the system. This requires that the utility provide for capital costs associated with renewing the current investment in the utility system on an annual basis. To avoid large deferred maintenance costs that often can only be funded via large bond issues, provision for renewal and replacement of the existing system should be undertaken. In this manner, future customers will not be burdened with the cost of replacing lines and plants for which they (or their predecessors) have already paid.

Annual renewal and replacement fund allocations should be calculated not only on the treatment plants and transmission system, but also the cost for replacement of older subdivision infrastructure, including old water mains and sewer lines, small pump stations, equipment, and other minor, capital expenditures that occur on a continuing basis, including service lines and taps. The asset management information system described in chapter 4 is the basis of R&R calculations (defined in chapters 9 and 10). This may mean that a sizable amount of money is collected and held in the repair and replacement fund for such things as major improvements in a wastewater treatment plant or upgrades to a major transmission line or pump station. Historical data may not represent a realistic value or amount of money required to be set aside, particularly if maintenance of the system has been neglected or deferred over time.

Provision for continuing renewal and replacement accounts for major facilities must be incorporated into the rate structure. The accumulation of funds for large projects requiring a significant commitment of financial resources will minimize funding from external sources, thereby minimizing any effect on rates to the customer. Over time, all customers will benefit because accumulated funds will be available for the renewal and replacement of infrastructure. In many older communities the use of services to slip-line sewer lines reduces infiltration to a point that plant expansions can be deferred. However, replacement funds should only be used to repair and replace existing infrastructure, not for growth-based facility needs. Growth-based customer needs should be met with impact fees.

REFERENCES

Kotlin, J. 2010. Demographics and Destiny. *Governing*, 23(5):38–39.

Nickerson, G.H. 1995. Issuing Bonds—Who Are These People and What Are They Doing Here? In *Proc. American Water Works Association Annual Conference and Exposition*. Denver, Colo.: American Water Works Association.

Vogt, A.J., D.R. Duven, S.H. Owen, and L.A. Cole. 1983. *A Guide to Municipal Leasing*. Chicago, Ill.: Municipal Finance Officers Association.

CHAPTER 12

Cost of Service Delivery

COST OF SERVICE

Previous chapters in this book have outlined the ongoing operations and capital components of utilities. Utility operations must be supported with budgets. As Figure 1-1 in chapter 1 indicates, finance and budgeting are major *support* services to the utility. Budgets are planning tools for the expenditure of funds. Determining the budget needs is an annual process, but there should be an understanding that the actual numbers may deviate, in some cases significantly, from the plan.

There is also a tendency among elected officials and finance and management personnel to compare the operation of their utility to that of utilities surrounding them. This can be problematic for a number of reasons, among them that different utilities may have vast differences in operations costs, and economies of scale always show that larger utilities are more cost-effective to operate than smaller ones per unit of production. The best way to demonstrate the methodology to compare utilities and the expected results is to provide examples. Fortunately there are two overlapping examples that illustrate the basic principles.

Example—1997 South Broward County Study

The first example is a comparative analysis of the operations of all southeast Broward County, Fla., utilities that was undertaken in the summer of 1997 (Bloetscher, 1998). South Broward County was used because all of its utilities are similarly situated, i.e., most are owned by a local government with full responsibility for providing water and wastewater services to their residents. The area is relatively uniform with respect to age, income, and ethnic mix of the population, and has similar environmental and business climates. Because no systems had received complaints about services or encountered significant problems with regulatory agencies over the previous four years, it was assumed that the services delivered met the needs of the residents. No previous financial comparison had been conducted. At the time there were two small private systems in the service area, both significantly smaller than the public systems. The study had, as a goal, a review of the argument for economies of scale by large, more regional, service providers.

To make comparisons between systems, the following information was secured from the utilities:

- Audits for fiscal years ending September 30, 1994 and 1995
- Budgets for fiscal year ending September 30, 1996, and proposed budget for 1997
- Consumptive use permits and backup data
- Wastewater treatment plant information, where applicable
- Rate schedule

- Most recent rate study, summary of customer information, usage, and customer type
- Maps of water and sewer systems showing pipelines, pump stations, tanks, and treatment facilities, where applicable
- Recent bond documents (official statements) for utility issues (later than 1990 only), if applicable
- Debt service schedule
- Capital improvement plans
- Organization chart with number of employees

These data were expected to provide significant insight into the following:

- Cost of service delivery from operations for water treatment, wastewater treatment, water distribution, wastewater collection, and administrative costs, excluding capital construction
- Rate comparisons
- Analysis of system condition, age, and deferred maintenance needs
- System inventories
- A set of comparative statistics that could be graphically compared

One utility in the city of Hollywood is nearly four times larger than the next-largest system, and over ten times larger than the smallest system. In addition, much of the wastewater treatment is performed at Hollywood's wastewater treatment plant, so wastewater comparisons had to be evaluated carefully. The age of the systems and their underlying maintenance needs were also considered, as opposed to strictly monetary comparisons. One would expect the newer systems to have fewer maintenance needs, and hence, fewer costs than the older systems. Based on these statistical parameters, the following information was evident.

- A graph comparing the average daily water and average daily sewer flows demonstrated consistency between water and sewer use, which confirmed that there were no extreme users in the service area (see Figure 12-1).
- The comparative statistic that provided the most useful information was the cost per thousand gallons.
- For water treatment, water distribution, sewer collection, and wastewater treatment, the graphics clearly demonstrated the economy of scale of the larger utility operations versus small-scale operations (see Figures 12-2 through 12-5).
- The administrative costs as a percentage of the total budget parameter also demonstrated the economy-of-scale argument that larger utilities can perform tasks at a lesser cost per unit than the smaller utilities (see Figure 12-6 and 12-7).

Having reviewed the operations costs, the next step was to review the existing rates. Given the economy of scale apparent in Figures 12-2 through 12-7, it was expected that there would be a tendency for smaller systems to have higher rates. Using the rate ordinances for each of the utilities, the cost for a 7,000-gal/mo user was derived. The 7,000-gal value was just above the average monthly use for a single-family user in south Broward County. Figure 12-8 outlines comparative single-family unit costs for the utilities. The

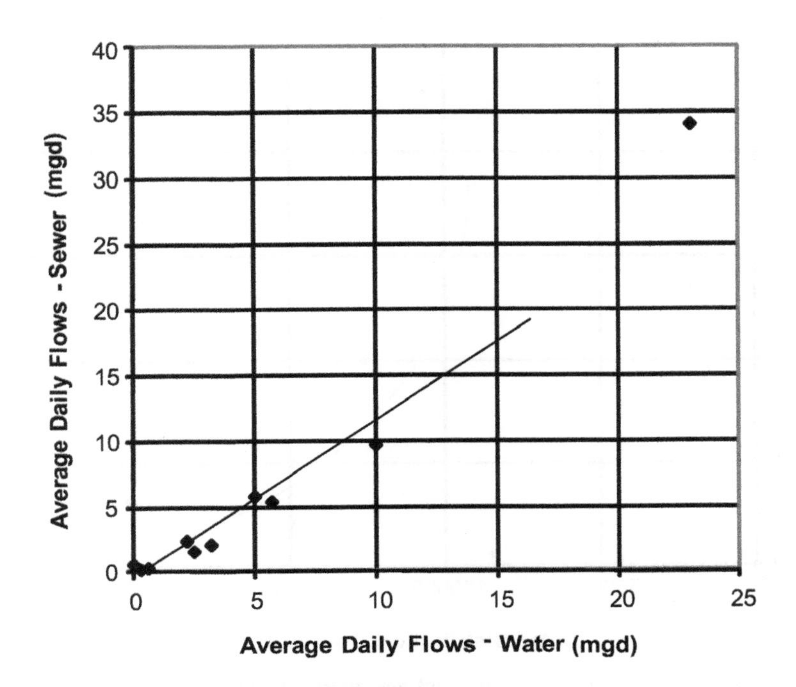

FIGURE 12-1 Average daily flow rates—water versus sewer

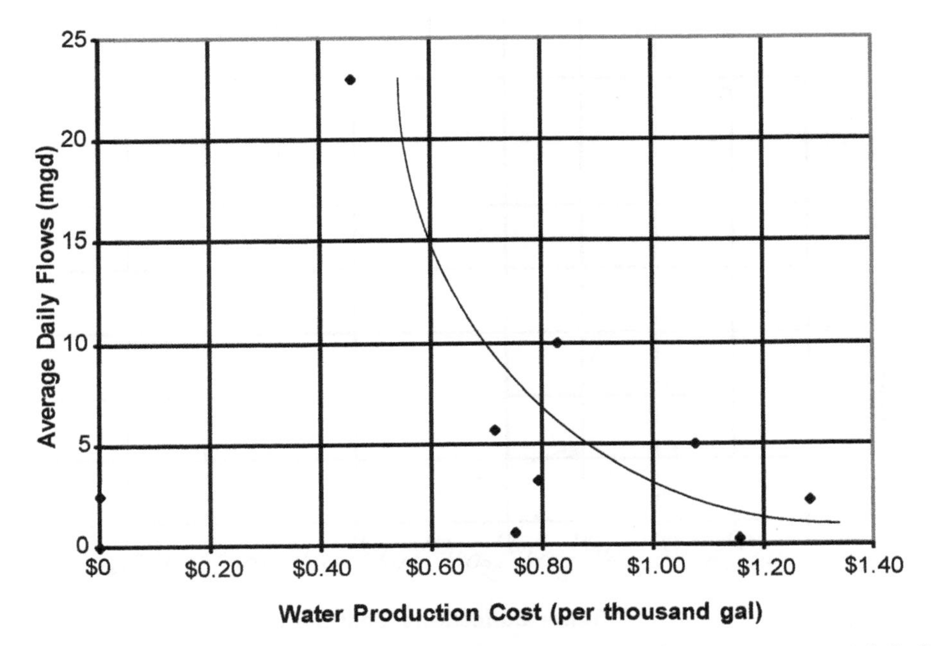

FIGURE 12-2 Water treatment plant operating cost per thousand gallons versus average daily flows

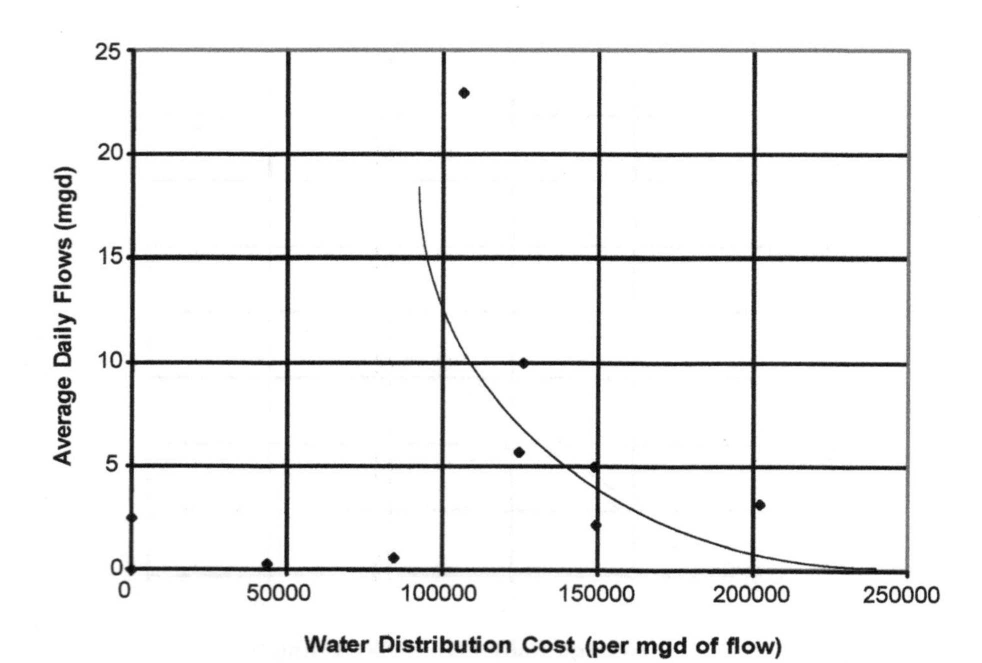

FIGURE 12-3 Water distribution operating cost per thousand gallons versus average daily flows

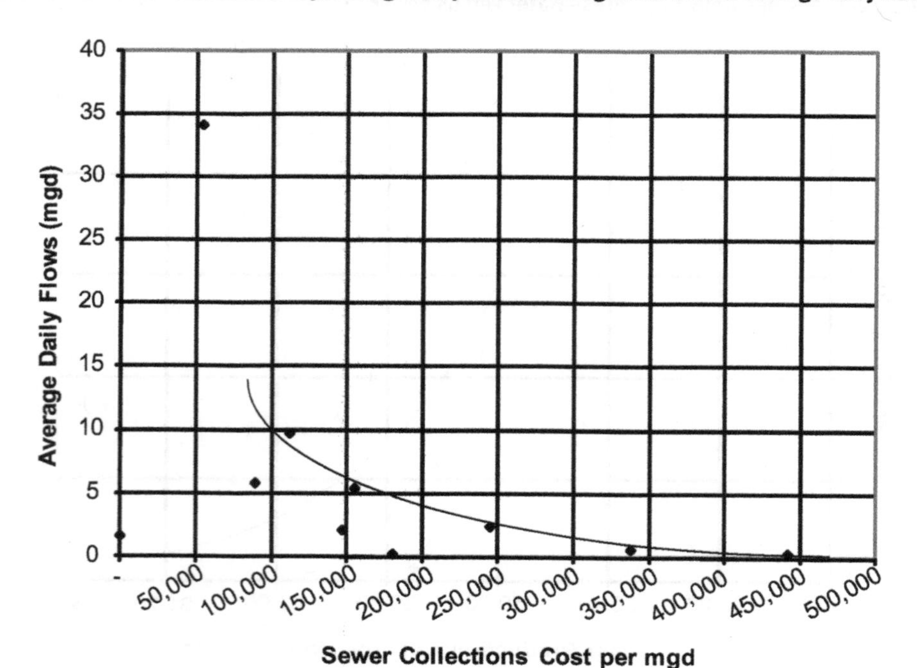

FIGURE 12-4 Sewer collection operating cost per thousand gallons versus average daily flows

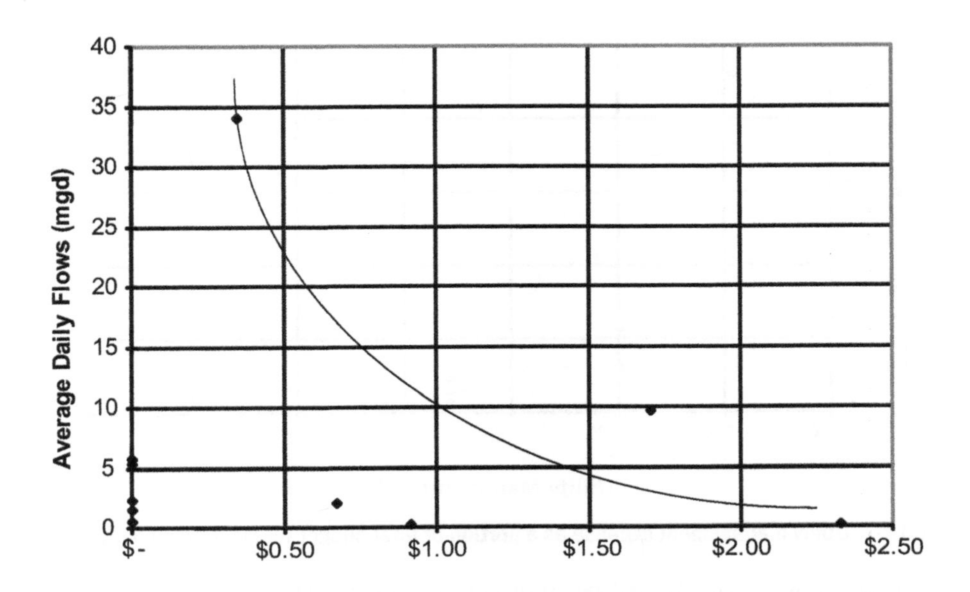

FIGURE 12-5 Wastewater treatment plant operating cost per thousand gallons versus average daily flows

25
20
15
10
5
0
Average Daily Flows (mgd)
$-
$0.50
$1.00
$1.50
$2.00
$2.50
Utility Administration Costs (per thousand gal)

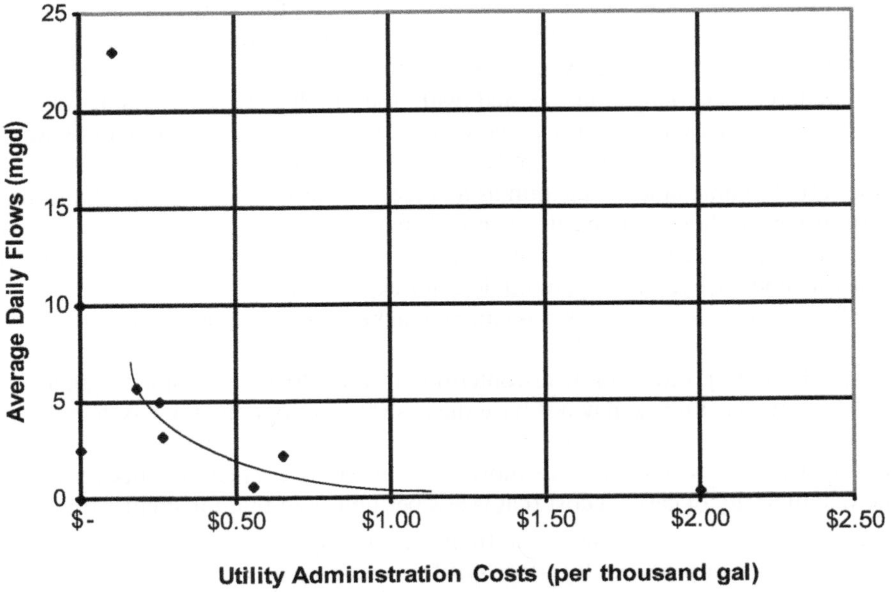

FIGURE 12-6 Utility administration operating cost per thousand gallons delivered versus average daily flows

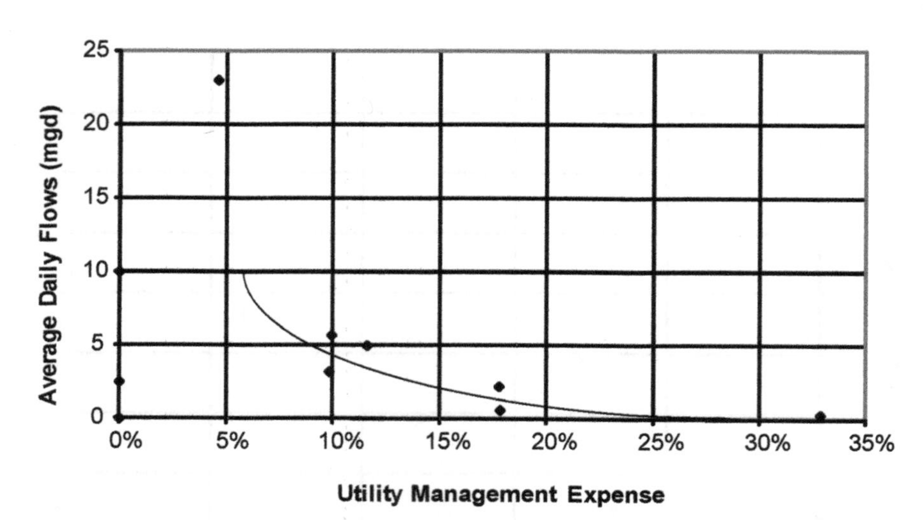

FIGURE 12-7 Utility management expense as a portion of total budget versus average daily water flows

surprising result was that the cost per thousand gallons for Hollywood, Hallandale, Pembroke Pines, and Dania all were within $2.00 of one another at the time of the analysis, despite Hollywood's much larger size. The Davie costs were significantly higher due to debt service on two utilities acquired in the past that have not seen a broad enough customer base over which to apply the debt.

If the rates are the same, but operations costs are higher for the smaller utilities, then one of two causes exists: (a) differences in the funds transferred to the general fund or (b) lesser amount of reinvestment in the system. An attempt was made to find out if substantial deferred maintenance obligations existed with the smaller systems. Using the concept explained in an earlier chapter for estimating deferred maintenance, an analysis was made. Based on the model developed, Figure 12-9 shows that the smaller utilities tend to have a greater deferred maintenance obligation as a percentage of total plant assets than the larger utility. In reviewing those systems, it was noted that indeed, they do have old infrastructure that is mostly depreciated. The two major outliers are newer cities that do not have very old infrastructure and have recently spent millions for expansion.

The following general conclusions can be reached for the systems evaluated.

- The study supports the long-held contention of the USEPA and other regulatory agencies that smaller utilities may not have the cash flow to operate their systems efficiently. Because operations tasks are similar, larger systems can develop the appropriate resources to deal with operations more efficiently than smaller systems, resulting in, in the vast majority of cases, a cost that is less per unit than that for smaller utilities.
- The newer systems had a lower cost than the older systems, as measured in the cost to maintain miles of pipe, but this does not translate to plant operations costs.
- The smaller utilities usually are accumulating enough funds to handle their operating requirements but insufficient funds for reinvestment in the system.
- The larger systems raise significant revenues for the construction of improvements on the system (via repair and replacement funds in the current rates, or borrowing).

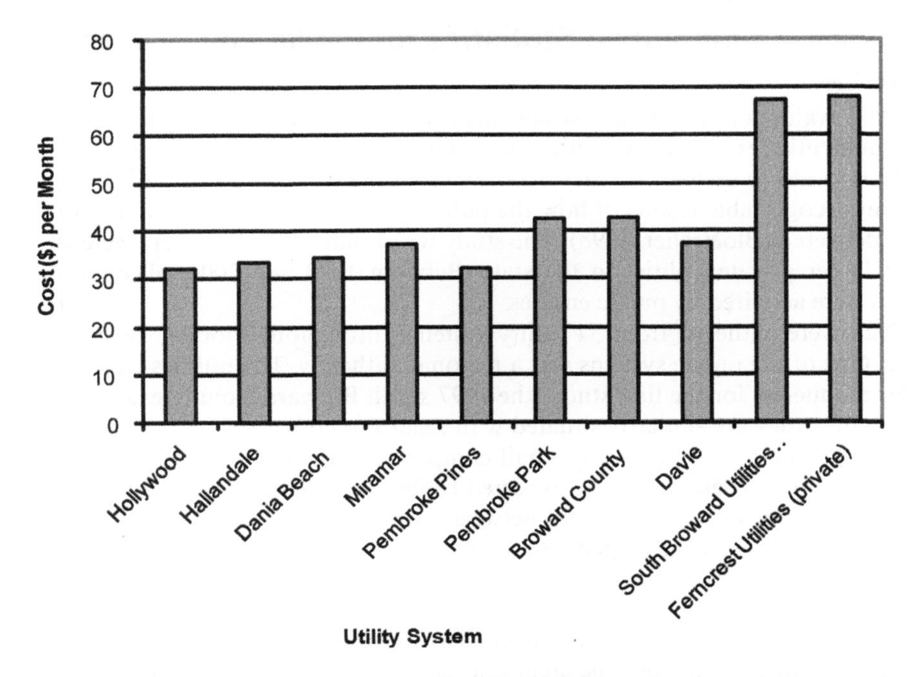

FIGURE 12-8 Comparison of water and sewer rates for south Broward County utilities (1997) for a 7,000-gpm customer

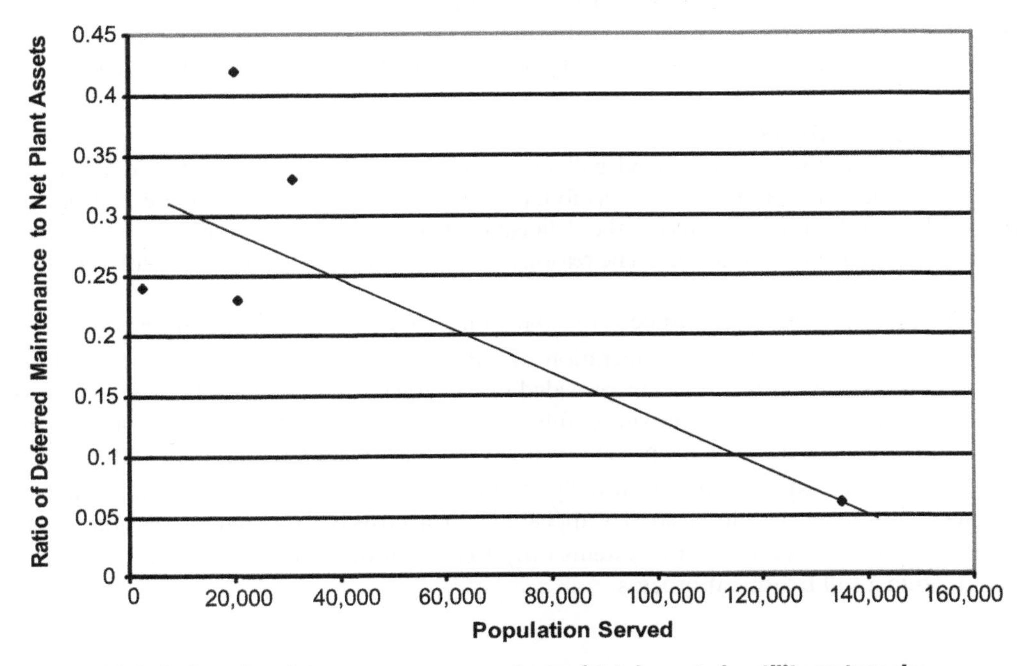

FIGURE 12-9 Deferred maintenance as a percentage of total assets by utility system size

Example—Comparative Statistics for Public and Private Systems

The 1997 evaluation mentioned earlier was conducted to identify trends in similarly situated public utilities. A second study was conducted in 1998 to compare a series of public and private utility systems of varying size across the state of Florida to discern if there were any recognizable results of how the public and private sectors may vary on how service is delivered (Bloetscher, 1998). The study was conducted when there were still a number of larger private utilities in the state. Between 1999 and 2005, all of those private utilities were acquired by public entities.

Data were gathered from 34 utility systems throughout Florida, 20 of which were public. One of the public systems was a regional authority. The utilities were selected in a similar manner as for the first study (the 1997 south Broward County example, presented earlier); thus they are similarly situated with regard to customers and treatment. The private utility systems were typically small compared to the public systems, which was the norm in Florida. These data were expected to find similar results to the first study, while providing insight on the differences between public- and private-sector costs for service delivery. Statistics were developed based on the prior analysis.

Like the first analysis, the data clearly demonstrated economy-of-scale advantages of the larger utility operations for water treatment, water distribution, sewer collection, and wastewater treatment. Comparing the administrative costs as a portion of the total budget parameter also demonstrates the economy-of-scale argument that larger utilities can perform tasks at a lower cost per unit than the smaller utilities. However, an obvious difference appeared, i.e., plotting the above information for all systems, and then plotting the same statistics for public and private systems individually permitted a second curve to be drawn. The data points lying below the curve represent private systems that appear to perform better than public systems of the same size. Further analysis was required to understand why these systems appear to be better performers.

Water treatment costs

Water treatment costs per thousand gallons, when plotted against the average daily flows, show the expected economy of scale as found in the earlier analysis (see Figure 12-10). The dashed line on Figure 12-10 shows the delineation between the public and private systems, which is a rather stark occurrence. The reasons for this differential include those listed below.

- The private systems tend to be very small compared to their public counterparts.
- Regulations concerning the operation of small water plants do not require full-time operator coverage as is typically provided in the larger public systems. Public-sector systems take minimal risks, providing full-time staffing to ensure that their systems have continuous operation, while private-sector systems are willing to minimize staffing.
- Maintenance expenditures are usually minimal in the private systems because maintenance costs are capitalized, so they appear as capital costs in a different expenditure code.
- Several of the small private systems only treat with disinfection as opposed to conducting a full treatment process.

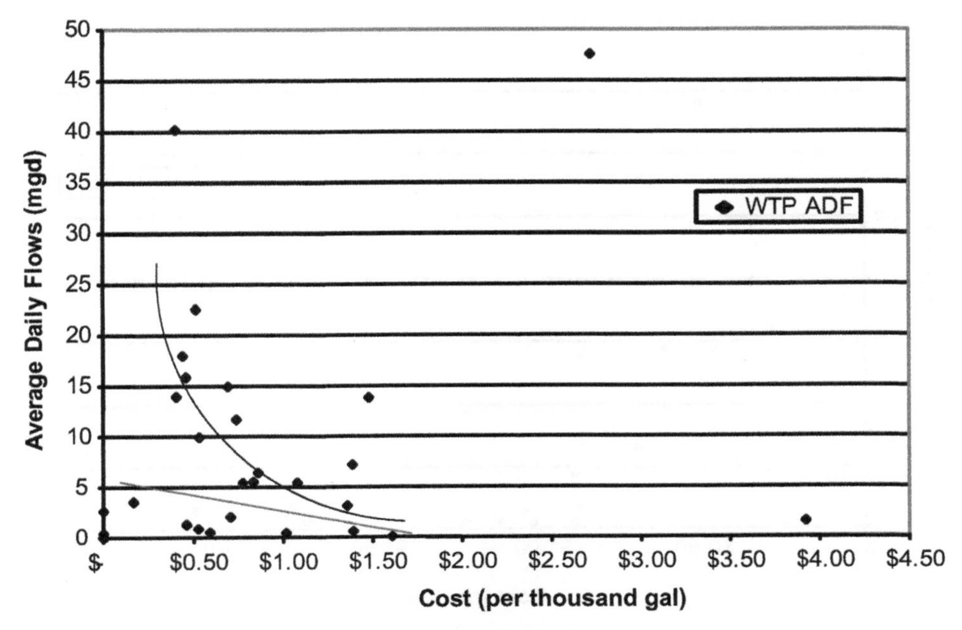

FIGURE 12-10 Water treatment costs per thousand gallons, public versus private

Wastewater treatment costs

Wastewater treatment costs per thousand gallons, when plotted against the average daily flows, show the expected economy of scale found previously (see Figure 12-11). The dashed line on Figure 12-11 shows the delineation between the public and private systems. The reasons for this differential are the same as for water treatment.

Water distribution

Water distribution costs per thousand gallons, when plotted against the average daily flows, show the expected economy of scale found previously (see Figure 12-12), but there are a series of data points huddled around the axes. This indicates that the private systems had no significant costs for their water distribution systems. In looking at the actual operation of these water distribution systems, nearly all expenses are associated with pipe and valve repairs—in other words, maintenance. All of these near-zero data points are private systems. The public systems track the economyof-scale plot. The dashed line on Figure 12-12 shows the delineation between the public and private systems. The reason for this is that private systems capitalize all repairs and maintenance work to increase the investment in their system, while public-sector systems have no such need. Even personnel costs associated with pipeline repairs are capitalized. Public-sector systems take minimal risks, so they keep personnel on staff to address pipeline maintenance and emergencies, and charge them as operating costs.

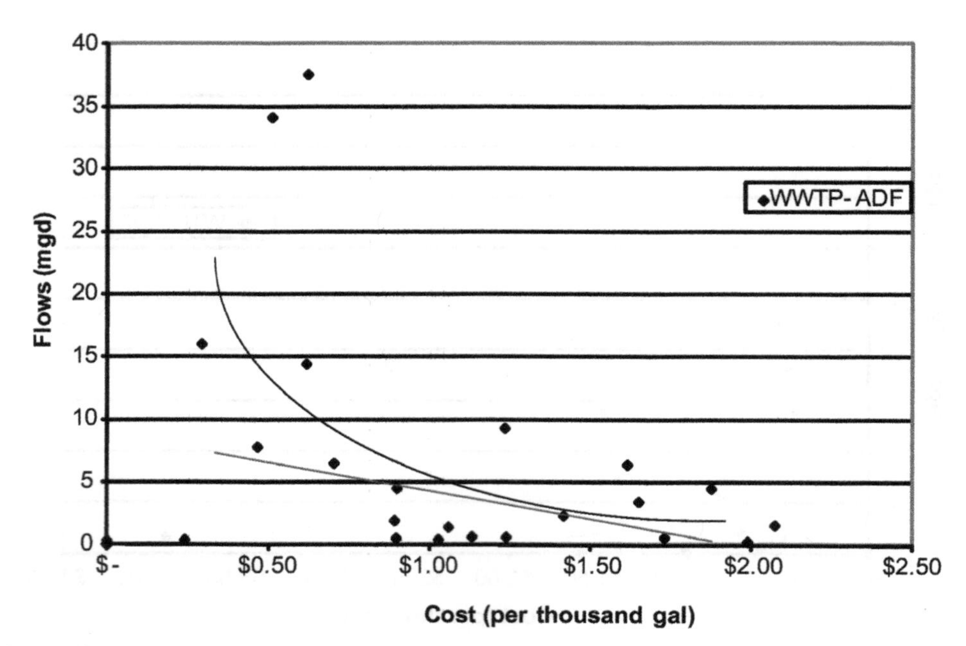

FIGURE 12-11 Wastewater treatment costs per thousand gallons, public versus private

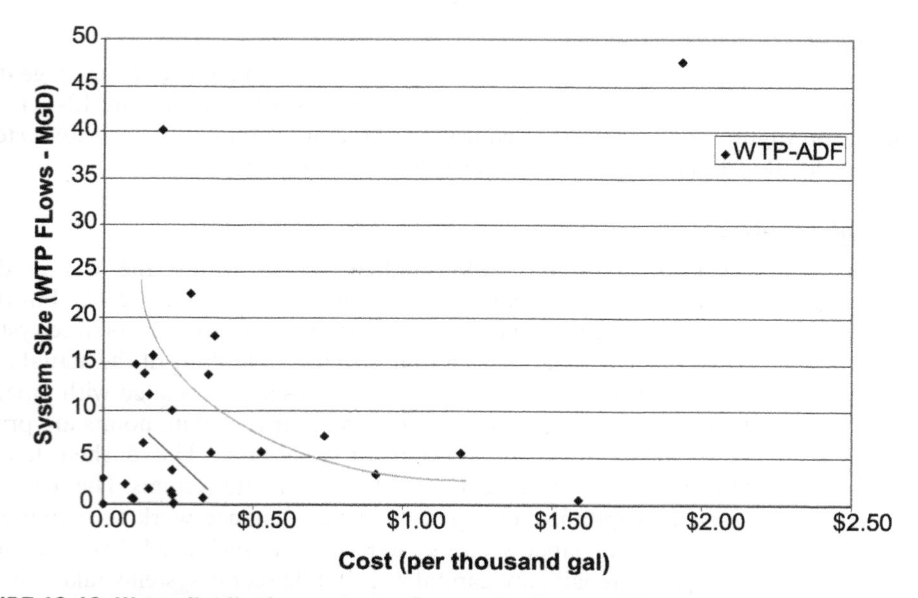

FIGURE 12-12 Water distribution costs per thousand gallons, public versus private

Sewer collection

Sewer collection costs per thousand gallons, when plotted against the average daily flows, show the expected economy of scale found previously (see Figure 12-13). This analysis shows the same trends as water distribution.

Billing costs

Figure 12-14 shows the billing costs for public, private, and both systems. There is no discernible pattern in any of these plots. The cause of this is likely the allocation of administrative, computer, or customer service costs to different sections of the budget and more variations in service delivery than the other sectors. For instance, Hollywood, with one of the higher costs, contracts with a private provider for most of the services, but scatters the remaining responsibility among multiple departments. The full cost is shown in Figure 12-14 for this system.

Administrative costs

An analysis of administrative costs shows a more consistent economy-of-scale argument than some of the other analyses (see Figure 12-15). A review of the private systems indicates that as a group, they have significantly higher administrative costs as a percentage of total budget than public-sector systems. This is due to management fees, overhead, profit sharing, and other private-sector charges made to the utility systems by their shareholding companies.

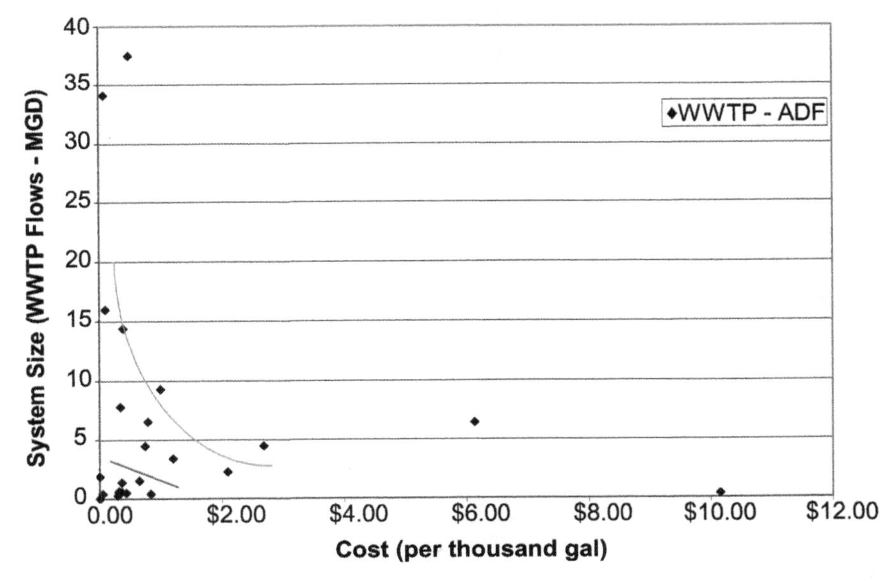

FIGURE 12-13 Sewer collection costs per thousand gallons, public versus private

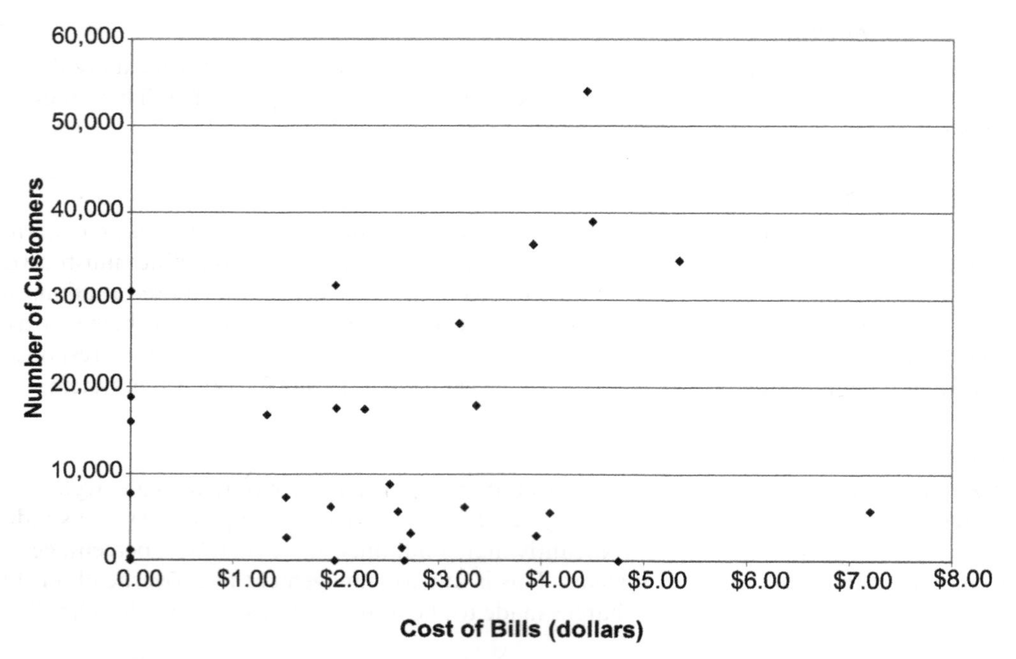

FIGURE 12-14 Billing costs per thousand gallons, public versus private

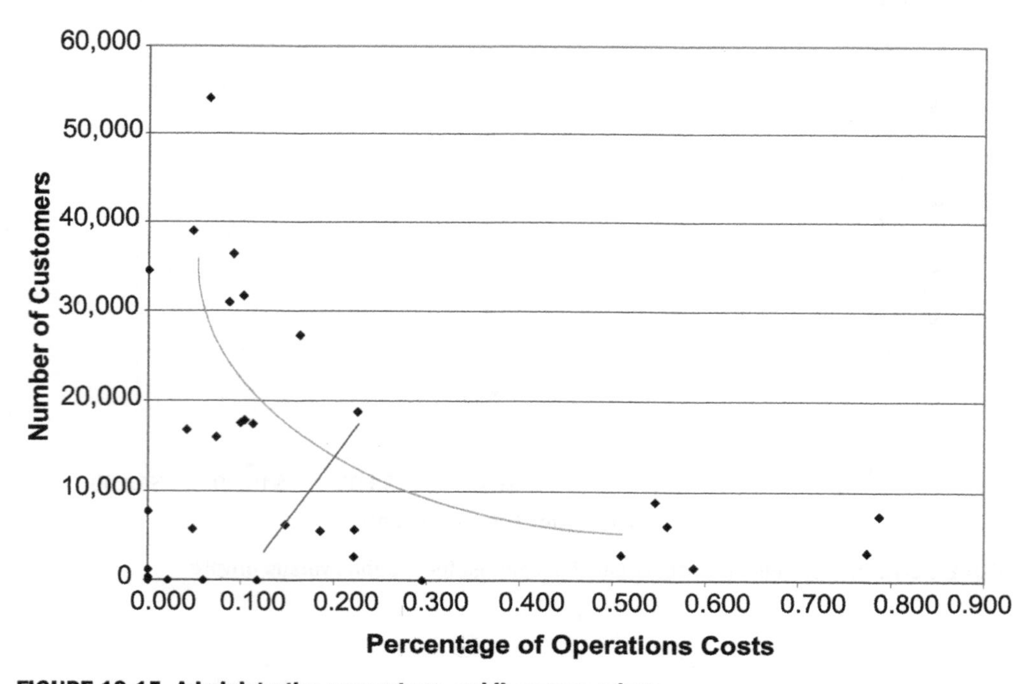

FIGURE 12-15 Administrative percentage, public versus private

The following general conclusions can be reached for the systems evaluated.

- The economy-of-scale arguments for utility operations are realized.
- Private utilities do not budget for maintenance—they budget for capital projects, which impacts how they record operations and maintenance costs. They are more likely to permit the asset to fail.
- Management fees, overhead, profit sharing, and other costs imposed on private-sector utility systems cause the overhead costs to be higher than for public-sector utilities.
- Rates were generally higher for the private systems. In part this is because the operating costs for private systems are shifted to capital, which earns a rate of return.

COMPARING SERVICE DELIVERY MECHANISMS

The conclusion for cost of service is that utilities will follow an economy of scale. It should not be unexpected that small systems have higher per-thousand-gallon costs than larger systems, nor should it be a surprise that larger systems may have lower rates. These analyses also indicate private systems do all their maintenance as capital construction, unlike public systems, which provides a distorted picture of their true costs of operations. This is an issue to note when negotiating any private operations contract.

These results provide an opportunity to discuss the four alternative service delivery mechanisms: local governments, regional authorities, developers, and private systems. Two are private, while two are public, and the operating environments of the two sectors are markedly different. Understanding this difference and the different priorities of each is helpful in understanding how each operates and the benefits and disadvantages to local residents.

Government-Owned Utility Systems

More than 90 percent of large utility systems within the United States are operated by governmental entities. Many of these large systems originally developed from the central cities and may serve much, if not all, of a metropolitan area. In these areas, government-operated systems have developed over time, based on bonds sold in anticipation of customers, and in some cases, with contributions from the general funds of the government meant to encourage growth or attract industry.

The benefit of publicly operated systems is that often the system is operated by a staff that reports to an elected body, and therefore users of the system have input into the policies and practices of the system. In addition, governments can coordinate their utility policies and expansions in a manner conducive to growth, and focus on long-term community needs in concert with transportation corridors. Thus, when new development is set to come online, utility and transportation networks are in place to serve them. The expertise and experience that exists within these systems often transcends jurisdictional boundaries, so these tools can be used to encourage growth in corridors where these services are available and can be required as a prerequisite to development. This allows expansion of the system by contributions-in-aid-of-construction at no cost to the users, taxpayers, or the local governments.

There is also an economic development benefit of a government-owned utility system in that the rates and charges for new connections are regulated by an elected body in conformance with the clear confines of case law and extensive litigation involving public bodies, so as not to unfairly penalize either new or existing users. Public ownership provides an avenue to eliminate health and safety hazards resulting from improperly operated water and sewer systems through police powers derived under statutes and case law, and a broader base to derive revenues to correct such problems. These same statutes allow local governments to require connection to water and sewer systems when the systems are available, and to enforce these connections, while protecting the current ratepayers by guaranteeing that the expenses will be recouped via property liens.

Government-owned utility systems pay no state, federal, or local taxes. There is no investor profit to be earned, no management fees or dividends, and the cost for borrowing for capital improvements is lower due to more favorable interest rates caused by their tax-exempt status. Loan monies may be available from state and federal sources to subsidize regional government-owned systems. Government-owned systems can benefit from the receipt of state and federal grants and impose assessments that nonpublic systems cannot.

One of the most important benefits derived from government-operated utilities is the ability of the local government to spread the cost of the utility in an equitable manner to all those who benefit from it. The existence of a water and sewer utility not only benefits online users, but also benefits developers who own raw land to be developed.

Public systems are normally very strong in public relations aspects due to the visibility of the rates, charges, and ongoing construction, and their impact on elected officials. This causes public-sector systems to be very responsive to requests from the public.

However, government-owned systems often suffer in part from the regulations put in place by elected officials and the public to protect the public. Examples of this include extensive personnel and purchasing regulations that make it difficult to respond to emergencies, replace employees quickly, or address equipment breakdowns, and budget limitations resulting from general fund shortfalls that are separate and distinct from the utility enterprise fund. (The utility may get caught up in general fund financial issues; e.g., across-the-board cuts of 10 percent of the budget to meet a shortfall, despite the fact that the utility is an enterprise fund of its own.) All are operational risk issues; thus public utility officials tend to build in excess redundancy to minimize the potential for failures that might adversely affect the utility or elected officials (whose response may be to terminate utility employees).

Politically charged issues brought before elected bodies often create unintended consequences and new issues that interfere with the objectives of the utility system. System maintenance and increases in rates, assessments, and impact fees are often difficult to get approved despite their relative importance in the continued growth of the utility system. Political issues, along with economies of scale, the need to focus on water and sewer issues more closely, and availability of resources (such as staffing and staff expertise) are reasons that local governments may participate in utility authorities.

Regional Authorities/Special Districts

Regional authorities are a vehicle to provide service within large urban areas or in rural counties. While local officials may not be fully responsible for operations and maintenance, the authority board, which is either appointed by local officials or consists of elected officials, wields control over the direction the authority will take. Coordination between the authority and the underlying local governments is necessary for development of the area so that concurrent availability of service will exist when the local governments require it.

While local governments may exercise significant input into the authority's operations and its long-term direction, the authority, not the local governments, is responsible for debt service, operations, and maintenance of the system. Because the focus of the authority is only on the utility system, routine maintenance of the system is not as difficult to "sell," nor is the need for long-term plans for expansion of facilities to meet the needs of growth. As a result, the coordination and planning necessary in intergovernmental cooperation facilitates the growth, maintenance, and reliability of the system. System reliability is probably the highest goal of authorities because of the interaction between governmental entities and the likely existence of agreements between those entities. As a result, capital projects may not be put off, and both system reliability and repair/replacement programs will sustain the system's value. Deferred maintenance obligations should be minimized.

Developer Systems

Of all of the delivery options, the one with the shortest-term view is the developer system. Developer systems are typically interim utility systems operated by land developers for the sole purpose of allowing land to be developed where publicly operated utility systems are not present. Land served by water and sewer gains significant value. Most developers expect their system to be absorbed into a larger centralized system at some point. The cost to provide an interim system, versus the rewards for development taking place prior to the availability of service from a central system, must be determined by the developer to be worth the investment risk.

The goals and objectives of a developer utility owner diverge from the goals and objectives of the ratepayer. This is because a developer usually enters the utility business only out of necessity for development. Therefore it is in the best interest of developer utility owners to build the utility with the capital cost burden placed on the ratepayer in terms of high rates, while keeping the utility costs incurred by the development company (connection fees and impact fees) at a minimum. The result is a utility with a high rate base and operating costs and minimal connection fees.

Most developer systems are small and short-lived, and in the regulatory framework that exists today, are becoming increasingly difficult to operate. The systems are becoming progressively more costly and less cost-efficient. As a result, the focus of such systems is primarily acquisition by large private providers with an eye toward establishing a foothold in an area, or sale of the utility system to a publicly operated entity at a profit to the developer.

Private Sector Owned Systems

There are two kinds of private-sector utilities. First, there are small, investor-owned utilities that exist within a limited service area that usually corresponds with some prior development (i.e., they used to be developer systems). Second, there are those owned by holding companies. Since operating a utility is very fixed-cost intensive, the number of customers connected to a system is extremely important. Due to economies of scale, most stand-alone utilities are not profitable until at least 1,000 to 1,500 users are connected to the system.

Utilities owned by holding companies, which provide management expertise and support in operating a series of local utilities, are typically run by local operators, maintenance personnel, and a local manager. The holding companies are often part of larger, investor-owned private electric utilities that were allowed to invest in like industries to earn a rate of return as a part of the Public Utility Reform Act of 1992.

At the local level, a private utility system may theoretically have similar operating costs as a publicly owned system, but capital is accounted for differently. Borrowing is typically arranged through intracompany loans or bond proceeds on the open market, which occur at higher interest rates than do public offerings (unless industrial development bonds or some similar public financing can be arranged).

Private-sector utility systems are monopolies that typically have their rates regulated by a public service commission or other public entity to ensure that the private provider is not taking advantage of its customers through its government-approved monopoly. The basis of the rates is developed from a Florida court case (*Keystone Water Company, Inc. v. Bevis*) that states that the rates must be just, reasonable, and compensatory, and balance the right of the public to be served at a reasonable cost with the right of the utility to earn a fair return on the value of the property used to provide the service.

There are three primary factors utilized in the development of rates: expenses, rate of return, and rate base on which the rate of return is calculated. The rate base is the depreciated value of the infrastructure. The methods used for determining rate base are the original costs, fair value costs, and reproduction costs. The US Supreme Court does not permit fair value as the sole basis of rate making, and as a result, the average investment of utilities during a given test year is normally used to determine the rate base rather than the value of the property at the end of a test year (*City of Miami v. the Public Service Commission*).

The operating expenses include the full gamut of potential expenditures, including salary and wages, maintenance, repair, costs of collection, distribution, extending services, and so forth as would be the case for a publicly operated utility, as well as franchise payments made to municipalities, depreciation, and the cost of filing for rate increases. Multilayer ownership of private systems creates special problems for government regulators in trying to discern the subtleties that exist between the parent company and the local utility, when looking at the local system only. This is because the local system typically pays management fees, a portion of the parent's dividends, and other monies as a part of the process, the accuracy of which is not easy to verify in rate records.

The reason for this manner of operation is that most private systems are concerned about the bottom-line profits as an indication of how well the system is operating. This view tends to make the private-sector utilities very short-term, profit-oriented agencies, which leads to very short utility planning windows. Short-term planning creates a problem in that most private utility improvements are designed and built in small increments on short timelines. These come at a higher cost than larger improvements that are planned and built over a three- to five-year time frame (based on economies of scale), in contrast to the longer-term needs of most growth management legislation. Ratepayers are penalized for these investments because the regulatory agencies evaluate the actual investment costs of used and useful components and provide a rate of return on these higher-cost improvements. Small improvements may be fully used, but the regulating agencies do not evaluate the efficiency of constructing them in this manner.

By definition, local officials have little or no say in the operation of the utility. They also have no responsibility for debt obligations, or for maintenance and operation. Routine maintenance of the system may not occur to the degree necessary, depending on the sophistication of the providers, the size of the system, and the diligence in recovering costs from the public service commission. Private companies are willing to risk system failure, as the only penalty is bad publicity. There is no penalty except from regulatory agencies, or condemnation from local governments that will have to compensate the private provider, usually at a price higher than the investment the private company has in the system.

Given the formula used for determining the rate of return for private-sector systems, private systems tend to make incremental, small improvements to their systems to keep the rate base constant and the rate of return the same each year. If the system grows or revenues begin to fall, the rate base is affected and adjustments are applied for, usually causing rates to climb. In turn, management fees and dividends are also increased, increasing the parent company's bottom line. In light of this, it is not surprising that many private-sector system operators focus on maximizing their return on investments, at the expense of ratepayers. This practice essentially becomes a disincentive to expand the system to meet longer-term growth needs; the system will expand where growth has already occurred and customers are readily available.

REFERENCES

Bloetscher, F. 1998. Comparative Analysis of South Broward County Utility System Operations. In Proc. *FSAWWA Specialty Conference–Clearwater, Fla.*

CHAPTER 13

Methods for the Establishment of Rates, Fees, and Charges

Utility systems charge a variety of rates, fees, and charges for service. These include monthly service charges, impact fees, assessments, and miscellaneous fees such as meter rereads, connection fees, late payments, and backflow testing. Each of these fees should have a basis for the charge generally consistent with the financial policy of the system. Only two fees have major legal constraints—impact fees and assessments.

The case law defining the employment of user fees varies from state to state, but is underlain by the basic concept of fairness. A utility's rates not only must be reasonable, but nondiscriminatory, although different user classes can be charged differently provided a valid rationale exists for the difference. A utility is entitled to make a reasonable profit from its utility operations and use the proceeds thus derived for other valid municipal purposes (*Rosalind Holding Company v. Orlando Utility Commission*). This profit has been held to be valid at levels less than 30 percent of total operations. As a result, municipal governments must exercise caution when using water and sewer fees to subsidize general fund operations, which could be construed as a means to transfer profit from the water fund to the general fund, where ratepayers outside the city have no voice in how the monies may be spent.

As mentioned, different user classes may be charged different rates if the rates can be justified. For example, a distinction can be made in some instances between user classes, i.e., residential customers being charged differently than industrial or commercial customers. Also, users of a new wastewater treatment facility can be charged differently than those using an older facility if there is an obvious separation and difference in costs of operation. Typically, however, impact fees and other revenue collection methods are used to absorb this difference and provide for consistent rates across all user classes.

As a result of the established statutory and case law, there are a number of potential revenue sources that can be designated within a financing plan of a utility. Capital recovery fees typically meet the legal test for impact fees, which also may be termed *system development charges, system capacity fees, reserve capacity charges*, or a variety of other names. These fees are collected from new customers who require increases in capital outlay. Repair and replacement (R&R) funds are used to collect capital to replace or upgrade existing infrastructure from existing rates. Repair and replacement funds are collected from existing customers to repair and/or replace the existing infrastructure at the requisite time. Periodic service charges, broken down in availability and volumetric portions, are used to collect the operation and debt service from customers receiving the service.

PERIODIC CHARGES FOR SERVICE

Periodic charges for service are the costs collected on a regular basis from existing customers for the amount of service they receive. Billing for service can occur at any interval, but typically occurs monthly, bimonthly, or quarterly, on a schedule determined by the utility. A case can be made for the use of monthly billings in preference to the others for the reasons listed below.

- The calculation of unaccounted-for water can be tracked more easily with more frequent readings, and reading errors are minimized.
- Frequent reading improves the ability of customers to detect leaks on their premises.
- Frequent reading improves the ability of utilities to track age/wear on meters to create more precise testing and change-out periods.
- Frequent reading reduces the amount of late or uncollectable payments due from customers who fail to pay their bills promptly (usually a monthly bill causes turn offs to occur no later than 45 days after the first water used in the period, while for quarterly bills it may be 115 or more days).
- Frequent reading lessens customer payments, which is of benefit where there is a population of fixed- or limited-income people.
- Frequent reading allows for more frequent inspection of utility facilities (e.g., meters) by utility personnel.
- Frequent reading provides earlier and faster price signals to the utility for water consumption trends and a clearer link to consumption billing.

Most utilities bill on a monthly basis. Reasons given for monthly billing not being performed include cost concerns (such as mailing, reading, and billing) and customer preference. In general, the benefits of monthly billing outweigh these concerns. Estimated (or averaged) bills of any type should be avoided at all costs due to the confusion they create for customers and billing personnel, especially when significant corrections, changes, or negative readings occur on subsequent bills.

Periodic service charges are usually broken down into two portions: availability charges and volumetric charges. Availability charges are the fixed portion of the bill that is usually based on equivalent residential (or dwelling) units (ERUs or EDUs), meter size, or some mixture of the two. The volumetric charge is based on the amount of water consumed by the customer as determined from meter reading. Due care must be exercised to avoid undercollection of fees with the imposition of any rate collection method.

Availability Charges

The fixed fee portion of the service charge is collected from every customer regardless of whether or not there is any usage at the address. This practice is intended to allow the utility to bill customers where service (e.g., a meter/service line) is available, because there is a cost for having the service available to the customer's property regardless of whether the customer uses the service or not. Two obvious and consistent charges encountered are those of meter reading and sending out the water bills. As a result, these costs should always be included in the fixed portion of the bill; likewise, debt service continues to occur whether or not the customer uses the system. Because the repayment of debt is important to protect the financial position of the utility, debt is often the highest priority in the budgeting process and, as a result, revenues to cover debt are typically included in the availability charge. This practice is also a safeguard in case there is catastrophic facility damage due to storms or other natural disasters; the availability charges continue to accumulate on the system to enable the utility to pay its debt, even though the service is not being used.

Inclusion of debt in the availability charge is something the bond rating agencies look for, and utilities can save money on long-term borrowing by providing assurance that the debt will be collected. The problem with this rationale is that in areas with large numbers of people on fixed incomes or who have limited economic ability, or where conservation is a goal, the associated higher fixed fee may not be politically palatable. However, all these issues can be addressed without compromising the system's stability or financial objectives through guaranteed minimum base amounts of water or other methods developed by creative rate consultants.

Two methods are used for creating the availability charge: a meter sizing system and an ERU-based system. A rational argument can be made for each, although each has its drawbacks. Keep in mind the fixed fee portion of the bill is for availability of service and does not consider if the service is used. This is sometimes a problem in areas with a significant percentage of part-time residents who want to have their water turned off to avoid a bill when they are away. The utility still must keep the service in operation ("available") even when the customer is not there.

Meter sizing method of allocation

The argument for using a meter size methodology is that the benefits to all users with similar-size meters is the same since a given meter size has an average and maximum flow capacity. The availability charge under this method is based on the fact that regardless of the actual usage of the property, assuming it is within the design parameters of the meter (refer to AWWA standards on this; see Table 13-1), each similarly sized meter should be billed the same fixed amount because their access to water is the same. In this manner, all customers who have identically sized meters would be charged the same amount and would have the same ability to draw water from the system. The fact that they may not use the full available supply is a matter of choice and in many cases is dictated by economic and conservation attitudes.

A typical, single-family residential unit uses a $5/8 - 3/4$ in. meter. All other meter sizes can be calculated to be a multiple of this size (see Table 13-1). Once all the equivalent meter sizes are accumulated, the debt and meter reading costs can be apportioned appropriately based on the $5/8 - 3/4$ in. meter size. However, there are two precautions to consider when using this system. First, the existing meters must be sized correctly. This is often a problem with existing customers, who will resist making the upgrades to their meter and system (and paying increased amounts for service) and developers who deliberately undersize meters to avoid impact fees. Residential users are typically subject to peaks (morning and evening) and virtually no use during the day (a diurnal curve). As a result, when sizing multifamily-unit meters, the peaking factor amount must be used, not the average flow, which may dramatically increase the meter size.

TABLE 13-1 Meter size chart

Water Meter Size, in.	Meter Type	Maximum Fixture Units	Peak Maximum Volume—AWWA
$^{5}/_{8}$ – $^{3}/_{4}$	Displacement	25	20 gpm
1	Displacement	45	50
1 ½	Displacement	100	100
2	Compound	225	160
3	Compound	500	320
4	Compound	750	500
6	Compound		1,000
8	Compound		1,600
10	Turbine		2,900
12	Turbine		4,300

ERU-based method of allocation

The ERU-based method resolves the issue with sizing multifamily residential meters noted with the meter size method. It also addresses historical undersizing of meters, which causes utilities to undercollect from trailer parks, condominiums, apartments, and the like under the meter size system where the political will does not permit the forced upgrading of substandard systems. This method assumes some control is exercised over the demands imposed by the ERU, whether it is defined as a typical residential single-family unit, average use (often 250 to 350 gpd for water), or some other system. This system is usually used in residential communities, and often condominiums, apartments, and mobile homes are deemed to be a partial ERU (0.7 is common).

The concern in using this system is the method of conversion to nonresidential or nonlike residential units, such as travel trailers and extensively landscaped homes, when their usage can vary widely. Table 13-1 can be referenced, but there is no clear demarcation that justifies a given commercial use as a multiple of an ERU. As a result, most utilities pursue a hybrid approach of the two systems, i.e., ERUs for all residential users, and meter sizing for all other uses. While the line of demarcation between residential and nonresidential is not always clear, such as in the case of RV parks and retirement homes, the hybrid approach mitigates concerns when using both systems.

In calculating the availability charge, the number of meters read should be considered. Where there is a single large meter servicing many units, the debt associated with each ERU should be assigned under the ERU approach, but not the reading costs. As a result, when calculating availability charges, there should be two considerations, one for the cost of reading the meter, and one for the cost of debt. A large meter should have only one reading charge.

Example—Determining the availability charge

Utility A has 10,000 meters (customers) serving 15,000 ERUs. The financial policy of the utility is to collect all meter reading and collection charges, and all debt as a part of the availability charge each month. The utility determines that its cost to read the meters monthly, send the bills, and collect the monies owed is $240,000 per year or $20,000 per month. The debt service the utility pays each year is $1.8 million. The utility needs to determine the appropriate availability charge for a single-family home and a 10-unit condominium if each condo unit is determined to be 1 ERU.

Based on this information, the utility determines that the cost per meter is $2 per month ($20,000 per 10,000 meters). The debt cost is $150,000 per month. Therefore the cost per ERU is $10 per month ($150,000 per 15,000 ERUs). The result is that the utility determines that the single-family customer's availability charge should be $12 per month, and the 10-unit condo should be $102 per month (10 × $10/month + $2 for reading the meter).

Volumetric Rates

With the exception of debt, meter reading, and billing costs, all other costs of the utility's operation are typically assigned to the volumetric portion of the bill. Most of the remaining costs are, in general, dependent on the amount of water or wastewater that is treated and pumped. The larger the utility, the more water or wastewater is being treated, the larger the piping system is to maintain, and usually, more staffing is needed. However, as shown in chapter 12, economy-of-scale arguments indicate that larger facilities will have fewer operators per gallon of water treated.

Within the volumetric charge for the utility are the planned expenditures in the operating budget, including water treatment, water pumping and transmission, water line maintenance, wastewater treatment, wastewater pumping and transmission, and wastewater system maintenance and repair and replacement funding. Within each of these accounts, there are a number of specific allocations that are charged, including but not limited to:

- **Water treatment**

  Chemicals

  Electricity

  Disposals of residuals

  Chemical testing

  Equipment repairs

  Records on maintenance performed and water pumped or treated

  Contracted maintenance and improvements

  Personnel

  Vehicles and vehicle maintenance

- **Water distribution pipeline maintenance, pumping, and transmission**

  Electrical costs

  Pump and line repair costs

  Contracted maintenance and improvements

  Record keeping

  Personnel

  Vehicles and vehicle maintenance

- **Wastewater treatment**
  Chemicals
  Electricity
  Disposal of residuals
  Chemical testing
  Equipment repairs
  Records on maintenance performed and water pumped
  Contracted maintenance and improvements
  Personnel
  Vehicles and vehicle maintenance

- **Sewer collection system pipelines maintenance, pumping, and transmission**
  Electrical and pumping costs
  Equipment repair
  Pipeline repair
  Contracted maintenance and improvements
  Personnel
  Vehicles and vehicle maintenance
  Record keeping

Nonoperations and support service costs include administrative costs, which would include purchasing of materials, legal fees, insurance, budgeting and accounting, human resources, engineering support, public information, technical information, dissemination of operational needs, and general management. Allocation of these nonoperating costs should be prorated over the operating budgets or personnel. The following example demonstrates this policy.

Example—Volumetric charge calculation

Utility B has 10,000 water customers who use 8,333 gal per month (gpm) and 5,000 wastewater customers who use a similar amount. The utility wants to determine the appropriate volumetric charge for operations (assume utility B has a financial policy to derive debt, and billing costs are derived from the base fee). Repair and replacement allocations are included in the operating budgets. The cost for operations is as follows:

Allocation	Population Served	Budget	Percent	Budget w/Admin
Water plant	10,000	$960,000	40	$1,000,000
Water distribution	10,000	$480,000	20	$500,000
Wastewater plant	5,000	$480,000	20	$500,000
Sewer collection	5,000	$480,000	20	$500,000
Administration	10,000	$100,000		

The utility finds that each customer uses 100,000 gal of service each year. The cost for the water system is $1.5 million. The cost per customer is $150 per year. As a result, the utility finds that the appropriate charge per thousand gallons used is as follows:

Allocation	Cost	Customers	Gallons Used per Year/Customer	Cost per Thousand Gallons
Water	$1,500,000	10,000	100,000	$1.50
Sewer	$1,000,000	5,000	100,000	$2.00

Therefore, based on the use per customer for the utility, the cost per thousand gallons of water is $1.50, and for sewer is $2.00.

FORECASTING RATES

Proper fiscal planning involves comparing projected utility revenue sources with the revenue requirements. A multiyear projection will identify the need for additional sources of funding, alternative financing, and steps to be taken to minimize impacts on the utility rates and charges to the customers. Projections for over five years are speculative, but where large expenditures may occur, projections are necessary. A computer model should be developed specifically for the utility, using budget information and the allocation method defined. From the model, adjustments for different rate structures (with elasticity effects), changing usage and population patterns, and other variables can be tested.

Once projections of the revenue components and the expenditure needs have been completed, the cash flow analysis and forecast should be completed to compare the expected annual revenue and expenditures for the planning period. Capital programming should be a portion of the forecasts. Cash flow analysis should include a beginning fund balance (cash on hand minus funds committed for current liabilities), be segregated into separate trust funds (such as for repair and replacement funds and impact fee trust funds), and define the work currently in progress or encumbered so expected carry-forward funds are known. As an initial plan, the cash flow analysis should identify any historical methodologies used by the utility to project revenues and expenditures, demonstrate the utility can meet all debt and infrastructure needs, and set minimum reserve levels.

A proper cash flow analysis should identify areas where the annual revenue from the sales of water or wastewater are adequate or inadequate to meet the operation and maintenance expenditures, and the impact that may have on subsequent years. As an example, if a shortfall might occur in year five, a minor increase in year two or year three may alleviate that concern, as opposed to a large increase in year five. In addition, where shortfalls may occur one year, but not in the subsequent or prior years, the fund balance can be used to balance the budget. Long-term changes in service demands, revenues, socioeconomic factors, and changes in state and federal revenue programs should be taken into account.

Cost Allocations

A rate study should allocate costs to those benefiting from the service and develop pricing strategies that can be clearly explained to the public. A lot of case law precedents come from Florida. Most of the Florida case law is based on the legal precedent that a utility may charge for the service and products it provides to the customers, but the basis for the charge must be reasonably related to the cost of service or the product (*New Smyrna Beach v. Fish*, and *City of Pompano Beach v. Oltman*). Aside from being just and equitable (the landmark case of *Contractors and Builders Association v. City of Dunedin*), Florida case law indicates that user fees and utility charges may include not only the cost of operating the utility, but anticipated future outlays. Cost allocations are hard to define as they are specific to a utility. However, the following example outlines many of the issues involved in creating a cost allocation methodology.

Example—Hollywood, Fla., reuse system rates

This case study outlines a process the city of Hollywood, Fla., went through to derive reuse rates a number of years ago. Reuse has not been actively pursued by south Florida utilities because of the cost for residential reuse, the density of development, a shallow water table, and the lack of large customers. The beneficiaries are also not always as clear as for water or sewer customers. Therefore a new method to allocate costs needed to be created. Typically south Florida homes are built on small lots, with driveways and buildings covering most of the land. The result is that residential reuse is not conducive to disposing of significant quantities of reclaimed water. In Hollywood, little additional development can occur, so partnering with developers to provide dual distribution systems is also limited. Differences of opinion over the allocation of costs for the service have further inhibited the growth of reuse in the area.

However, Hollywood was able to pursue and implement a slow-rate land application reuse system, targeting water users competing with the city for its raw water supplies. These competitors included golf courses, large industrial and commercial customers, and open space. The first six users were golf courses, all of which had consumptive use permits to withdraw water within the cone of influence of the city's well field. Their total demands were 4 mgd. Permitting of the program proceeded smoothly with the Florida Department of Environmental Protection (FDEP) and allowed uprating of the wastewater plant by 4 mgd. All were withdrawals in the cone of influence of the city's potable water supply wells (competing users).

The city applied for and received state revolving fund (SRF) loan monies to construct the pipeline portions of the reuse system. The loans were for 20 years at interest rates below 3 percent. Over 10 mi of pipelines were constructed in city rights-of-way with these funds. A portion of the city's recent $135 million bond issue was designated for upgrades at the wastewater plant, including filters, chlorination equipment, effluent pumps, and aeration basins. These facilities are necessary to implement the reclaimed water system.

The city encountered resistance when it began to negotiate with private golf courses to secure reuse agreements. One of the first issues the city faced was how to properly allocate the costs to install and treat the reclaimed water. From the city's perspective, the reclaimed water was a disposal mechanism for the wastewater treatment plant. A high reclaimed water rate would negatively affect the goal to dispose of the water. However, the cost to treat reclaimed water is significantly higher than the cost for using groundwater. As a result, no incentive exists for irrigation and industrial users to take a high-cost reclaimed water source, when much cheaper options exist.

An effort by south Florida utilities to have legislation passed for mandatory connections to reclaimed water (when available) was successful. However, it came with the caveat that if the reuse recipients were charged exorbitant fees, the legislature would revoke the law or subject the utilities to oversight regulation by the Public Service Commission. The fee of $0.10 to $0.25/thous gal was deemed appropriate, because the legislature wanted the polluters, not the reclaimed water recipients, to pay the cost of the cleanup.

Reuse customers have argued that the polluters, who are the wastewater customers, should pay all the costs, and the reuse customers should receive the water at little or no cost because reuse is a disposal option. However, self-supplied groundwater users incur costs for their wells, permits, and other items that the reuse water would replace. The cost of their on-site service is avoided when they convert to reclaimed water. Therefore an argument can be made that the reclaimed water cost, if it approximates the present resource cost, is an appropriate charge. Others have argued that the public water system should pay all or a portion of the costs for reclaimed water because competing groundwater users are eliminated when reclaimed water is delivered and used on a site that previously had a consumptive (water) use permit.

In designing the reclaimed water rates, the city identified both benefits and beneficiaries, plus the logical cost allocations for the reclaimed water improvements. Three beneficiaries are the wastewater utility, the reclaimed water users, and the water utility. The wastewater utility benefits because, by reusing 4 mgd, the utility was able to rerate the plant. The reclaimed water users benefit because using reclaimed water improves water supply availability—they are not subject to water use restrictions during drought periods when groundwater withdrawals might contribute to salt-water intrusion. The water treatment plant benefits by eliminating competing users.

From a wastewater treatment plant perspective, since reclaimed water (while a resource to some) is principally a disposal method, it requires pricing incentives to encourage new customers. This means that the reclaimed water price should be kept low. As a result, the bulk of the operations cost to filter the effluent is also paid by wastewater customers. A portion of the reclaimed water piping costs, because of the limited customers, was also included.

The pumping, transport, and storage costs benefited the reclaimed water customers more than anyone else. This operating cost was determined to be about $0.13/thous gal. The rate was set at $0.10 to encourage connections as the avoidance cost for on-site wells varied between $0.07 and $0.10/thous gal. Additional costs for on-site improvements necessary to accommodate the reclaimed water on the golf courses were assigned to the golf course users directly.

Because of the reduction in competing aquifer withdrawals from the golf course recipients, an annual potable water supply allocation was secured for the city's water system. The theory was that the water customers benefit as much as reclaimed water customers, so a matching amount was contributed to the reuse system annually for operations.

The cost allocation was that 75 percent of the costs were paid for by the wastewater customers. The remaining amount was split equally between water customers and the reuse recipients. To date, no one has complained about this breakdown being unfair to any party.

Water Rate Structures

The cost of supplying utility services increased significantly over the period from 1970 to 2005. This occurred for many reasons, among them stricter regulations established through passage of the Safe Drinking Water Act and the Clean Water Act, the need for many utilities to develop increasingly more costly water supplies, and rapid economic development in certain areas. With the onset of droughts in some areas and higher cost for development of new water sources throughout the country, water rates and pricing have been used to play a role in demand management.

Traditionally, utilities have used water pricing as a means to recover costs by charging users of a specific type in accordance with the cost of serving that type of user. This water pricing practice is both effective and equitable. However, pricing can also work to reduce demand by providing an incentive for customers to manage water use more carefully. The customers' sensitivity to pricing is enhanced by rate structures that increase the cost of water at a constant or increasing rate as usage increases. Since there is significant flexibility in structuring rates to achieve this objective, it should be the goal of the utility to select the rate structure that achieves cost-of-service principles while meeting community needs. Among the considerations in developing a rate structure are those listed below.

- **Financial sufficiency**—generating sufficient revenues to recover operating and capital costs
- **Conservation**—encouraging customers to make efficient use of scarce water resources
- **Equity**—charging customers or customer classes in proportion to the costs of providing service to customer groups
- **Ease of implementation**—having the capability to implement the rate structure efficiently without incurring unreasonable costs associated with reprogramming, procedures modifications, and redesigning of forms
- **Compliance with appropriate legal authorities**—being consistent with existing local, state, and federal ordinances, laws, and regulations
- **Effect on customer classes**—minimizing negative financial effects on utility customers
- **Long-term rate stability**—producing rates that are reasonably constant from year to year (i.e., so that the methodology does not produce rates that fluctuate widely from one period to another)

As noted, it is important to project operating and capital costs over an extended period so that fluctuations in potential rates can be evaluated. It is essential that revenue requirements are sufficient to provide for adequate facilities, to allow for proper replacement and maintenance, and to ensure that the utility is operating on a self-sustaining basis.

After costs have been allocated to user classes, it is then necessary to design a rate structure for appropriate charges to customers. Precedent and reasoning used in developing water rates must be sound so that conclusions reached can withstand scrutiny in the event of litigation. The development of water rates to achieve political and social objectives should be minimized in the rate design to avoid the appearance of discrimination.

When selecting a rate design, a utility should consider its operations and economic environment, the community's objectives, and its customers. To achieve its objectives, a utility should adopt a comprehensive process for evaluating the appropriateness of alternative rate structures. Factors to consider in revising existing rate schedules are

- Impact on customer groups or individual customers
- Compliance with local, state, and federal laws and regulations
- General public reaction to changes in rates
- Impact of shifts in the cost burdens from a group of customers that has been overcharged to a group that has been subsidized under the existing rates
- Reluctance to depart from the existing rate form because of tradition
- Pressure from special interest groups

- Ease of implementation
- Equity
- Simplicity and understandability by customers
- Water conservation

The development of a rate schedule to meet costs of service will have to take into consideration local practices and conditions and should be in the best interests of both the community and the utility.

The application of rates to maintain a sound utility system is a policy decision. This decision can be accomplished by reviewing past practices, considering present and future impacts on customers, and providing rate modifications that will accomplish the necessary results with the least controversy. It may be necessary in the application of a rate structure, such as one for conservation rates, to spread the modifications over an extended period to accomplish the results desired.

Designing rate structures that conserve water are a requirement of consumptive use permits granted by some regulatory agencies and have become an integral part of the water supply planning process. The following paragraphs outline the goals, benefits, and problems of the more common rate structures. Most of these rate structures have more applicability to water systems than wastewater systems, due to the need to conserve the former (Bloetscher, 2009).

Declining block rates

For many years, a single schedule of declining block rates applicable to all customer classes was the predominant water rate structure in the United States. The declining block rate provides a means of recovering costs from the customer classes under a single rate schedule, recognizing the different water demands and costs associated with each customer class. Under this rate methodology, economies of scale are recognized since the price per unit declines as the customer consumes more water.

This methodology provides no incentive to conserve water; in fact, it encourages just the opposite. Recent national rate surveys indicate that the traditional, declining block rate schedule applicable to all customer classes, once popular in industrial cities, may no longer be the primary method of charging customers. This is a result of shifts in industrial uses and the need for conservation of water resources in many parts of the country. Table 13-2 is an example of a declining block rate (Bloetscher, 2009).

TABLE 13-2 Declining block rate structure

User Class	Residential	Nonresidential
Availability charge per ERU	$10.00	$10.00
Volumetric charge (per thous. gallons)		
0–6 thous gal/mo	$2.50	$2.50
6–10 thous gal/mo	$2.25	$2.25
10–25 thous gal/mo	$2.00	$2.00
25–250 thous gal/mo	$2.00	$1.75
Over 250 thous gal/mo	n/a	$1.50

TABLE 13-3 Inverted block rate structure

User Class	Residential	Nonresidential
Availability charge per ERU	$10.00	$10.00
Volumetric charge (per thous gal)		
0–6 thous gal/mo	$1.50	$1.50
6–10 thous gal/mo	$2.50	$2.50
Over 10 thous gal/mo	$4.00	$2.50

Uniform volumetric rates

A uniform volumetric rate is one in which all water (and/or sewer) use is charged at the same volumetric rate to all units. This is the rate structure typically used with wastewater systems, although many systems will cap the sewer usage at some point. There is no change in the unit rate for water used during the billing period. The uniform rate may encourage water conservation by eliminating the lower-priced water rates under a declining block rate structure, and it may be perceived as being more equitable in some communities because all customers pay the same rate per thousand gallons regardless of water use volumes. The amount of conservation to be expected when using a uniform rate structure is minimal.

Inverted block rates

Inverted block rates are the opposite of the declining block rate structure. Under this alternative, rates increase for progressively larger volumes of water use. As a result, larger-volume customers pay a progressively higher average rate for increased water use. The reasons for using an inverted block rate structure are to offer financial incentives for reducing water use, as well as to price the cost of increased capital facilities necessary to meet peak demands (i.e., there is a high correlation of peak demand coincidence with water use by large users). Table 13-3 shows an example of an inverted block rate structure.

Some utilities apply one inverted block rate system that is applicable to all customers. Other utilities adopt separate inverted block rates for different customer classes. For example, the residential class may be charged at a higher unit rate for water use in excess of an established amount per billing period (although if a sewer cap is used, the block rate should be higher than the total of water plus sewer at the cap limit). Typically, the higher-priced block(s) are established at amounts that would typically include nonsanitary usage, such as irrigation, car washing, and filling swimming pools. Under this alternative, these uses are charged at a higher rate to discourage excessive usage. A variation of the inverted block rate structure could include a mechanism whereby all water use is charged at higher rates if usage exceeds a certain preset amount. This element can provide greater conservation incentives to the customer since excess water use will cause all purchases to be charged at higher rates, not only the usage in the higher blocks. Consideration should be given to the amount of increase between blocks if sewer charges are capped. For example, the following would create a disincentive to conserve because the sewer volume is capped:

Block, gal	Rate/1,000 gal Water	Rate/1,000 gal Sewer*	Rate/1,000 gal Combined
<10,000	$1.50	$2.50	$4.00
10–20,000	$2.00	$—	$3.00
20–50,000	$2.50	$—	$2.50
>50,000	$3.50	$—	$3.50

*Sewer capped at 10,000 gal.

A solution to this problem would be:

Block, gal	Rate/1,000 gal Water	Rate/1,000 gal Sewer*	Rate/1,000 gal Combined
<10,000	$1.50	$2.50	$4.00
10–20,000	$4.50	$—	$4.50
20–50,000	$5.50	$—	$5.50
>50,000	$7.00	$—	$7.00

*Sewer capped at 10,000 gal.

The ability of a price increase to affect consumption is termed *price elasticity.* Price elasticity of demand is a measure of the relative influence that a change in price of a commodity (water) has on the demand for that commodity. Water savings due to this type of increase in cost depend on the price elasticity of the demand. Price elasticity can affect long-term savings of inverted block methodologies. Several variables have significant effects on price elasticity: for example, elasticity will vary with the type of customer, income of the customer, and the amount of water used. Although it is clear that in most cases, a price increase will reduce consumption, the magnitude of the reduction depends on how the customers react to the financial increase. Many utility systems have adopted inverted rate structures for conservation purposes. The structure appears to yield some overall reduction in water, at least in the first few years.

Off-peak rates

Off-peak rates apply to water service provided during periods when the utility is not providing water service at its daily or hourly peak rates of flow (in much the same manner as power companies). This rate reflects lower utility costs during these off-peak periods. Off-peak rates can encourage customers to take larger portions of their water needs during off-peak periods. By shifting some demand to off-peak periods, the existing facilities are less likely to be overtaxed during peak periods and the subsequent need to construct new facilities to meet load growth may be reduced. This rate structure is used widely in electric utilities, but is nearly impossible to implement without advanced metering technology in the water/wastewater industry.

Seasonal rates

A variation of off-peak rates is seasonal rates. Seasonal pricing to affect peak use is probably the most effective and equitable means to manage demand. Outdoor usage, which is responsible for most of these peaks, provides an opportunity for generating the greatest reduction with the least burdensome pricing strategy. Seasonal rates establish a higher rate for water use during the utility's peak season, reflecting limitations in supply during those periods. Seasonal rate structures indicate to the consumer the importance of efficient use of resources. The method is becoming a popular and effective rate structure in areas where seasonal peak uses are high (for example, in beach communities where most of the water is used for irrigation and tourism).

Several variations are possible with this method, but each charges higher rates for use during the peak seasons than during nonpeak seasons. Water use during peak-demand months can be either surcharged or simply charged at higher rates. Alternatively, all water uses that exceed predetermined usage levels during base months (such as the winter period) can be charged at higher rates, with the established base usage charged at a lower rate. Table 13-4 is an example of a seasonal rate structure.

Flat rates

A flat rate is one in which all water (and/or wastewater) use is charged at the same rate regardless of amount received. This rate structure is still used by some sewer systems. Meters are not required. The practice encourages (even rewards) waste. Most utilities with flat rates have changed this system to accommodate requirements in grants or loans received for past construction. The rate system is not equitable except where all users will use the same amount of water; an example could be condominiums. This rate structure is not favored by regulatory agencies and, in general, is viewed as unfair to many sectors of the customer base. Some SRF programs and bond rating agencies will insist on volumetric rate conversion before loans are granted.

Schedules by customer class

Another option is to establish a separate rate structure or schedule of charges for each customer category served by the utility. However, a rate structure applicable to all classes of customers cannot reflect the cost of service for any particular customer group. By establishing rates by class of user, there is a more direct recovery of cost from each customer category. Since the rates can better reflect cost differences among the various classes, the customers in each are made aware of the cost of each unit of water consumed. The major difficulty in establishing a rate schedule by class is identification of the various classes and assignment of each customer to the appropriate category. Nearly everyone can make an argument that they belong to a special user class.

Impact of alternative rates

Many alternative rate structures discussed in the prior paragraphs have not yet been adopted extensively across the United States. A utility using any given rate structure could find itself with rate inconsistencies and/or competitive problems with other utilities, or it could be challenged in the courts by customers over the fairness of the rate structure. For example, some alternative rate structures could impact larger water users heavily and discourage economic

TABLE 13-4 Seasonal rate structure

User Class	Residential October–April	Residential May–Sept.
Availability charge per ERU	$10.00	$10.00
Volumetric charge (per thous gal)		
0–6 thous gal/mo	$1.50	$2.00
6–10 thous gal/mo	$2.50	$3.50
Over 10 thous gal/mo	$3.50	$6.50

development within a particular area. As a result, the utility adopting an alternative rate structure should be aware of the pricing approaches of similar, adjacent utilities.

The rate structure implemented by the utility must meet its needs and policies, but may be significantly different than the utility's current rate structure. Large differences can have major impacts on the utility and its customers. An implementation plan should be developed that considers and addresses these effects. The implementation plan should include provisions for phasing in and testing the impacts of the proposed alternative prior to its full implementation.

Any increase in periodic availability or volumetric rates will have a corresponding percentage decrease in actual water usage. In performing an analysis of the expected revenues, lower usage should be taken into account, regardless of the rate structure used, as each has its benefits, disadvantages, and a different economic response to differing price elasticity.

Any time there is a significant change in required revenues, the effects on the utility customers must be gauged, which may require an updated cost of service to identify areas where cost savings can be realized. Where these changes affect different customer categories, it may be beneficial to review a rate structure that will charge the differing customer groups accordingly, so that one class is not unduly subsidizing another.

A comparison of typical periodic bills is the most common way of determining rates and their effects on consumers. However, in doing this evaluation, a variety of meter sizes and water use ranges should be used to provide a broader picture. Too often, rate consultants will compare only an average bill, failing to look at large users and small users, who may be affected in drastically different ways. Elected officials may express concern if their constituency is among those paying the higher rates. They will generally push to cut costs, with capital and maintenance costs being the easiest to delete. The tendency to defer capital projects or reduce contributions to capital funds should be avoided.

Example—Impact of changing rates on customers

Utility C has a large number of single-family residential customers, some of whom are in an exclusive area with high water use. The utility recognizes that its water supplies may be limited; thus conservation is a policy the utility wants to pursue. Currently the utility has a uniform rate system with an availability charge of $10 per month and a volumetric charge of $2.00 per thous gal, regardless of use. The utility noted that many of its residents are on fixed incomes. These same customers use an average of only 3,000 gpm. The city council has directed that these fixed income and commercial users should not be impacted by the rate change.

The utility proposed an inverted block rate that penalizes high-volume water users, defined as those consuming over 10,000 gpm, and separates residential and nonresidential users into two user classes. The goal is to be revenue neutral. The rate structure recommended is as follows:

User Class	Residential	Nonresidential
Availability charge	$10.00	$10.00
Volumetric charge (per thous gal)		
0–6 thous gal/mo	$1.50	n/a
6–10 thous gal/mo	$2.50	n/a
Over 10 thous gal/mo	$4.00	n/a
All users	n/a	$2.00

The following outlines the impact on 3,000- and 15,000-gpm single-family users, and a 3,000-gpm commercial user:

Customer	Current	Proposed	Change
3,000 gal/mo single-family user	$16.00	$14.50	$(1.50)
15,000 gal/mo single-family user	$40.00	$49.00	$9.00
3,000 gal/mo commercial user	$16.00	$16.00	$—

The major concern for this utility is if the conservation is too successful, the utility may lose revenue as a result of the changes for low-end users. As a result, reducing current user fees should be considered carefully before implementation.

IMPACT FEES

Impact fees are charges imposed against new development or connections to provide the cost of capital facilities made necessary by that growth. In general, the "capital facilities" are deemed to be treatment facilities, wells, surface intakes, and regional transmission systems (infrastructure installed in front of individual houses is not an appropriate use of impact fees because they serve no regional value). The use of impact fees is based on Florida case law derived from *City of Dunedin v. Contractors and Builders Association of Pinellas County*, where a utility's "water and sewer facilities would be adequate to serve its present inhabitants were it not for drastic growth, it seems unfair to make the existing inhabitants pay for new systems when they have already been paying for the old ones." This case is the basis for much of the impact fee law that currently exists both in Florida and nationwide (Bloetscher, 2009).

Impact fees have been extensively litigated within the state of Florida, less so in other locales. The Florida case law is cited in impact fee cases throughout the nation and its basic tenets are upheld. As developed under this case law, impact fees must meet the "dual rational nexus test." The first prong of this test requires that there be a reasonable connection between the anticipated need for additional facilities and anticipated growth (*Hollywood, Inc. v. Broward County*). The second prong requires that there be a reasonable connection between the expenditure of impact fee revenues and the benefits derived by new connections (*Hollywood, Inc. v. Broward County*). In addition, case law requires that these fees be just and equitable. As a result, a profit cannot be earned on impact fees; they must be related to the actual cost of providing the service as defined in the second prong of the dual rational nexus test.

Utilities have instituted impact fees as a method of generating revenue from new customers to finance major facility construction made necessary by the addition of those new customers. To meet the dual rational nexus test, these charges are typically based on the incremental or marginal costs of providing the service, an average cost to provide an incremental portion, or an estimate of the cost of the construction to be provided. Because facility planning timelines may be extensive, and because of the geographical variance in growth demands, a multiyear estimate is used to forecast needed expenditures and proper impact fee amounts.

The driving force behind impact fees is the sentiment to have growth pay for growth. The magnitude of impact fees varies throughout the country, depending on the municipality or the utility's desire to encourage growth. For utilities in Florida, impact fees gained considerable favor after passage of the 1985 Growth Management Act, which requires localities to have capital infrastructure (including water and sewer service) available at the time development actually occurs. Similar laws exist in some other states. These large facilities are typically financed with both bonds and impact fees.

In establishing impact fees for water and sewer services, the findings typically made by governing bodies contemplating the use of impact fees include the following:

- That the land regulations and policies require owners of land to connect to regional facilities when they become available
- That those requiring additional demands on the system from growth should contribute their fair share to the cost of improvements and additions to the regional system
- That these contributions are an integral and vital element of the regulatory and growth management plan
- That capital improvement planning is an evolving process defined by a level of service adopted by the governing body
- That the impact fees will protect the interests of the citizens currently served or intended to be served by the utility system, which enhances the health, safety, and general welfare of the residents and landowners within the utility's service area
- That the impact fees are an important source of revenue
- That the deficiencies that exist between the existing system and the adopted level of service cannot be funded through impact fees

All properties that are connecting to the utility system are subject to payment of impact fees at the time of connection to the system (usually when building permits are applied for) in addition to any costs for installation of subdivision infrastructure. Normally both are paid as a part of new lot costs.

During growth cycles, collection of impact fees can be considerable. However, since they are tied to growth, significant fluctuations may occur in impact fee receipts from year to year as a result of local and national economic conditions. Consequently, the revenues are not always predictable, and make pledges toward debt service of these funds difficult without supplemental revenue pledges. High impact fees may discourage the growth for which the impact fees are intended to pay. In areas that are trying to grow in order to continue the growth of the tax base and services, high impact fees are a problem. However, having a subsidy by current ratepayers to encourage growth may be equally unsatisfactory. In other areas, where growth is too rapid, impact fees charged at the full cost of providing the facilities (not subsidized) may help to control growth.

In determining the value of an impact fee, an important consideration for any defense in the event of a challenge is that the impact fee should reflect the incremental costs to provide the treatment and transmission capacity to the consumer. As such, the present worth of any debt service amounts that would be paid for during the life of a customer being connected to this system on a current debt should be deducted from the value of the impact fee. For example, assuming that a single-family home requires an average of 250 gpd of water service, 250 gal of treatment plant capacity must be set aside for the house. Assume this cost is determined to be $850, based on the cost of expanding the treatment plant, divided into 250-gal increments.

Next, assume the transmission and pumping costs for the storage tanks, high-service pumps, and major transmission system to get the water to the local area is $500. The impact fee value would then be $1,350. However, if there is an outstanding bond issue that the new customer will pay on as part of his monthly service charge, the present-worth value of that bond issue should be deducted from the $1,350. Otherwise the customer is paying twice, both for expansion needs and for infrastructure already in place, and the impact fee can be invalidated under challenge. Typically the present worth of debt paid as a part of periodic water bills over a 20-year period amounts to between $150 and $250. Therefore, case law would limit the value of the impact fee in this case to about $1,100.

Customer impact fees need to be determined by a similar methodology. For commercial users, meter sizing is an appropriate method to calculate impact fees since meter size represents the average and maximum available water supply at that address. Since the use at the property may change over time, unless the meters are changed, there is a certain maximum amount of water that can be utilized and remain within the design parameters of the meter. This is the rational nexus for establishing impact fee rates using meter size, although proper sizing of the meter must be ensured.

In developing the appropriate funding levels for impact fees, the options for funding the capital projects anticipated to meet future demands must be established. This would include separating repair and replacement projects, deficiencies in the current system, and future growth into the appropriate funding mechanisms.

Example—Simple Impact Fee Calculation

Utility D currently serves 1,000 houses. Assume no substantial commercial/industrial usage exists. The existing water treatment plant (WTP) capacity is 250,000 gpd, with two wells, each 0.250 mgd. The average household use is 250 gpd of water and 200 gpd of sewer. A new development is proposed in a newly annexed area. The developer will build water and sewer lines in the development and connect to the utility system. However, the utility's registered professional engineer notes that both the water and sewer treatment systems must be expanded. Based on reasonable increments, the engineer recommends the following:

- 0.10-mgd wastewater treatment plant (WWTP) expansion is $500,000,
- 0.10-mgd water treatment plant expansion is $250,000, and
- 1- to 0.25-mgd well is $250,000.

Money will be borrowed to construct these improvements. The net present worth of the debt service over the life of the bonds is anticipated to be $100 for water and $100 for sewer. The utility needs to find the appropriate impact fees it should be charging.

At 250 gpd for water, the plant expansion will serve 400 units. The initial impact fee calculation for the water plant is $250,000 divided by 400, or $625 per unit. But the well is also required. For the well, 250 gpd at 250 mgd means the well can serve 1,000 units. Therefore the portion of the well associated with each unit is $250,000 divided by 1,000 units, or $250. The total initial impact fee is then $875. Deduct the $100 debt credits and the appropriate impact fee is $775 per unit for water.

At 200 gpd for wastewater, the plant expansion will serve 500 units. Therefore the initial impact fee calculation is $500,000 divided by 500, or $1,000 per unit. Deduct the $100 debt credits and the appropriate impact fee is $900 per unit for sewer.

Example—Impact Fee Calculation from Capital Improvement Plan

Utility E has a capital improvement plan as shown here. The board of commissioners directs staff to develop a revised impact fee. The utility's financial policy requires growth to pay for growth.

Item	Capacity	Cost (millions)
WTP expansion	7 mgd	$14.0
Wells	7 mgd	$3.5
Transmission pipeline		$2.1
WWTP expansion	2.5 mgd	$10.0
Two major lift stations		$1.0
Transmission pipeline		$2.0
Retrofit area with sewer		$10.0
Infiltration/inflow correction		$2.0

The utility notes that the infiltration and retrofit sewer projects are not appropriately funded by impact fees (the latter should be an assessment district). As a result, the utility calculates the impact fee as follows:

Item	Capacity	Cost (millions)	ERUs Served	Cost/ERU
WTP expansion	7 mgd	$14.0	20,000	$700.00
Wells	7 mgd	$3.5	20,000	$175.00
Transmission pipeline		$2.1	40,000	$52.50
TOTAL				$927.50
Debt PW				$127.50

(continued)

Item	Capacity	Cost (millions)	ERUs Served	Cost/ERU
Impact fee water				$800.00
WWTP expansion	2.5 mgd	$10.0	10,000	$1,000.00
Two major lift stations		$1.0	20,000	$250.00
Transmission pipeline		$2.0	20,000	$250.00
Retrofit area with sewer		$10.0	n/a	0
Infiltration/inflow		$2.0	n/a	0
TOTAL				$1,500.00
Debt PW				$200.00
Impact fee sewer				$1,300.00

The recommended impact fees are $800 per ERU for water and $1,300 per ERU for wastewater.

SPECIAL PROJECTS AND ASSESSMENT

There are a number of instances where impact fees logically do not apply because they have no regional benefits. Such facilities would include the following:

- Small gravity sewer lines
- Local water lines
- Neighborhood pump stations and attendant force mains
- Interconnecting transmission lines and other facilities typically installed and dedicated to the utility at the time of construction of subdivisions or developments by developers, by other assessment districts, municipal services taxing or benefits units, or like similarly or specially funded projects in areas determined to need new installations or retrofits
- Connections to the utility system

These improvements serve a limited geographical area. They are typically termed *subdivision infrastructure*. This would include the installation of water or wastewater service where it is currently not available, or replacement of outdated or older infrastructure on the system. In many cases, the appropriate way to do this is through assessments. In addition, residents may request that water or wastewater service be replaced or supplied to their neighborhood, and they will petition the utility's governing board to undertake the project. Assessments are an appropriate means to accomplish this also.

Assessments are collected to meet special benefits for a sector of the population, and must represent a fair and reasonable portion of the cost to each of the projects subject to the improvements and the assessment (the assessment version of the rational nexus test). Payment of the assessment bill may be enforced through a lien against the property, most easily accomplished by placement of the assessment on the property tax bill so that failure to pay the property tax bill and assessment (which cannot be separated) will cause the tax collector to pursue liens on the property. While a detailed discussion of assessments in all jurisdictions is beyond the scope of this document, there are fairly strict requirements established for assessments in state statutes.

CONNECTIONS AND MISCELLANEOUS CHARGES

The category "connection and miscellaneous charges" pertains to a variety of services provided to specific customers as desired or required. For instance, when someone wants to connect to the system there is a charge to install the service tap, to turn the meter on, and so forth. Each of these items should be listed on an individual bill to the customer in the amount associated with the actual cost to provide the service (rarely this is the periodic user charge bill). An analysis should be undertaken to determine appropriate charges for these services, including the personnel, overhead, vehicles, and so forth. Such things as connections, turn-ons, turnoffs, meter rereads at the customer's request, backflow tests, meter tests, meter change-outs, meter installations, repair of damage caused by the customer, and water audits can all have fees assigned to them. These costs should be paid by the only person benefiting from the services—the customer.

Appropriate allocation and charging for these services can be a significant source of revenue for a utility system in given year. Where problems exist in regard to too many charges for too many services, it is usually caused by attempting to track these billings. A good computer system will alleviate the problem, although any of these services can be included within the general rate for monthly service. An example ordinance is included as appendix D.

BULK USER RATES

Bulk users are a specific subset of recipients of water or wastewater services. Unlike service provided by a utility to its own customers, the typical bulk service arrangement is with a neighboring utility or other user that would cause one of the parties to receive the service on a wholesale basis, bypassing much of the distribution and collection system costs of the wholesaling utility. As a result, the cost to maintain the extensive piping systems and the connections to individual customers is borne by the bulk user, not the wholesaler, so the cost allocations may differ significantly from retail customers.

In trying to determine the appropriate rates for bulk users, it is recommended that the availability and volumetric charge system be implemented. The debt service portion of the cost must always be recovered, often on the basis of capacity reserved. This is best done up front through collection of an availability charge. Secondly, the treatment and transmission costs need to be determined. While treatment cost is readily available from treatment plant budgets, the cost of transmission is not as easily determined because transmission cost is often included as part of water distribution or sewage collection budgets, which will include both the transmission system and local subdivision infrastructure. As a result, when allocating these costs, carefully consider maintenance costs per mile for large-diameter pipelines and force mains, and the cost to maintain large lift stations. Where maintenance records exist, it may be beneficial to allocate the large pump station costs or tank costs based on past experience. Also consider any allocations of costs associated with risk on the system, such as guarantees of water quality in the bulk user's system after it is passed through the meters, and any growth needs or pretreatment limitations.

A rational basis for demonstrating the costs of operations and maintenance to the bulk users will alleviate a great many potential problems. In trying to determine the formula for a bulk user, establishing a per-thousand-gallon charge along with the availability charge treats the bulk user like a regular retail customer and alleviates the need to do follow-up corrections or adjustments on an annual basis. The cost per thousand gallons to the bulk user usually will be significantly less than that to the retail user.

In many cases, a portion of the bulk user agreement should include a payment of impact fees, collected by the bulk user and paid directly to the other utility. This alleviates the need to calculate reserved capacity in the availability portion of the bill. However, where reserved treatment plant and/or transmission capacity is desired, a cost to provide and reserve that capacity should be included in the bulk user rate, ideally as a part of the availability charge.

REFERENCES

Bloetscher, F. (2009). *Water Basics for Decision Makers: What Local Officials Need to Know About Water and Wastewater Systems*. Denver, Colo.: American Water Works Association.

Rosalind Holding Company v. Orlando Utility Commission. District Court of Appeal of Florida, Fifth District, Nos. 79-1298/T4-596; 80–89.

CHAPTER 14

Financial Policies and Evaluations of Financial Health of the Utility

The top priority of water and wastewater systems is to provide a level of service that meets state and federal regulatory requirements, and the demands and expectations of its customers. The long-term goal of a financial policy is to develop a utility system that is stable, requiring limited increases in the cost of services, minimizing those increases that are required, and minimizing the acquisition of long-term debt, which in a stable utility system will be the driving force for significant rate increases. Because the ratepayers bear the ultimate cost of service, there needs to be a financial plan that will permit the utility to meet its priorities at an affordable and stable cost over the long term.

To accomplish these goals, most utilities are set up as enterprise funds, operating like a business. The water and wastewater customers are not subsidized by the general fund taxpayers of the local government, who may not be the same group as the ratepayers. As an enterprise, the utility must derive its own revenues and use those revenues to fund operations, capital, debt, and reserves to meet customer and regulatory demands. As federal funding sources have been phased out, revenue-backed borrowing has increased. As a result, water and sewer systems face additional pressures to control and maintain historical debt service coverage levels and liquidity. A utility's ability to finance projects to satisfy environmental regulations has become a critical bond rating determinant, along with its ability to meet growing customer demands.

Prior chapters focused on the operation and capital needs of a utility system and the funding sources to meet those operational and capital requirements. To determine the appropriate levels for these revenue sources, a more encompassing picture of the financial stability of the utility system is needed. The overall view of the fiscal strength of the utility system extends beyond simply meeting operating and capital needs—a set of guiding principles is also needed. The focus of this chapter is to provide (a) a basic understanding of the need for a fiscal policy, (b) the theory behind the allocation of costs to revenue sources, (c) legal or policy provisions that should be identified and included in the overall financial picture, and (d) a discussion of forecasting revenues and expenditures.

FINANCIAL POLICIES

Financial policies are a set of long-range business principles, goals, and guidelines approved by the governing board to provide direction to staff on the courses of action to take with regard to the finances of the utility, establishing policy guidelines for developing budgets, and providing standards on which to measure performance (Kavanaugh and Williams, 2004).

Financial policies serve a number of purposes for both management and the governing board of the utility. Among these purposes are (Kavanaugh and Williams, 2004):

- Allowing elected officials and managers to approach financial questions from an overall, long-range vantage point (the financial policy should include the capital improvement plan)
- Presenting an overall financial picture of the entity, rather than receiving financial decisions on a case-by-case basis, which saves time and provides direction and focus on long-range policy objectives
- Improving financial stability of the utility by allowing the utility to plan and prepare for financial emergencies through the establishment of long-term reserve funds and mitigation of future increases in the expenditures or reductions of revenues

A formal adopted policy provides a written set of policies from which the governing board can base decisions that will avoid the conflicts resulting from inconsistent financial decisions. This, in turn, prevents problems that occur as a result of case-by-case decisions that inevitably lead to conflicts with current, past, or future policy alternatives that may cost more in the long run (Kavanaugh and Williams, 2004).

Formal adopted policies, such as the operating budget, usually involve open discussion within the community. Formal policies promote continuity regardless of changes in personnel or governing board members, providing a long-term view of fiscal procedures. Formal policies can increase efficiency by standardizing fiscal procedures, while informing new employees and officials of the expected courses of action. Bond rating agencies look favorably on utilities with adopted financial policies in place and followed.

Informal policies develop with time and a limited paper trail. They are more related to the culture of the organization and the peculiarities of the staff. Informal policies rely on past practices or specific employee preferences. Whether specific or vague, they are not formally approved by the governing body. However, they have the benefit of being flexible and may have more support internally than certain formal policies. Whether formal or informal policy documents, they should be considered "living" documents that change as conditions require.

Many states provide some guidelines for fiscal management, but in most cases, these are not rigid rules. Figure 14-1 shows the types of financial policies and the flexibility involved. Guidelines are useful and rarely will limit flexibility unless they attempt to micromanage the fiscal operations. Such policies are appropriate for all entities, regardless of size. Figure 14-2 shows the ramifications of financial policies.

A financial policy should provide direction to (Kavanaugh and Williams, 2004):

- Allocate appropriate resources in accordance with expenditure programs and revenue needs
- Project revenues and expenditures on a multiyear basis and verify the veracity of the projections from current and prior years
- Act as a planning tool from which the utility can plan expenditures in a systematic or rational manner over time, rather than large amounts at one time
- Determine when revenue shortfalls may appear so plans can be made in advance to avoid or address them

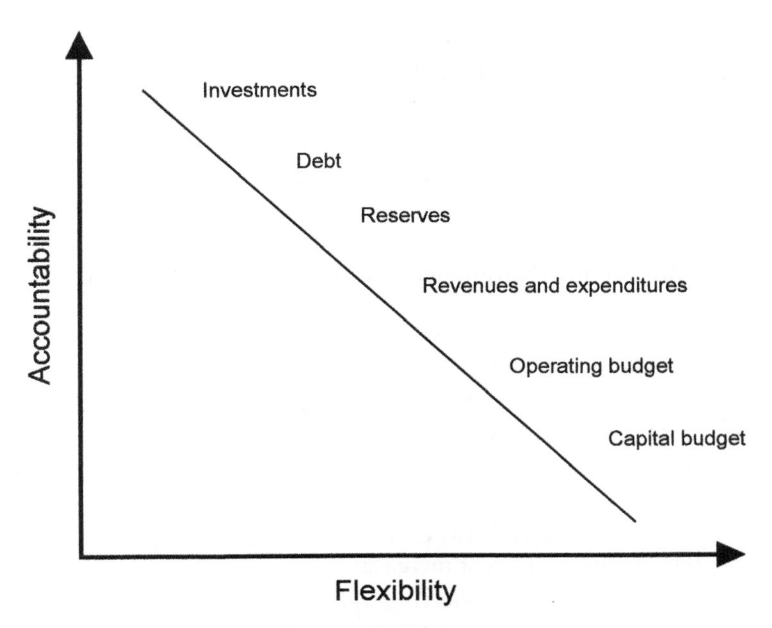

FIGURE 14-1 Accountability versus flexibility in financial policies
Adapted from: Kavanaugh and Williams (2004).

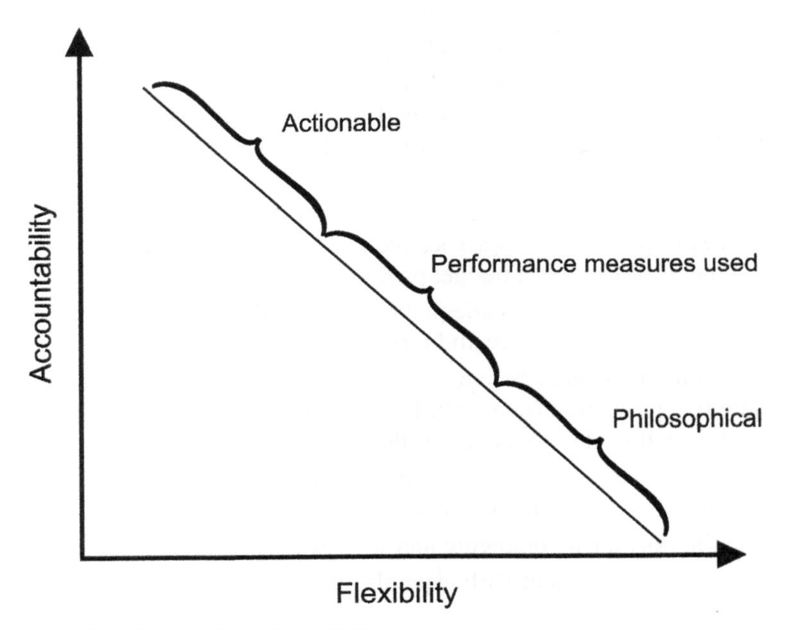

FIGURE 14-2 Ramifications of financial policies
Adapted from: Kavanaugh and Williams (2004).

- Simplify the annual budgeting process
- Plan multiyear programs appropriately
- Develop capital programs to meet growth and maintenance needs, including the access to debt at a reasonable rate
- Anticipate changes in service demands and revenue needs
- Facilitate coordination among departments within the local government

A financial policy should (Kavanaugh and Williams, 2004):

- Maintain a diversified and stable revenue system to shelter it from short-term fluctuations in any one revenue source.
- Set fees and user charges for each water and sewer enterprise fund at a level that fully supports the total direct and indirect cost of the activity.
- Establish all user charges and fees at a level related to the cost of providing those services in conformance with laws of the state.
- Require calculations of the full cost of activities supported by user fees to identify the impact of inflation and other cost increases.
- Follow an aggressive policy of collecting water and sewer revenues.
- Estimate its annual revenues by an objective, analytical process.
- Project revenues for the next five years and update this projection annually, examining existing and potential revenue sources.
- Maintain a multiyear plan for capital improvements and update it annually.
- Enact an annual capital budget based on the multiyear capital improvement plan. (Future capital expenditures necessitated by changes in population, changes in real estate development, or changes in economic base will be calculated and included in capital budget projections.)
- Coordinate development of both the capital improvement budget and the operating budget. Future operating costs associated with new capital improvement will be projected and included in capital budget projections.
- Maintain all its assets at a level adequate to protect the utility's capital investment and to minimize future maintenance and replacement costs.
- Project its equipment replacement and maintenance needs for the next several years and update this projection each year. For this projection, a maintenance and replacement schedule will be developed and followed.
- Identify the estimated costs and potential funding sources for each capital project proposal before it is submitted to the governing board for approval.
- Determine the least costly financing method for all new projects.
- Allocate resources consistent with expenditure needs, state and local ordinances, and proper funding policies.
- Maintain adequate repair and replacement funding to be appropriated for water and wastewater capital needs to minimize future debt obligations.
- Maintain a separate accounting for reserve capacity fees, repair and replacement funds, telemetry contributions, inspection reimbursements from developers, and contributions toward capital construction by project.

- Levy all fees and charges in a fair and equitable manner, reflective of the actual cost of providing the service for which the fee was rendered.
- Project regulatory changes and budget capital expenditures.
- Develop pricing strategies that will protect the overall utility system.

Governmental entities create financial policies for a number of functional areas, including the following (Kavanaugh and Williams, 2004).

- Operating budget
- Revenues
- Expenditures
- Capital improvements
- Debt
- Procurement
- Investment of assets
- Risk management
- Human resources (from a compensation, pension, and classification perspective)

Effective policies are those that are (Kavanaugh and Williams, 2004):

- Explicit, yet brief so that they can be agreed on by all parties
- Literal in meaning and therefore easily interpreted by laypeople
- Comprehensive in addressing all salient issues, but not complex
- Maintained current through periodic review and amendment
- Available for reference

Table 14-1 shows typical budget policy components. Table 14-2 shows typical revenue and expenditure policy components. The benefit of defined policies is that they lend themselves to measurable policy components (Table 14-3).

Development of financial policies should not be delayed until a crisis erupts. This limits the potential options that can be considered and may create a rush toward incomplete or misguided policies without the necessary public discourse.

Rate Stabilization

Utilities should develop a policy to guide staff on the maintenance and use of rate stabilization reserves. Because debt drives rates, it is often helpful to begin increasing rates before the debt service begins. In the intervening period, the excess cash can be used for rate stabilization purposes, which extends the time for implementation of large rate increases. It also is a buffer in the event some emergency disrupts cash flows. However, rate stabilization funds are a one-time expenditure and, therefore, once spent cannot be reallocated unless the rate stabilization fund is replenished.

TABLE 14-1 Operating budget policy components

Operating Budget Links	GFOA* Strength of Recommendation		
	High	Medium	Optional
Links between financial and strategic or other plans	X		
Scope	X		
Comprehensive		X	
Budget format	X		
Basis of budgeting defined	X		
Budget calendar established	X		
Responsibilities assigned	X		
Budget and control system		X	
Process of budget amendments	X		
Balanced budget	X		
When reserves can be tapped		X	
Performance measurement			X

*Government Finance Officers Association.
Adapted from: Kavanaugh and Williams (2004).

TABLE 14-2 Revenue and expenditure policy components

Revenue and Expenditure Policy Components	GFOA* Strength of Recommendation		
	High	Medium	Optional
Diversification/stabilization			
Prevent fluctuations	X		
One-time revenues	X		
New revenues		X	
Estimate of revenues			
Conservative, objective, and reasonable	X		
Multiyear forecasts			X
Revenue manual			X
User fees			
Cost recovery	X		
Review of fees and charges	X		
Changes in fees and charges		X	
New fees		X	
Expenditures			
Maintaining capital assets		X	
Program review		X	

*Government Finance Officers Association.
Adapted from: Kavanaugh and Williams (2004).

TABLE 14-3 Performance measures for policy components

Policy Component	Performance Measure
Operating budget	
Budget calendar	Number of deadlines met
Revenues and expenditures	
Estimates of revenues	Accuracy of forecast
Cost recovery	Cost:recovery ratio
Maintenance of assets	Physical condition
Fund balance	
Appropriate size of unreserved fund balance	Percent of operating fund
Capital improvement	
Minimum level of spending	Appropriation > minimum
Pay-as-you-go	High percent of PAYGO
Debt instruments	Percent of revenues assigned to capital projects
Debt management	
Refunding bonds	Present-worth savings
Debt service limitations	Annual debt/capita
Debt outstanding limitations	Debt vs. assessed valuation
Repayment provisions	Percent of principal paid in 5 years
Bond at rating goals	Meets rating goal
Investments	
Yield	Percent return vs. market
Portfolio diversification	Percent of portfolio invested

Adapted from: Kavanaugh and Williams (2004).

Contingency Policies

Each utility should develop guidelines for maintaining contingency funds for use in the event of emergencies, natural disasters, unforeseen failures, and unexpected events. A catastrophic event can seriously hamper the utility's ability to provide service to its customers and can seriously impact its financial condition. The vulnerability assessment and emergency response plan discussed in chapter 4 will identify those potential areas where emergencies might occur and their potential cost.

Certain reserve funds, equipment inventories parts, and contracts for services can be arranged for the most common problems. An analysis of past budgets will indicate the likelihood of large, unanticipated expenses with time. This can be used as a guide for the utility for allocating a contingency amount in the annual budget. Having a contingency amount in the budget will permit the utility to fund such emergencies without having to amend the budget each time a problem occurs. If the money is not spent, then it can be reappropriated in following years. However, no utility can bank enough funds for every emergency. A line of credit through a local bank would be prudent for high-cost items.

Fund Balance

Fund balances are those monies not encumbered for other purposes. Utilities maintain fund balances to cover those potentially volatile expenses like power, chemicals, unexpected overtime costs, and emergencies. When significant amounts of fund balance are collected over a period of years, they can be appropriated for capital projects. Fund balance should not be budgeted to balance operations expenses unless some unforeseen revenue shortfall occurs. Fund balance should be invested to earn interest, which is an added revenue for the utility. The appropriate level of fund balance should be based on the volatility of revenues, the variability of expenses, the need to replenish debt, rate stabilization, and unreserved fund balance funds.

Unreserved Fund Balance

Unreserved fund balances are separated from reserve funds as they provide a different opportunity. Revenues will generally lag expenditures simply because utilities bill their customers after they have received service. As a result, the utility must have an unreserved fund balance to meet this discrepancy.

The Government Finance Officers Association (GFOA) recommends that the unreserved fund balance be a minimum of 15 percent of the annual user fee revenues. A month and a half to three months of revenues are other suggestions, depending on the frequency of billing. If the unreserved fund balance falls below the policy level, a mechanism should exist to restore this amount. Figure 14-3 shows the total and unreserved fund balance. Table 14-4 shows typical fund balance policy components. Unreserved fund balance monies should be highly liquid investments.

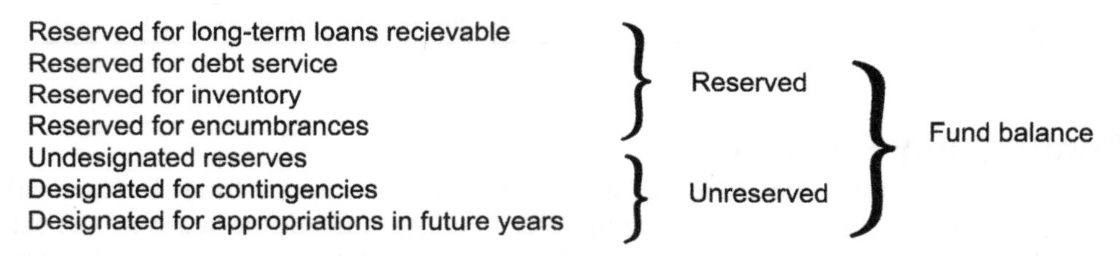

FIGURE 14-3 Graphical representation of fund balance

TABLE 14-4 Fund balance policy components

Fund Balance Policy Components	GFOA* Strength of Recommendation		
	High	Medium	Optional
How reserves are established for unreserved fund balance	X		
Determining appropriate size of fund balance	X		
Methods for utilizing unreserved fund balance	X		

*Government Finance Officers Association

Adapted from: Kavanaugh and Williams (2004).

Debt Policies

Utilities intending to issue debt are advised to develop a debt policy. The GFOA recommends that a debt policy should address the following (Miranda and Picur, 2000):

- Types of debt permitted to be issued (e.g., revenue bonds, line of credit, and COPs)
- Method of sale of debt instruments (negotiated sale, competitive bids, or private placement)
- Selection procedure for consultants to help with issuance of debt
- Disclosure to investors
- Use of debt proceeds
- Debt capacity limitations
- Integration of debt and capital planning activities in the capital improvement program
- Structure of the debt issuance (e.g., terms and redemption polices)
- Investment of debt proceeds
- Maintenance responsibilities
- Credit policies and compliance with existing laws
- Policy of refunding debt

Table 14-5 outlines debt management policy components recommended by GFOA. Debt issuance can come in a variety of forms, depending on the size of the issue, the expediency required for the proceeds, and whether the obligation is intended to be long or short term. The types of debt a utility can use have been discussed previously as has the use of debt proceeds. A financial analyst can make a recommendation as to the appropriate method of sale for any given issuance, but the utility should not limit its options.

TABLE 14-5 Debt management policy components

Debt Management Policy Components	GFOA* Strength of Recommendation		
	High	Medium	Optional
Purpose of policy defined		X	
Oversight responsibilities defined		X	
Conditions for debt issuance	X		
Purposes and use of debt	X		
Project life impact	X		
Types of debt permitted	X		
Refunding of debt	X		
Restrictions to debt utilization	X		
When debt will not be issued		X	
Minimum size of issuance			X
Maturity limitations	X		
Statutory limitations	X		

(continued)

TABLE 14-5 Debt management policy components (Continued)

Debt Management Policy Components	GFOA* Strength of Recommendation		
	High	Medium	Optional
Debt service limitations	X		
General fund contribution	X		
Per capita income limits	X		
Expenditure percentage limits	X		
Limitations of outstanding debt	X		
Market value limitations	X		
Assessed value limitations	X		
Per capita income limits	X		
Misc. limitations	X		
Overlapping debt			X
Characteristics of debt structure	X		
Repayment provisions	X		
Maturity guidelines	X		
Debt service fund requirements	X		
Use of letters of credit/insurances, etc.			X
Investment of bond proceeds		X	
Debt issuance process	X		
Sale process	X		
Professional service usage and solicitation	X		
Bond rating goals			X

*Government Finance Officers Association
Adapted from: Kavanaugh and Williams (2004).

The capacity for debt issuance should result from a study of the financial condition of the utility and the community. The intended audience of the report includes the media, rating agencies, investors, and the governing body. It should include how the affordability might impact the capital program, the impact the results may have on current policies, and new policies that might be required. Measures of debt capacity can take many forms.

Unfortunately, there is no standard set of debt capacity indicators, but the collection of information about the community is suggested. Figure 14-4 shows the framework for analyzing debt capacity, including the collection of data for the community and for neighboring communities for comparison purposes, development of debt issuance scenarios, determination of the amount of debt that can be issued, and development of or revision to existing debt policies. The following are suggestions:

- The amount of debt service in comparison to the margin between revenues and expenditures (the amount that could be spent for debt service)
- Total outstanding debt (all types)
- Total annual debt service (all types)
- Bonded debt (excluding short-term notes and liabilities)

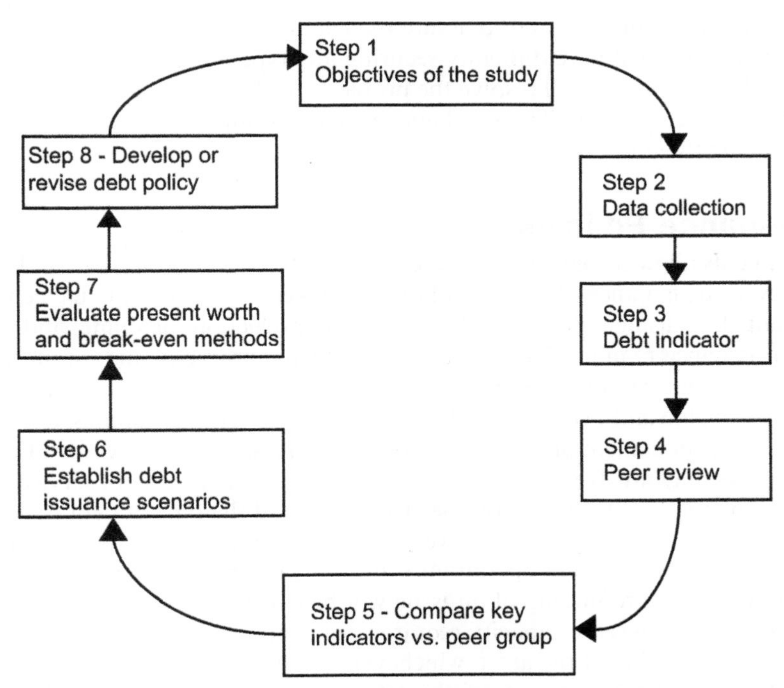

FIGURE 14-4 Debt capacity analysis framework that shows how determining the ability to issue debt is an iterative, multistage process
Adapted from: Kavanaugh and Williams (2004).

- Debt as a percentage of the total property values (which measures the total debt obligations as a portion of all property values)
- Debt per capita
- Debt as a percentage of per capita income (which measures the ability to pay)
- Debt service as a percentage of per capita income (which also measures the ability to pay)
- Debt service as a percentage of total system revenues
- Debt service as a percentage of operating revenues

Use of debt indicators in comparison to other communities provides information on whether or not the debt to be issued will create any potential hardship to the utility customers.

Using the comprehensive annual financial report (CAFR), the data are easy to obtain, but different reporting requirements make the comparisons difficult and may confuse refunding and debt proceeds.

The following list includes characteristics of the debt issue.

- Repayment provisions
- Maturity guidelines
- Debt service funds
- Credit enhancements
- Investment limitations of bond proceeds

Credit enhancements are used to assure the lender that the utility will repay the loan. Small utility systems will have difficulty securing any type of rating. Credit enhancement measures such as insurance will resolve the problem. Such measures may include letters of credit and insurance policies. These enhancements will improve the rating of the utility, thereby reducing the interest.

Debt Issuance Policies

Policy documents allow the governing board and staff to make decisions about how to issue debt. However, the issuance of debt should not be taken lightly. Debt issuance policies are project-oriented, focusing on capital infrastructure construction, planning, and land acquisition. Debt issuance requires definitive purposes for the proceeds, and, in most cases, the debt cannot be issued without specificity about the proposed expenditures. The expenditures should be reasonable and necessary and related to some form of planning document. The lack of a planning document or study will raise questions about the necessity of the debt.

The policies may involve philosophical issues, such as that the debt service should not constitute an unreasonable burden on the customers, and the type of debt issued. Restrictions may be placed on the size of any given issue to avoid rate shock to the customers, the use of debt (debt to supplement operations is usually prohibited for obvious reasons), and the terms under which refunding of an issue will be considered. A limit of the term debt can be issued and the reserve capacity required are common (usually borrowing does not exceed 30 years or the life of the asset, whichever is lesser).

Refunding may be appropriate if the refunding issue will substantially reduce the interest rate without extending the payback period or if the opportunity exists to remove restrictive bond covenants. In general, for refunding to be appropriate with regard to interest rates, the interest rate should be at least 1.5 percent less than the current bond issue (e.g., 4.5 percent versus 6.0 percent per year in interest rates).

Debt service is considered to be outside the operating budget. As a result, a series of policies may be adopted that relate to the amount of debt that can be issued in relation to the operating budget. Caution should be used when setting such a policy, as new systems or those undergoing significant rehabilitation may have debt in percentages that are significant when compared to operating costs. In all cases, the operating revenues should cover operating costs plus debt, with some reserve capacity. Most debt issues require a debt reserve that is a sizable percentage of the debt service.

Debt Service Reserves

The rating agencies look for a fully funded debt reserve because utility systems can incur fluctuations in revenues and expenditures for a variety of reasons, some of which include facility damage, sudden changes in chemical and power costs, and new environmental regulations that impact the utility's ability to provide services. Debt service reserves are most important for utility systems or projects that exhibit a concentration of assets or customer base, shallow service area economy, cash flow constraints, lack of operating history, or competition for services. A utility with these characteristics will need a fully funded reserve to obtain the highest potential bond rating.

Investing Borrowed Proceeds

Once a bond issue is closed, the funds are transferred to the account of the utility. In most cases, the funds arrive prior to the obligation to pay occurs because most governmental entities must have the funds prior to executing a contract. The total of the funds may be significant, and rather than allowing the funds to sit idle, the utility will invest the proceeds. A financial advisor can help with investment decisions. Most important is that the investment decisions must protect the bond principal.

The bond proceeds cannot be commingled with operating and other funds of the utility. To do so would interfere with the ability of auditors to ensure compliance with federal arbitrage provisions. Arbitrage, the extended earning of interest in an amount greater than interest paid by governmental entities, is prohibited. Table 14-6 outlines the GFOA-recommended policies for debt management.

TABLE 14-6 Investment policy components

Investment Policy Components	GFOA* Strength of Recommendation		
	High	Medium	Optional
Scope	X		
Objectives	X		
Standards of care			
Authority to invest	X		
Prudence			X
Conflicts of interest/ethics		X	
Oversight			X
Investment portfolio			
Authorized investments	X		
Diversification		X	
Maturities		X	
Portfolio valuation		X	
Safekeeping and custody of investments			
Eligible institutions	X		
Safekeeping	X		
Collateral	X		
Internal controls	X		
Reporting			
Frequency and format		X	
Performance measurement			X
Other components			
Policy changes			X
Review and adoption		X	

*Government Finance Officers Association

Adapted from: Kavanaugh and Williams (2004).

Transfers out of the Utility Fund

A major concern for many local government utility managers is general fund transfers, which happen frequently in local government operations. A well-documented cost allocation study will provide support for transferring the cost of purchased services by the utility from the local government. Local governments should be careful with "payments in lieu of taxes." The utility is tax-exempt if it is a public entity. Attempting to value these assets can be a challenge as depreciation and backup of value should be included. Payments in lieu of taxes would be designed based on applying the local tax rate to the net plant asset value of the utility. It is not appropriate to use arbitrary sums to balance the general fund. Franchise fees may be appropriate, but typically yield limited amounts of money and would be duplicative of payments in lieu of taxes, so charging for both is inappropriate. An example of another poor option would be a "return-on-investment" charge, which would seem unsupportable for most utilities since the general fund usually has not spent money on the utility in recent memory. Since most utilities are enterprise funds, they invest in themselves and are therefore entitled to keep the "rate of return."

Consumer inquiries indicate that unjustified subsidies are losing voter support, and with good reason. Excessive subsidies may not comport with existing case law and may penalize ratepayers, while providing others benefits for which they do not pay full value. Taking money from the utility has some impact on the operation of the utility—usually deferring capital obligations (the norm) or increasing rates. This also creates difficulties for utilities that try to measure their performance through comparisons with others, as noted in chapter 12.

MEASURING FINANCIAL CAPACITY

Fiscal Factors

The amendments to the Safe Drinking Water Act (SDWA) in 1996 included a section on the federal match to state revolving loan funds that stated that funds could not be disbursed to any utility system that lacks the technical, managerial, or financial capacity to maintain SDWA compliance unless the state determines that the loan will help the utility obtain long-term compliance. Hence many state revolving loan programs began to look at utility finances and managerial capacity in a manner similar to bond rating agencies.

The Environmental Finance Center at Boise State University has developed tools to help measure financial capacity. The tools involve the following (Boise State University, 1998):

- Total user fee revenue
- Total expenditures
- Cash flow
- Rate-setting frequency
- Affordability to the customer

The financial factors are less clear-cut than technical issues, which either meet standards or do not. The concern with financial matters can only be viewed in comparison to other, similarly situated utilities. Following are guidelines that provide insight into the utility's financial condition.

- Revenues should meet or exceed expenditures. The expenditures should include all relevant expenses based on full cost accounting of system operations. A ratio of the revenues to expenses should be determined. The ratio should be greater than 1:1. Most lenders want net revenues to exceed debt amounts by at least 15 percent.
- There should be no subsidies from other funds.
- The affordability index indicates the amount of money current customers are expected to pay for water and sewer service as a proportion of their income. This is measured by dividing the average bill by the median income of the service area. Most utility bills fall between 1.25 and 1.75 percent of the median income. Utilities with rates below 1.25 percent are well under the average and should have no problems securing the rate revenues to fund debt service. If the ratio is above 2 percent, further investigation is warranted. Current and proposed rates should be analyzed.
- The utility should maintain cash reserves at a minimum of 1.5 times the monthly operational expenses for emergencies, payment delinquencies, and cash flow purposes.
- The ratio of current assets to current liabilities should be well in excess of 2:1. Low ratios indicate either that the utility has significant amounts of debt and may not have the capacity to incur more, or that the system is dilapidated.
- The ratio of total sales to receivables should be in excess of 10:1.
- The ratio of sales to working capital should be greater than 1:1. Where the ratio is close to 1:1, one of two things is happening—the utility has very low liabilities and maintains a high cash balance, which is good, or the opposite, which may be an indication of financial stress that would concern potential lenders. The typical ratio is 6:1 (Boise State University, 1998).
- The ratio of total sales to net fixed assets should also be considered. Medium-size utilities have a ratio of 0.3:1 (Boise State University, 1998). If the ratio is high, it is likely that the utility is in deteriorated condition. If the ratio is low, the utility may have just completed a series of capital assets that distort the picture, or may have insufficient revenues being generated on the system. Those reviewing the fiscal condition of the utility will delve into this factor in more detail to determine if the underlying utility system assets are in poor condition.
- The ratio of sales to total assets should also be considered in the same light as the ratio of sales to net fixed assets, and for the same reasons.

From a financial management perspective, the following questions are raised:

- Does the utility produce and follow an annual budget? Most governmental entities are required to budget annually and have that budget approved by resolution or ordinance by the governing body after due consideration, advertisement, and public hearing.
- Does the utility have an audit of the finances conducted annually? Most governmental utilities are required to have an annual audit of finances conducted by outside entities. This audit is the CAFR.
- Are rates reviewed at least every two years, and if so, what basis or guidelines are used for rate-setting purposes? Use of a professional rate analyst is preferred to staff rate analyses for this type of rate review.
- Is there a capital budget? If so, is it followed and how is debt used to fund it?
- Does the utility have a bond rating?

In conjunction with the previous points, written polices and procedures are considered desirable, and their absence may lead to concerns about the preparedness of management in the event of emergency. Policies that utilities should have on hand include

- Personnel and human resources
- Customer service
- Operations safety
- Risk management
- Emergency response
- Operations and maintenance manuals for the treatment and transmission facilities

Other Fiscal Health Factors

A strong financial position entails efficient and effective use of financial resources and adequate contingency and reserve fund levels. A strong fiscal condition will increase the bonding capacity and bond marketability and will lower interest rates, which is of benefit to the public. There are three rating agencies in New York that evaluate the creditworthiness of utility systems (and local governments). As a result, their analysts are fully familiar with indicators of the strength of the utility. Their evaluations translate to ratings applied to the bonds, which drive the interest rate for the bonds. Using the rating agency criteria, the utility can measure its financial condition to yield insight on its available resources; taxing, revenue, and bonding capacity; and financial position.

In addition to the factors discussed previously, the municipal financial evaluation criteria documents include four other areas that are evaluated by bond rating firms in determining financial health and establishing a rating of debt risk. These include

- Administrative factors within the utility, including political and managerial stability and knowledge
- Local economic factors
- Legal provisions
- Debt

A brief discussion of each of these issues is warranted.

Administrative factors

Bond rating firms analyze the financial policies of a utility as a reflection of financial performance. Of concern are not only the financial controls and policies, but also the quality of management. Financial policies can work favorably when management achieves noteworthy financial performance despite a mediocre economic base. The form of government (e.g., professional managers are preferred to elected officials as the chief executive officer), political stability, and the ability to implement plans (i.e., quality of staff) and fulfill legal requirements are important considerations. Also included in the firms' analysis is an evaluation of personnel turnover ratios, history of labor–management relations, and legal/political constraints.

The following are questions the rating agencies try to answer to discern the utility's managerial capabilities:

- Are there institutionalized management systems for controlling cash, debt, and budgets?
- What does the organization do to attract and maintain qualified managers and middle managers?
- How are costs controlled? The agencies' question seeks to identify the extent to which alternative service delivery methods have been examined that might promote economies and efficiencies in providing services to the customers.

Reliability and continuity of information are important elements of management, as is exhibiting a willingness to make hard choices.

Economic factors

An economically diverse service base, creation of jobs, and adequate income levels ultimately translate to the ability to repay debt. All are important economic factors considered by the rating agencies. The economic environment is scanned from several strategic viewpoints, including the community's growth prospects, wealth levels of its population, and employment stability. Moderate growth is valued since it makes planning for future capacity needs more manageable. Rapid population increases or decreases can create operational and financial stress for a utility system. The rating firms also compare growth expectations for a community with the utility's supply capacity and expansion plans.

The rating agencies' focus on employment includes not only overall trends and size of the workforce, but also its distribution across employment sectors. Having a diverse economy will allow the utility to compensate if a sector of the economy declines. The goal is to evaluate whether the local ratepayers can endure economic disruptions, or if such disruptions would create additional loss of employment, reduced wealth levels, and slower growth in the service area. In many "company towns," this has proven to be a problem.

Legal provisions

An otherwise highly rated utility can be rated lower because of weak legal provisions in the bond documents, or in the legislation that confers powers to the utility (or municipality). Assurance that revenues will cover debt service fully is essential. There must be some legal affirmation that funds will be available to support efficient operations and capital investment because the rate covenant for repayment is linked fundamentally to the flow of funds. A strong bonds test requires that the new debt service resulting from the issuance of additional bonds be covered by the prior year's net value revenues.

The rating firms also assign ratings to water and wastewater revenue bonds after consideration of the regulatory environment of the water and wastewater systems. Compliance with federal and state quality standards is of particular concern. The utility's ability to finance projects mandated by external regulatory requirements is also pertinent in light of reduced availability of grant and loan funds from federal and state sources.

Debt factors

The type of security being pledged to debt repayment, overall debt burden, debt history, and trends are the factors reviewed for debt purposes. Debt burden is measured by the ability to repay debt by comparing revenues and the total budget resources to expenditures and existing debt. A utility desiring an optimum debt rating must demonstrate an effective planning program for capital improvements. Questions to be answered include

- What additional debt assurances are included in the capital plan?
- How realistic is the capital improvement plan?
- Does the utility's capital plan address growth demands and regulatory requirements in a timely, fiscally prudent manner?

The concern is that if the plan is not realistic or reasonable, higher construction costs could significantly increase a facility's ultimate costs, necessitating further borrowing or inability to bring facilities on-line to generate revenues.

The rating agencies look for a fully funded debt reserve because utility systems can incur fluctuations in revenues and expenditures, including those due to facility damage, sudden changes in chemical and power costs, and new environmental regulations that impact the utility's ability to provide services. Debt service reserves are most important for utility systems or projects that exhibit a concentration of assets or customer base, shallow service area economy, cash flow constraints, lack of operating history, or competition for services. A utility with these characteristics will need a fully funded reserve to obtain the highest potential bond rating.

Signs of Fiscal Stress

The objective of assigning ratings to bond issues is to identify significant economic strengths and weaknesses that would affect pledged revenues for the debt under review. Through this analysis, signs of fiscal stress that a rating agency will try to identify are listed below.

- National economic slowdowns
- Declines in revenues or deficits on the balance sheet
- Substantial court awards
- User charges to support an increased debt burden that may rise significantly over short periods of time
- Deferred rate increases that could have a detrimental impact on future financial performance
- Substandard fiscal policies and procedures
- Loss of a primary revenue source or limitations on expenditures

The result is a perception that the cumulative effects can impact the system's ability to maintain regulatory compliance, repay bonds, and provide capacity, which may lead to moratoriums on new hookups, and violations of state and federal discharge permits and contaminant levels, which may lead to costly fines.

FISCAL PLANNING PROCESS

Having reviewed what the rating and regulatory agencies look for from the utility, the financial plan can be developed. In developing a financial plan, a multistep process must be initiated that includes the following:

- Establishment of business principles
- Service area evaluation and capital improvement program development
- Development of capital financing strategy
- Customer and usage forecast
- Principles for the design of utility rates for service
- Public information program
- Post-implementation monitoring

Many of these items have been outlined in prior chapters or sections. The business principles are developed from the outline of the evaluation of the utility's current financial status, including its deficiencies. Business policies deal with who pays the bills, when, and how. The fundamental principle for a utility system is fairness; i.e., customers, whether existing or new, should pay their fair share of the cost of construction, financing, and operations of the utility's assets. While staff and consultants may be used to make these decisions, the ultimate responsibility for local financial policy lies with management and the governing board. Therefore the financial policy should be developed to bring out the short- and long-term consequences of each element, and to facilitate better understanding of the ramifications of each decision.

Service Area Evaluation

Once a formal financial policy is in place, the next step is to identify the service area, or market, characteristics. The market analysis includes existing and future customers to be served, an evaluation of competing suppliers (e.g., small private utility systems or other municipalities), a review of customers that can use wastewater effluent, and an evaluation of operational constraints that can affect development in the future. This evaluation should be made in concert with a reevaluation of the capital program as a result of updates in the planning processes outlined in chapters 8, 9, and 10.

Appropriate phasing of major capital additions, customer additions due to utility extension, and acquisition programs all are associated with the projected capital needs and long-term system revenues. Once the utility market is identified, the development of extension and expansion policies can be linked to the available financial strategies to equitably fund the capital improvement needs of the utility.

Once a customer and usage forecast is prepared that recognizes the existing customer needs and normal and incremental growth due to the acquisition and construction programs, then the amount of funds that will be generated from capital charges and utility rates can be identified. Details on development of rates and fees are discussed in chapter 13. The financial plan is intended to provide the framework to develop these rates and fees.

Development of Utility Financial Plan

Once the capital funding strategy and customer forecast are completed, the incremental changes in the utility operations expenditures should be recognized. These changes result from a variety of factors, including incremental cost increases due to facility additions and changes in regulations, acquisition of utilities, changes in the way business is conducted (e.g., implementation of a comprehensive meter replacement program), and recognition of target reserves and replacement programs.

As previously mentioned, maintenance of operating and replacement reserves is a major concern of the rating agencies as this indicates the ability of the utility to respond to significant emergencies. The implementation of an ongoing repair and replacement program funded by existing users for utility assets at the end of their service lives is strongly recommended to avoid excessive future borrowing. Where such policies bring increased rates, a public education effort is required. An informed public understands the purpose of the capital improvement program and the need for increased utility rates and capital facility fees to support the program.

REFERENCES

Boise State University. 1998. *DWSRF Capacity Assessment, City of Twin Falls, ID.* Boise, Idaho: Boise State University, Environmental Finance Center.

Kavanaugh, S., and W.A. Williams. 2004. *Financial Policies: Design and Implementation.* Chicago, Ill.: Government Finance Officers Association.

Miranda, R.A., and R.D. Picur. 2000. *Benchmarking and Measuring Debt Capacity.* Chicago, Ill.: Government Finance Officers Association.

CHAPTER 15

Customer Service and Public Relations

CUSTOMER SERVICE

Customer service (shown on the organization chart in Figure 1-1 in chapter 1) is the interface mode for most utility customers and the utility itself. Most customers neither talk to the utility staff, nor attend its meetings. But they all pay bills. To collect the fees and charges due to the utility, a set of policies, ordinances, and procedures must be developed, and the personnel to perform the work must be hired. The ordinances should not only include the rates for various fees and charges, but the conditions for services being delivered. A typical rate ordinance would include sections on

- Definitions
- Applications for service
- Connection fees
- Frequency of billing for service
- Meter sizing parameters
- Treatment of multiunit developments
- Master metering
- Location of meters
- Meter tampering
- Monthly availability and volumetric charges, including any associated conservation, peak, or other alternative rate structure
- Miscellaneous service charges
- Payment requirements and deadlines
- Delinquent bills
- Turnoffs and turn-ons in connection with nonpayment
- Customer service policies
- Billing dispute resolution
- No free service
- Penalties for violations
- Reserve levels and use
- Other services particular to a specific utility

An example rate ordinance is presented in appendix D. A procedure for billing and dispute resolution should be developed to guide the staff in implementing the ordinance and performing the billing function. In addition, impact fee collection procedures should be outlined, preferably in a separate policy or ordinance because the funds must be treated differently and kept separate from the rest of the revenues. A typical impact fee ordinance should include the items listed below.

- Findings
- Definitions used in the ordinance
- Defining the regional utility system, service area, and facilities to be constructed using impact fees in accordance with statutory and case law
- Due process appeals
- Impact fee rates
- Other information particular to the utility, but including such ideas as financing impact fees for existing homes now connected to the system, time of payment for new construction, public–private partnering arrangements, and interutility service area transfers

Standard forms should be developed for service applications, work orders to track services provided, and, if contemplated for financing, arrangements to promote consistency and fairness for all customers and to ease the staff burden. Excellent customer service defines a successful utility. Customer service is the first point of entry for the public (e.g., utility bill or payment thereof). Customers assume they are the highest priority—so make them feel like it. Customers expect things like online billing, online payments, and online bill review. Customer service functions include

- Reading meters
- Calculating and sending bills
- Processing customer payments
- Opening and closing accounts
- Responding to customer accounts
- Investigating billing issues
- Turn-ons/turnoffs
- New service

Needed customer service policies include

- Application for new service
- Connection
- Protocols and fees
- Billing adjustments
- Customer responsibilities
- Theft of water

- Collection mechanisms
- Termination for nonpayment
- Payment options
- Appeals of issues

Collection of Charges on the Utility Bill

As noted previously, the collection of charges for water and sewer services is usually done via a periodic utility bill. Failure to pay the bill will subject the customer to termination of service until the bill is paid. Penalties and fees are usually imposed to regain service. The legal issues associated with the collection of a charge on a utility bill are concerned with whether or not payment can be enforced by the threat of termination of service to the customer. Termination is achieved by turning off a valve in the meter box.

A question that sometimes arises is whether the failure to pay one charge (e.g., sewer) can be enforced by termination of all the water services. In *State v. the City of Miami, Florida*, the court upheld the utility's authority to terminate service if the bill is not paid in full. In this case the sewer charges on a water bill were challenged under due process. The court found that these two services were essentially equal to one another and thus failure to pay one would constitute failure to pay both and therefore service could be terminated.

In the typical utility system, an interlocking relationship exists for water, sewer, and perhaps reclaimed water irrigation service. Since there is a logical relationship between the services, failure to pay one would constitute a reason to terminate all the water-related services. However, many local governments use the utility bill to cover not only water and sewer services, but other services such as stormwater and garbage fees. Care should be exercised when placing these other charges on the water and sewer bill. In addition to significantly increasing the perception of the cost of water, termination of service for failure to pay the full amount of the bill may not be enforceable. In *Edris v. Sebring Utilities Commission*, the court found no interdependence between electrical services and water services. Lacking an identifiable mechanism to tie these services together, the court found against the utility. The court said that for the authority to use the enforcement mechanism depends on whether there is a financial, interlocking relationship between the two programs, or a logical and essential relationship between the two.

METER READING

It is recommended that utilities meter all water use. Metering allows the utility to track customer use patterns and account for water piped out of the treatment plant. Meters also provide utilities with a means to recover costs by charging user classes in accordance with the cost of serving that type of user and their actual consumption, which is both effective and equitable. Therefore, meters can be viewed as the utility's "cash register."

For metering to work in the most equitable manner, the appropriate meter sizing is important. Meters tend to underregister both low and high flows. As a result, the appropriate meter size should be designed to register the low flows and provide adequate water pressure and supply for normal use. Where demands vary widely, the use of a compound meter (one that measures both high and low flows) is recommended.

Meter sizing and type should be based on the previous Table 13-1, which was derived from the National Association of Plumbing, Heating and Cooling Contractors Water Supply Calculator, as derived from the National Bureau of Standards BMS 66 and 79 as amended. This table requires that the meter size be based on the actual number of fixture units as derived on calculation, based on an 8-gal-per-second (gps) flow rate. The code suggests that no meter be permitted to have more than the number of fixture units shown in Table 13-1, nor be allowed to exceed the peak flow volume as established by AWWA Manual M22. Therefore, in designing the appropriate meter size, there are three tests the meter must pass. The meter must be upsized until all three tests are met.

For instance, a proposed commercial use has 90 fixture units and a peak volume of 150 gpm. Based on the meter sizing chart, a 2-in. compound meter is the appropriate size. Correctly sizing the meter is important because the impact fees and monthly water bill are usually based on the meter sizing decision.

Any meter found to be undersized based on this table should be upsized to the appropriate meter size. All costs, charges, and additional fees relating to replacement of the meter should be paid by the property owner in conformance with applicable policies and ordinances. Impact fees may be affected by the increase in meter size since larger meters are assumed to require additional water demands. Developers and residents tend to request smaller meters than needed to reduce their periodic charges and their impact fees.

When a customer who has a water meter makes application for installation of a larger meter, the cost should be borne by the customer and credit should not be given for the tapping charges paid on the smaller meter, although credit for the prior impact fee should be considered. There should be no refunds or credits given to any customer requesting a smaller meter.

Meters must be left accessible to utility employees at all times. Reading meters is often the only time the customer and utility interact on a routine basis. As a result, public perception is important. Periodically a customer may dispute a billed amount, which requires a reread of the meter prior to the delinquency date of the bill. If possible, the reread should be done in the presence of the customer during normal working hours of the customer service representatives. During the reread, a customer service representative should check for indications of leaks, such as wet areas on the lawn or high meter revolution (or leak detector), and report any anomalies on the reread work order so the customer can be notified. The customer should turn off all spigots and other water uses during the time the customer service representative is on-site. If the meter is found to have been read wrong, the utility must adjust the bill accordingly to reflect the actual usage and issue a new bill.

If the customer wishes to pursue the matter further, then he or she may request a meter test. In general, utilities do not make adjustments unless the meter tests higher than standards set by AWWA for meter accuracy. If the customer remains unsatisfied and demands that his/her meter be replaced, the customer should be advised that meters tend to underregister low-volume water use, which is the most common household use, and the customer will likely receive a higher bill the month after the meter is replaced.

Meters should be protected at all costs because they are the means by which utilities obtain their revenue, and therefore accurate readings are important. Meters larger than 2 in. should be rebuilt at least every two years as accuracy may decrease noticeably over this period. Because there are usually a limited number of large meters, staff can accomplish this task. Small meters wear out after 7 to 10 years, depending on local conditions. Rarely is it cost-effective to repair these meters, so a routine change-out program should be budgeted as a part of the capital budget. Monitoring for unaccounted-for water will help identify when leaks or meters may be interfering with revenue collection. An example is shown in Table 15-1. Because of changes in usage patterns over the year, it is recommended that an annual running average be used to measure percentages of unaccounted-for water. For the utility shown in the table, the rise in unaccounted-for water was coincident with finding nearly 8 percent of the meter stock with zero readings for 90 days. A meter change-out program was initiated as a result.

TABLE 15-1 Unaccounted-for water for a utility

Month	Water Pumped (mil gal)	Estimated Fire and Similar Flows	Water Sold (mil gal)	Unaccounted-for Water Month (%)	Running Annual Avg. (%)
Oct-02	76.959	0.708	61.099	19.7%	
Nov-02	75.659	0.696	76.786	–2.4%	
Dec-02	75.031	0.690	65.716	11.5%	
Jan-03	75.549	0.695	74.273	0.8%	
Feb-03	71.411	0.657	69.335	2.0%	
Mar-03	80.19	0.738	71.412	10.0%	
Apr-03	77.834	0.716	72.831	5.5%	
May-03	78.811	0.725	72.426	7.2%	
Jun-03	75.074	0.691	67.369	9.3%	
Jul-03	80.107	0.737	67.131	15.3%	
Aug-03	73.837	0.679	67.188	8.1%	
Sep-03	72.507	0.667	67.725	5.7%	7.8%
Oct-03	81.047	0.746	60.781	24.1%	8.2%
Nov-03	79.918	0.735	71.766	9.3%	9.2%
Dec-03	83.139	0.765	62.844	23.5%	10.3%
Jan-04	87.843	0.808	83.129	4.4%	10.5%
Feb-04	77.062	0.709	74.14	2.9%	10.5%
Mar-04	88.959	0.818	68.938	21.6%	11.6%
Apr-04	84.704	0.779	82.525	1.7%	11.2%
May-04	83.368	0.767	65.423	20.6%	12.4%
Jun-04	91.907	0.846	66.339	26.9%	14.0%
Jul-04	93.27	0.858	82.147	11.0%	13.6%
Aug-04	82.356	0.758	75.638	7.2%	13.5%
Sep-04	79.395	0.730	67.391	14.2%	14.1%
Oct-04	84.243	0.775	75.506	9.5%	12.9%
Nov-04	86.16	0.793	73.762	13.5%	13.2%
Dec-04	92.7	0.853	80.075	12.7%	12.4%
Jan-05	91.143	0.839	67.755	24.7%	14.1%
Feb-05	83.826	0.771	74.119	10.7%	14.7%
Mar-05	89.893	0.827	70.464	20.7%	14.6%

Example—Utility Billing Issues

The city of Hollywood experienced a number of past issues with water billing functions. These issues arose from conversions of billing methodologies, new computer systems, changes in utility rates, and estimated billing. The issues included the perception that meters were not read, the bimonthly estimates (readings were bimonthly but bills were sent monthly) were incorrect, and customer service response was poor. To review the issues in a comprehensive manner (from meter reading to receipt of revenues), a task team was created to identify areas where productivity could be improved and processes streamlined.

The task team consisted of representatives from information services, commercial finance, the office of management and budget, water distribution, utilities administration, and the city manager's office. Each representative lent certain experience and expertise to portions of the process in an attempt to improve the following:

- Productivity improvement methods
- Accuracy of water bills
- Timeliness of water billing
- Customer relations

The task team was asked to develop

- Performance measures for meter reading and work orders to use as comparisons with other utilities
- Alternative mechanisms for delivery of billing services
- Operational and organizational procedures to smooth delivery of billing services
- Policy and ordinance changes to improve the billing and collection process

After several months of study, the task team discovered the most significant concern experienced by the billing staff was the routine bimonthly estimation of meter readings. This was because the estimates were based on the past four bills, two of which were estimates; thus errors compounded themselves. The result was the perception that many meter readings were incorrect and/or the meters were not read, despite the investigation indicating that *less than two tenths of 1 percent* of the meter readings or estimates provided by the meter readers were incorrect.

Other problems included

- An unusually high number of work orders, especially when a significant portion of those work orders were for account turn-ons and turnoffs.
- Lack of enforcement of delinquent accounts.
- Access problems. There remains great difficulty in reading meters in some areas due to fences, pools, buildings, dogs, and so forth.
- Lack of meter reader productivity. The meter readers worked for water distribution in the field, while the rest of the revenue function was in the finance department in city hall.

The meter readers were picked up at the water distribution center and transported to their routes, where they read meters in the morning. They were then picked up and transported to lunch (during which time they are also paid), transported back to their routes where they finished reading meters, and then waited to be picked up. This created between 2.5 and 3 unproductive hours each day per reader. A more inefficient system was unlikely.

The task team recommended that the city pursue the following:

- Conversion to monthly meter reading as soon as possible
- Contract meter reading because current staff could not read the meters monthly and additional staff was not a politically viable option
- Consolidation of the meter reading function, billing, and customer service into one division within one city department

The meter reading function was outsourced and monthly reading requirements were instituted by the contractor. The city also contracted the billing and customer service program. However, the cost to provide these services was significantly higher than the city staff's estimate with added meter readers. The justification for the contract was the lack of trust the customers had in the existing customer service and meter reading personnel.

PUBLIC RELATIONS

The issue of public trust brings the next topic to the fore. Perception is reality in local government. As a result, the activities of the utility will be scrutinized constantly. There will be times when failures of the water or sewer system will occur simply because underground conditions, lightning strikes, severe weather, and other natural occurrences cannot be predicted. Usually failures are small, e.g., blocked sewer lines, water main breaks, or pumps that short out, leaving people out of service for a few hours. But ice and wind storms put thousands of people out of service every year, sometimes for longer periods. Depending on the nature and severity of the failure, there may be an adverse effect on public confidence and perception of the utility, which is why it is important that there be an ongoing public relations and communication effort with the public that maintains public confidence in the utility prior to a failure. When a failure occurs, the response to the public should be carefully planned and factual.

The following suggestions for dealing with the public during failures can involve the press, television, and other outlets for publicity. Whether good or bad, in the information age in which we live, attention spans are short and perception often matters as much as or more than the facts. Since a failure is perceived as a negative, the press and public will expect negative issues. However, keeping the following suggestions in mind when speaking to the public or press will help with public relations efforts and in maintaining the public's confidence in the utility.

- Make sure the speaker is qualified to make the response. Confidence is lost when the speaker is not someone who really understands the issue.
- Keep the answers short and to the point. The press likes sound bites.
- Minimize jargon, which the public may perceive is only being used to confuse the issue.
- Avoid negative words and phrases, numbers, and dire predictions.

- Avoid discussions of risk, cost, and promises, all of which can be misinterpreted; instead focus on the utility's goals in providing service.
- Be genuine. Body language and the perception of the speaker may be as important as the words spoken.

This is not an exhaustive list, and no two communities will pursue the same strategies.

It is helpful if there is some planning for responsibilities of staff in the event of a failure. It may be helpful to provide training of specific employees who are identified as having a potential role during a problem. Likewise, employees who lack qualifications or the personality traits to represent the utility in a positive light should never be placed in the position of speaking to the public or press. A system to funnel information to a central point (such as the utility director) should also be instituted prior to a failure occurring. Evaluation of facilities may help identify potential areas where a failure could lead to significant impacts.

Other issues to keep in mind in dealing with the public include the need to avoid absolutes. Instead, focus on the goals of the utility, which is one reason strategic planning can be useful. The goals are to minimize risk and impacts to the public. Because of the technical and regulatory nature of utilities, the verbiage used may not be clear to the uninitiated, so pertinent commentary is important. Speculation, promises, costs, and humor should be avoided. Presentations should be 10 to 15 minutes at the most, and reserved only for critical issues if the press is involved. Otherwise the message will get lost in the details.

Example—Failure of a Utility to Protect Local Groundwater

A large city-run utility system in the mid-Atlantic region of the United States is considered to be well run with a high-quality staff. The system serves nearly 1 million people, and has been growing very quickly. Keeping up with wastewater demands through construction of pipes and treatment facilities has been a challenge.

In the late 1990s, the city was applying its wastewater sludge to a field outside the city. In 2002, high nitrates were found in 39 individual service wells in the area. The state decided that the city's sludge practices were at fault, i.e., too much sludge was applied to the field. The state fined the city more than $82,000, but residents were outraged that their wells could no longer be used.

The problem quickly became headlines in the local papers, one of which was that the mayor "apologizes to well owners." This was not the kind of publicity the city needed. Ultimately the city decided to spend additional money on the treatment system and to run pipelines to connect the residents to city water at no cost to the residents. The city's cost approached $5 million. The mayor indicated that the city was willing to spend as much as $30 million to solve the problem.

The city had been proactive in dealing with the problem. The city manager, mayor, and utility personnel were educated on the issues and, as a result, were able to deal with residents and the press. Yet, while the issue will be resolved, the reputation of the city remains tarnished, and affected residents do not trust the city and do not want to pay for the service. Unfortunately this is often the short-term outcome of utility failures.

Example—Risk Versus Cost of Failure

A North Carolina city had an emergency generator in its capital budget for years, but never funded it because the cost–benefit analysis never showed it was a worthwhile investment. An ice storm brought down power lines in 2002, which eliminated both feeds to the water plant. It was 48 hours before power was restored, leaving most of the city without water, freezing water mains and services, and creating significant damage to the pumping and piping system. Had the generator been in place, service would not have been lost except possibly in a few isolated areas. The city staff attempted to defend their decision. Later the mayor acknowledged that not installing the generator was an error, the implications of which were not well understood by decision makers. Too often the costs and benefits are not well understood by the person conducting the assessment. What was the "cost" of not having the generator to that community versus the "benefit"?

Example—Utilities Maintain Service After Hurricanes

Florida utilities have extensive experience dealing with weather-related impacts to their operations. As a result, most water and wastewater treatment plants have generators in place to avoid outages that occur routinely as a result of lightning strikes and hurricanes. During Hurricane Andrew and the seven storms of 2004 and 2005, most of the water systems remained in service, which allowed other emergency agencies to rely on the water and wastewater systems during cleanup in all but the most affected areas. The main reason for service to be out was broken pipes resulting from roots of uprooted trees planted in medians and rights-of-way. Numerous small breaks made it impossible to hold pressure in areas of the system, not a failure of the entire system itself. As a result, most people have confidence in the ability of their utility to maintain service during most weather events in Florida.

WATER CONSERVATION

Another area where the utility can engender significant, positive public relations for itself is through water conservation efforts. Experts have compiled large databases on the effectiveness and cost benefits of various conservation measures that incorporate a community's physical and environmental conditions necessary to yield a cost–benefit analysis of the most efficient courses of action to pursue. Analysis of cost–benefit ratios helps a community select the appropriate measures and make allowances for budgeting the program, and provides more flexibility in response to changing conditions.

The most effective programs are planned for five- or ten-year periods as a part of other planning efforts. The possibility of revenue shortfalls is always a concern when conservation programs are undertaken, so rate experts should develop rates that encourage conservation, yet generate the required revenues. Utilities should ease into water conservation efforts, or revenue problems will arise. Higher-usage categories should be targeted for using conservation measures before low-use categories as they yield the highest cost–benefit ratio to the utility. Because residential customers typically use about half of their water inside the home, the focus is usually on irrigation uses.

Water Conservation Measures

While there are numerous water conservation measures that can be employed to reduce water usage, depending on the community, some measures may be more beneficial to the system than others. For any water conservation measure under evaluation, the savings and costs in terms of capital, installation, and operation and maintenance of the measures are critical inputs to a cost–benefit analysis (Bloetscher and Meeroff, 2009).

Public information programs

Water conservation information programs increase customer awareness of habits, behaviors, and procedures that waste water. These programs also increase awareness of water scarcity, available sources of water, water source protection, distribution network capacity, and treatment issues. Public information programs are designed to promote understanding and dialogue in the community on water conservation topics, as well as to motivate customers to conserve water.

In general, there appears to be widespread public support for water conservation. When surveyed on various water conservation issues, respondents favored policies and programs, including increased prices for water, to improve water conservation. An interesting observation from a survey conducted by Tampa Bay Water is that while 87 percent of respondents agreed more should be done to conserve water, 93 percent also believed that they personally are already doing all they can to conserve water (Tampa Bay Water, 2006).

A public information program is normally directed by the water purveyor. Residential, commercial, and/or industrial customers are segmented and targeted as appropriate. Print media, bill stuffers, Internet sites, radio, and/or television can be used to disseminate the desired information, via free public service announcements, paid advertising, or actual news stories. Brochures and a speakers' bureau can also be used to spread the conservation message. Brochures and fliers are typically distributed as billing inserts in customers' water bills or provided for free distribution at schools, libraries, and community centers. Information displays can be placed in shopping areas. A school outreach program with a developed curriculum of materials and activities on water conservation for elementary, secondary, and high school groups can be used. The curricula should be designed to complement existing science and social studies programs, thereby increasing the likelihood that teachers would incorporate these materials into their regular curricula. Water conservation seminars could be provided for homeowners, architects, developers, and landscape/irrigation professionals and vendors. Examples are as follows (USEPA, 1998):

> ***Information available.*** Water systems should be prepared to provide information pamphlets to customers on request. Public information and education are important components of every water conservation plan. Consumers are often willing to participate in sound water management practices if provided with accurate information. Furthermore, providing information and educating the public may be the key to getting public support for a utility's water conservation efforts. An information and education program should explain to water users all of the costs involved in supplying drinking water and demonstrate how water conservation practices will provide water users with long-term savings.

Understandable water bill. Customers should be able to read and understand their water bills. An understandable water bill should identify volume of usage, rates and charges, and other relevant information.

Informative water bill. An informative water bill goes beyond the basic information used to calculate the bill based on usage and rates. Comparisons to previous bills with proactive warning for large increases in usage compared to previous year or month, and tips on water conservation can help consumers make informed choices about water use.

Water bill inserts. Systems can include inserts in their customers' water bills that can provide additional information on water use and costs. Inserts also can be used to disseminate tips for home water conservation.

School program. Systems can provide information on water conservation and encourage the use of water conservation practices through a variety of school programs. Contacts through schools can help show young people the value of water and conservation techniques, as well as help systems communicate with parents indirectly.

Public education program. Utilities can use a variety of methods to disseminate information and educate the public on water conservation. Outreach methods include speakers' bureaus, operating booths at public events, printed and video materials, and coordination with civic organizations.

Workshops. Utilities can hold workshops for industries that might be able to contribute to water conservation efforts. These might include workshops for plumbers, plumbing fixture suppliers, builders, and landscape and irrigation service providers.

Advisory committee. A water conservation advisory committee can involve the public in the conservation process. Potential committee members include elected officials, local business people, interested citizens, agency representatives, and representatives of concerned local groups. The committee can provide feedback to the utility concerning its conservation plan and develop new material and ideas about public information and support for conservation in the community. Of course, to be meaningful, the utility must be receptive to ideas offered by the committee.

It is widely believed that customers who are informed and involved are more likely to support the water system's conservation planning goals. Motivating customers to use water wisely focuses on encouraging customers to practice water conservation in their daily water use habits. Customers can be encouraged to discontinue wasteful water use habits, such as flushing tissues down the toilet, leaving the water running while brushing teeth, and so forth if they believe that the measure is either good stewardship or that a real need for conservation exists. Savings from such changes in behavior are included under the public information program.

A good public awareness program will help facilitate the message in times of crisis situation that accompanies implementation of emergency measures. Public information during a crisis can best be facilitated by having a prepared procedure developed for use before the need arises. There should be an exercise conducted, designed to test the operational aspects of the plan so as to identify unintended impacts that might result from its implementation.

The exercise should not be limited to the water utility but should include local government, members of the media, and other participants who would be impacted by the plan. The consequences of what might happen if it were actually necessary to implement specific provisions of the plan must be fully reviewed and examined.

The public information program is the glue that ties together all other conservation measures. A public information program must exist on a continuing basis to maintain community interest and continue conservation. It is usually helpful to assign one individual as the water conservation coordinator, or to hire a professional promotions agency. In addition, a good public awareness program will help get the message to the public in any crisis situation that requires implementation of emergency measures.

High-efficiency plumbing fixtures/measures

Significant opportunities exist for water conservation in plumbing fixtures. These fixtures include toilets, faucets, showerheads, and water-saving inserts. Many of these devices require local plumbing code or development of legislation, and some are already mandated, such as the use of high-efficiency fixtures in the 1992 Energy Policy Act and the various building codes. When legislation is involved, code enforcement is required. In most cases, this can be accomplished by working with local plumbing and building inspectors to clear permits and perform site inspections to verify compliance. Efforts to accelerate the water savings involve the request for installation of water-efficient toilets, showerheads, and faucets at the point of property transfer or lease change. The largest demand on the water system comes from the residential sector. The following measures are designed to reduce average and peak demands from indoor plumbing units.

Toilets. Toilets are by far the main source of water use in the home, accounting for nearly 30 percent of residential indoor water consumption. Toilets also happen to be a major source of wasted water due to leaks and/or inefficiency. Because toilets do not use water heating energy, no additional energy costs or energy savings are realized. A secondary benefit of increased toilet efficiency is reduced wastewater flows. Reductions in toilet water use equal a direct 1:1 reduction in wastewater flow because water used to flush a toilet enters the wastewater system. All of the reduction occurs in fluids, not in the solids content. Most of the wastewater treatment costs are associated with solids settling, aeration, handling, and disposal. Potential programs may exist if solid contents exceed 300 mg/L at the wastewater plant, as most wastewater treatment plants are designed to treat wastewater at concentrations below 300 mg/L.

Toilet conservation measures include high-efficiency toilets, ultra-high-efficiency toilets, toilet tank displacement dams, and toilet tank displacement bags. Historically, standard toilets used 5.5 gal (21 L) per flush, while high-efficiency toilets used 3.5 gal (13.2 L) per flush due to changes in the design. Low-flush toilets use special designs to reduce water used to about 1.6 gal (6 L) per flush.

Three main types of low-flush toilets are currently available. The first retains the gravity flush concept, but operates very efficiently because of improvements in the design. Toilets with this design typically use 1.0 to 1.5 gal (3.78 to 5.7 L) per flush. The second type eliminates the gravity flush concept. One model of this type of toilet features a pressurized flush tank in which water is forced into the bowl using pressure from the water system. A third type is a dual flush system where 0.8 gal (3 L) per flush is allotted for nonsolid waste and 1.6 gal (6 L) per flush for solid waste. Also available on the market are flushometer toilets, waterless urinals, and composting toilets (although the last would not be used on a utility's system).

WaterSense, a program sponsored by the USEPA, helps consumers identify high-performance, water-efficient toilets that can reduce water use in the home and help preserve the nation's water resources (USEPA, 2008). The USEPA estimates that over the course of a lifetime, a person will flush the toilet nearly 140,000 times. The WaterSense label is used on toilets that are certified by independent laboratory testing to meet rigorous criteria for both performance and efficiency. USEPA released a final specification on Jan. 24, 2007, for high-efficiency toilets (HETs). Manufacturers that produce HETs meeting WaterSense efficiency and performance criteria can apply to have their products certified. Only high-efficiency toilets that complete the third-party certification process can earn the WaterSense label. Design advances enable WaterSense–labeled toilets to save water with no trade-off in flushing power. The USEPA reports that many perform better than standard toilets in consumer testing. It should be noted that the entire toilet should be replaced because tanks and toilets are designed together. Replacing only the tank may provide unsatisfactory flushing, causing water use to increase.

Displacement systems. Displacement systems should be avoided as these devices generally encourage double flushing, and do not conform to USEPA's WaterSense standards. As a result they do not significantly save water.

High-efficiency showerheads. According to USEPA (1998), showering represents approximately 17 percent of residential indoor water use in the United States. Showers and baths account for about 30 percent of the water use in a typical bathroom (Vickers, 2001). A brief five-minute shower can consume 15 to 35 gal (57 to 132 L) of water with a conventional showerhead with a flow rate of 3 to 7 gpm (9 to 26.5 L/min). Since showerheads are found in all residences, much like toilets, they constitute a significant potential savings on a broad scale for a utility. Replacing showerheads that use 3.0 gpm (11.3 L/min) or more with more modern units that use 2.5 gpm (9 L/min) or less can make a significant difference in the amount of water used per room (16 percent). The payback period can be on the order of three to four years depending on the extent of the project (Alexander, 2002). Replacing showerheads will not only save water, but also reduce the cost of heating water.

Several manufacturers now offer ultra-high-efficiency showerheads that can achieve 1.0 to 1.5 gpm (3.8 to 5.7 L/min) with no observable decrease in flow to the user. This compares to 2.75 gpm (10.4 L/min) for modern high-efficiency showerheads (plus or minus 10 percent) and 5 to 8 gpm (20 to 30 L/min) for nonconserving showerheads. One design, with a flow rate of 1.5 gpm (5.7 L/min), draws air into the water to create a high-velocity spray.

Three types of incentive programs have been identified for existing customers: (1) provision of high-efficiency showerheads free on request, (2) mass mailing or delivery of these showerheads with no direct customer contact, and (3) an intensive door-to-door delivery with follow-up installation program, including direct customer contact. Providing showerheads free on request is likely to result in a customer acceptance rate of 50 percent for these normal replacements.

Regulations can also be an effective method of achieving installation of high-efficiency showerheads. The high-efficiency WaterSense draft specification was 1.5 gpm (5.7 L/min), which is appropriate. The code for new construction could require such showerheads in all new homes and buildings, resulting in a 100 percent customer acceptance rate for new construction. For existing construction, a regulation requiring all homes or buildings sold to be retrofitted with high-efficiency showerheads would result in a 100 percent customer acceptance rate for all homes or buildings sold.

Secondary benefits from shower flow restrictors and high-efficiency showerheads include reduced wastewater flow and energy use. Reductions in shower water use equal reductions in wastewater flow because all water used to take a shower enters the sewer system. However, all of the reduction occurs in fluids, not in the solids content, so wastewater treatment plants will need to plan for higher solids content (estimated to increase from 200 mg/L to 300–350 mg/L), with accompanying air demands.

High-efficiency faucet aerators. According to the USEPA (1998), faucets account for more than 15 percent of indoor household water use. High-efficiency faucet aerators save water by reducing the flow rate through restriction or aeration. Nonconserving faucets have rated flow rates between 2.75 and 5.0 gpm (10.5 and 20 L/min), while high-efficiency faucet aerators have rated flow rates between 1.5 and 2.5 gpm (5.7 and 9.5 L/min). Self-closing faucets save water by limiting the length of time the water flows per use. High-efficiency faucet aerators and self-closing faucet aerators both have rated flows of 2.75 gpm (10.5 L/min) or less.

Federal guidelines mandate that all lavatory and kitchen faucets and replacement aerators manufactured after Jan. 1, 1994, use no more than 2.5 gpm (9.5 L/min) of water measured at normal water pressure (typically 20 to 80 psi). Metered valve faucets manufactured after the same date are limited to 0.25 gal (1 L) per cycle. The USEPA WaterSense standard for high-efficiency lavatory faucets is 1.5 gpm (5.7 L/min). There are also 1.0 and 0.5 gpm (3.8 and 2 L/min) aerators on the market. These are commonly used in water kit programs distributed by vendors. These factors are often used to estimate potential savings of water conservation programs. Options for water savings in this category include fixture replacement, leak detection, and installation of aerators.

Kitchen faucet aerators. Faucets can waste large amounts of water, as they are one of the most heavily used water sources in the kitchen. One way to save water is to install pedal-operated faucet controllers to ensure that valves are closed when not in use. Commercial kitchen, low-volume, automatic shutoff nozzles typically cost $20 to $80. By installing a foot-actuated faucet, one food service facility in North Carolina reduced its monthly water usage by 3,700 gal (14,000 L), an annual savings of nearly $700 (NCDENR, 1999). Another way is to install infrared or ultrasonic sensors that activate water flow.

Rubber gaskets often wear out and deform because of the high volume of hot water use. By installing a brass gasket and an automatic shutoff nozzle, a facility could save as much as 21,000 gal (80,000 L) of water per year (NCDENR, 1999). Merely replacing spray nozzles with the newer 1.6-gpm (6-L/min) models (versus the older 3- to 4-gpm [11.3- to 15-L/min] nozzles) can save 50,000 gal (189,000 L) of water per year and nearly 2000 kW·h of electricity per year (White, 2004), while saving $50 to $70 per month on a typical 3-hour/day usage pattern (West, 2006). Water audits of commercial facilities have shown that 60 percent of identified water savings comes from simply installing faucet aerators in all kitchen sink outlets (NCDENR, 1999).

Efficient dishwashers. All dishwashing machines employ wash, rinse, and sanitizing cycles. There are four main types of dishwashing machines: undercounter, door, conveyor, and flight. Requirements for machine size can be calculated by estimating the amount of traffic that will be served in the food service area. Commercial dishwashers use approximately 1.0 to 1.5 gpm (3.8 to 5.7 L/min), while conventional rack washers use 9 to 12 gal (34 to 45 L) per cycle. Newer units use only 0.75 to 2.5 gal (3 to 9.5 L) per rack (NCDENR, 1999). Undercounter washers use the most water, and conveyor types use the least. Energy-efficient, low-flow conveyor washers can reduce water consumption by 43 percent (NCDENR, 1999). An Energy Star® dishwasher saves about $100 over its lifetime, mostly from using less hot water than conventional models. Energy guidelines and water consumption levels for dishwashers are continuing to tighten, and manufacturers are offering more water-saving models. Using an appropriately sized, water-efficient model will save a significant amount of water.

High-efficiency dishwashers reduce the amount of water required per load. A nonconserving dishwasher uses 14 gal (53 L) per load. Assuming a dishwasher is run twice per week, the average per user is 3 gal (11.3 L) per day. Since the efficiency of dishwashers is continuing to improve, the lower end of the range, 2 gal (7.6 L) per day, is used for water use by efficient dishwashers. The savings for an efficient dishwasher over a nonconserving dishwasher is 1 gal (3.8 L) per day.

An aggressive incentive program would have a large impact on the customer acceptance rate for efficient dishwashers. If an incentive payment or connection fee discount is high enough to make installation of efficient dishwashers markedly cost-effective to the customer, then a 70 percent acceptance rate could be achieved for new construction and for normal retirements in existing construction.

Secondary benefits from efficient dishwashers include reduced wastewater flow and energy savings. Reduction of water use in a dishwasher equals reductions in wastewater because all water used to wash dishes enters the sewer system.

Efficient clothes washers. Clothes washers are the newest area of study. Front-loading washers can be rated as low as 14 gal (53 L) per cycle. This is far below the 1980s average of 42 to 47.5 gal (160 to 180 L) per load, and below nonconserving clothes washers that were estimated to use 55 gal (207 L) per load (Brown and Caldwell, 1984). The Brown and Caldwell study also collected data on the number of loads of wash done per week. Based on these data, the study estimated 0.3 loads of wash per capita per day. This implies that the water use per clothes washer per day is 40 gal (151 L) for a nonconserving clothes washer, 31 to 35 gal (117 to 132 L) for a conserving clothes washer, and about 4 gal (15 L) for the front-loader. The savings for an efficient clothes washer over a nonconserving clothes washer is more than 40 gal (151 L) per clothes washer per day.

Large conventional washer-extractor machines use fresh water for each wash and rinse cycle without internal recycling. The capacity of these units ranges from 25 to 400 dry lb per load, requiring 2.5 to 3.5 gal of water per pound of laundry. Coin-operated machines (16-lb) are slightly larger than residential units (14-lb) and use 35 to 50 gal of water per load (Vickers, 2001).

An aggressive incentive program could have a large impact on the customer acceptance rate for efficient clothes washers. If the incentive payment is high enough to make installation of efficient clothes washers markedly cost-effective, then a 70 percent acceptance rate could be achieved for normal retirements in existing construction.

Leak detection tablets. Toilet leaks can be detected using leak detection tablets. The tablets are placed in the toilet tank, turning the water a bright color. If the water is leaking from the tank to the toilet bowl, the water in the toilet bowl will turn color. Leaky toilets can usually be repaired by replacing ball cocks and flapper valves. The *HUD Survey of Water Fixture Use* (HUD, 1988) found 20 percent of the toilets tested leaked; thus, this estimate is used as the percentage of all toilets that leak (HUD, 1988).

An estimate of the capital cost, installation cost, operations and maintenance (O&M) cost, and lifetime was obtained by contacting manufacturers and/or vendors of this measure. The retail cost of dye tablets is $0.25 and the wholesale cost is about $0.05. The labor cost of the toilet leak repair is $100.00 (1.0 hour × $100/hour). The lifetime of a fixed toilet leak, corrected as the result of using leak detection tablets, is expected to be five years.

Fixing leaky toilets requires two actions by the customer: identification of a leak and fixing the leak. Use of leak detection tablets is the first step in this process. A number of water agencies have supplied leak detection tablets to their customers. For mass mailing programs, approximately 20 percent of the households reported using the leak detection tablets, as discussed in Vickers (2001). The acceptance rate for leak detection tablets in Aurora, Colo., was 60 percent in its program, which distributed retrofit kits, including leak detection tablets, door to door in response to a direct mail offer. This estimate can be used for the customer acceptance rate for leak detection tablets using door-to-door delivery with follow-up installation programs.

An information program is assumed to result in 5 percent of the toilets being tested for leaks with leak detection tablets, while a program that provides leak detection tablets free on request would have a customer acceptance rate of 10 percent.

Fixture retrofit kits. This measure overlaps with low-flush toilets, high-efficiency toilets, leak detection tablets, high-efficiency showerheads, shower flow restrictors, faucet aerators, and leak detection, as well as with water pressure reduction, metering, pricing, public information, and in-school education.

A comprehensive kit is defined as one containing two high-efficiency showerheads, two toilet tank displacement dams, two toilet tank leak detection dye tablets, and a pamphlet containing instructions, household leak detection/repair tips, and information on efficient landscape irrigation. The tank displacement dams are intended for use in nonconserving toilets. A fixture retrofit kit is a collection of individual water conservation measures. Each of the measures included in the kit has been discussed separately.

For residential units, four methods to distribute retrofit kits are available: depots, mailings, door-to-door canvassing, and drop-off systems. The difference among the acceptance rates of the retrofit measures primarily results from the varying distribution methods and the level of the promotional effort.

Depot retrofit. The depot retrofit method encourages customers to pick up free retrofit kits at selected mini-depot locations such as the billing office, public/government buildings, schools, libraries, or supermarkets.

Mailings. A promotional campaign advertisement placed in the customers' bills that free retrofit kits are available would institute the retrofit mailings method. Interested customers would reply to request a retrofit kit. Administration of mailers, use of the consumer confidence reports, and solicitations would require administrative and clerical assistance.

Canvas and drop-off. These measures are not effective because the equipment is rarely installed.

Distribution events. These would include showerhead exchange and toilet exchange done as part of outreach programs.

Utility conservation measures

Meter reading. A number of studies have been conducted on the impact of metering on water use. These studies agree that metering water use reduces consumption and report savings in water consumption ranging from 13 to 45 percent after meters were installed (Mayer and Deoreo, 1999).

Water audits. Water surveys can bring one-on-one interaction between a customer and a water conservation specialist that can lead to significant reductions in water use both indoors and outdoors. Surveys take as little as an hour and can be directed toward specific service areas or targeted to customers with high water usage. Results vary according to household size, type and age of home, weather conditions, and the efficiency of interior and exterior water-using fixtures.

A residential water audit is conducted by an agency representative at the request of the homeowner. The specific design of an audit should be tailored to the community's needs. A water audit consists of five components. First, water uses are identified and discussed with the homeowner. Second, a walkthrough of the premises is conducted to identify potential water-saving strategies. Third, an offer is made to install high-efficiency showerheads, tank displacement dams, and faucet aerators, and to use leak detection tablets on the toilets. Fourth, leaks, if identified, are repaired. Finally, information on further actions that can be taken to conserve water, including a lawn watering guide, is provided.

A modified version of the water audit can be conducted as a multifamily water consultation measure to target large complexes (25 units or more) with high water use. The manager of the complex could be contacted and a site visit arranged. During the site visit, the water conservation specialist would conduct a walkthrough to check each unit for leaks, identify needs for high-efficiency fixtures, check the irrigation schedule, and advise on the use of drought-tolerant plant species. Low-flow devices will be supplied for the manager to install. Materials include high-efficiency showerheads, toilet tank displacement devices, faucet aerators, toilet tank leak detection tablets, toilet tank ball cocks and flapper valves if needed to repair toilet leaks, faucet washers to repair faucet leaks, and literature on efficient irrigation and irrigation systems, low-water-use plant material, and other water-saving tips.

Savings from a water audit depend on the equipment installed and actions taken by homeowners in response to the audit. Each of the measures included in the audit has been discussed separately in this report. Estimates are provided of the water savings per conservation measure in the appropriate sections.

A commercial water audit is conducted by an agency representative at the request of the commercial customer. The same basic procedure is used, although the specific design of the audit will vary. However, a typical water audit for large commercial customers involves inspection of plumbing fixtures, cooling equipment, and water-using processes pertinent to production (if applicable) to determine recycling opportunities and detect inefficient or malfunctioning fixtures and equipment responsible for leakage or other water waste. Instructions on efficient landscape irrigation, if appropriate, are given. Savings from a water audit depend on the equipment installed and actions taken by customers in response to the audit. Of all the conservation opportunities reported to all the customers receiving audits, 25 percent typically cooperate.

This measure overlaps with high-efficiency toilets, showerheads, and faucets; fixture retrofit kits; residential leak detection; shower flow restrictors; and landscape measures as well as with metering, pricing, public information, and in-school education.

Distribution system water audit. A distribution system water audit compares the amount of water produced from wells, surface supplies, and interagency ties by the water purveyor with the amount of water used by customers as reported by meter readings. The difference is unmetered water. After accounting for authorized unmetered uses, such as fire-fighting, main flushing, and public use, the remaining amount is termed *unaccounted-for water*. If the numerical percentage of unaccounted-for water is substantial (e.g., more than 15 percent), it may indicate inaccurate meter readings, malfunctioning valves, leakage, or theft. Unaccounted-for water reports should be compiled on a one-year rolling average to obtain true data for unaccounted-for water. This is because variations in meter reading dates and changes in use patterns due to weather or population (e.g., snowbirds) can affect month-to-month totals.

Water losses, whether due to leakage, theft, illegal connections, underbilling of customers, or faulty system controls, represent revenue losses to the water utility. This is water that the agency has already paid to obtain, treat, and pressurize; however, because it is lost, the water produces no revenue. For example, in 2005, 15 percent of the water provided by city of Dania Beach, Fla., was unaccounted for. A review of water billing records indicated that 7 percent of the meters on the system were not functioning at all. These were all replaced, and unaccounted-for water dropped to 7–8 percent. There was an associated revenue increase as well.

Water-saving opportunities can be identified and quantified from continuing reviews of the distribution system. The actual water savings are derived from taking action to curb the water losses identified. These actions include meter calibration, leak detection/repair, valve repair, control system corrections, and corrosion control, which are all programs the utility should pursue.

A water audit of the water agency's storage, transmission, and distribution system is recommended as the first step in system water conservation. Once a utility has conducted a comprehensive water audit, annual updates provide data to help managers decide how to adjust priorities and monitor progress made on system maintenance. Equally important, the update can identify new areas of system losses to establish new annual maintenance goals. Updating a water audit will usually be less expensive than the original baseline audit.

Water savings from leak detection and repair are included as a separate measure. The water savings coming from the combined actions of meter calibration, valve repair, control system corrections, and corrosion control can vary from 3 to 30 percent of the water agency's total production.

System leak detection program. Brigham Young University, Department of Civil Engineering performed a study for the American Water Works Association in 1987 called *Water and Revenue Losses: Unaccounted-for-water*. This report summarized research performed by the Department of Civil Engineering on unaccounted-for water and revenue loss problems of water utilities. The objectives of the effort were to

- Assess the value of various methods used to measure quantities of water that are lost or unaccounted for
- Provide an appraisal of the techniques available to monitor such losses
- Suggest standardized definitions of the terms used to describe the types and sources of water and revenue losses
- Identify solutions available to utilities to control such losses of water or revenue

Simply stated, the problem is twofold: (1) all of the water purchased or produced by a water utility does not reach its intended destination, and (2) some of the water that does arrive at its intended destination is never paid for.

A leak detection program is designed to determine where water losses occur on a water system. This measure overlaps with water audit and leak detection programs. Known and authorized water uses that do not produce revenue and are often not measured, such as fire-fighting and firefighter training, sewer and street cleaning, hydrant and water main flushing, freeze protection, water quality testing, backwash water, and similar operational uses are discussed. Other authorized water uses that sometimes do not produce revenue, such as public fountains, public buildings, parks, cemeteries, public golf courses, highway landscaping, public swimming pools, and other such beneficial uses are also covered in this section.

To determine water loss quantities, water volumes must either be measured or estimated. This can be done via meter calibration/accuracy checks and its effect on water loss quantities, inspection of water use records, inspection of water loss records, and water audits. Methods of leak detection range from simple, passive processes to extremely complicated, active methods using electronic correlators and requiring specially trained personnel. Leak detection programs can be conducted in house or by consulting experts in the field. There are enough existing water utilities running efficient, well-maintained operations with minimal water losses (i.e., less than 10 percent) to indicate that leak detection and water audits are successful.

Leak detection programs will focus on visible and invisible leakage. Visible leaks are those that can be seen emerging from the ground or pavement and bubbling up to the surface. The source of the leak may be a considerable distance away from the area where it is observed. Many visible leaks are reported by water customers and can be readily localized. Invisible leaks include those that percolate into the surrounding ground and those that enter other conveyance facilities, such as storm drains, sewers, stream channels, or old abandoned pipes.

There are three methods to find leaks:

Audible leak detection. This method uses electronic listening equipment for detection. Pressurized water that is forced out through a leak loses energy to the pipe wall and to the surrounding soil area. This energy creates sound waves in the audible range, which can be sensed and amplified by electronic transducers or, in some cases, by simple mechanical devices. The sound waves are then evaluated by an individual trained in leak detection who can determine the exact location of the leak, based on the pitch or frequency. This specialist conducts an initial listening survey of the entire distribution system and records all suspect sounds. If these sounds are heard again when rechecked later, leaks are pinpointed.

Zone flow measurements. This method can reasonably be undertaken as an extension of the water audit or, in some cases, directly as a leak detection method. Its purpose is to determine whether or not a sector or zone of a water system is suffering major leakage. To effectively conduct a zone flow measurement, a utility must maintain good maps, have valves located at zone control points, and provide a tap in the main for the recording pitometer.

Normal course of operation. This method involves discovering leaks accidentally during the normal course of operations and maintenance, for example, in a valve exercising program. Meter readers have the opportunity to check for visible meter box leaks when reading the meters.

Water pressure reductions. Water pressure reduction lowers water consumption and water system leakage by reducing the system pressure. This can take two forms: systemwide reductions done by the utility or individual reductions conducted on consumer systems. Systemwide pressure reductions can be used during high-demand periods to reduce usage and/or leakage on the system.

Typically, water pressure varies anywhere from 40 to 80 psi on most systems. However, reducing pressure in developed areas can negatively impact the level of service provided to customers by reducing the performance of irrigation systems and fire protection systems substantially. In some cases, pressure reductions will necessitate the replacement of the customer's service lines as well. These drawbacks may outweigh the conservation benefit that could be achieved for existing construction, primarily with landscape irrigation. This is an especially useful measure in areas with significant irrigation or seasonal use.

The current saturation of water pressure reduction systems in new construction is very low, so the current saturation is zero. The eligible areas for pressure reduction are limited. The most effective means of achieving water pressure reduction in high-pressure areas would be through a regulation that requires all new residential developments to be designed and built with pressure below a fixed maximum. A 100 percent acceptance rate would be achieved for a regulatory program in new residential developments. However, this measure is not appropriate for new residential construction in an already developed area because it would create additional administrative problems.

Water reuse (reclaimed water system). Reclaimed wastewater (reuse water) is designed to increase conservation by serving as a substitute for potable or well water in irrigation and other nonpotable applications. Benefits of this option are far-reaching despite the potential costs, and include the following:

- Centralization of facilities would lower costs of operation and likely construction due to economies of scale
- Potential elimination of surface water outfalls
- Reduction in the number of injection wells
- Increased ecosystem restoration
- Increased efficiency in the use and retention of local and regional water resources

Water savings for this measure would theoretically be all potable water used for irrigation within the reclaimed system service area. However, the use of reclaimed water often requires retrofitting existing neighborhoods, which is expensive and, in some cases, impractical. Significant users are more practical and save more water use (e.g., golf courses).

Irrigation systems. The volume of water used for lawn and landscape irrigation is not well documented. Outdoor water use in warm climates can be on the order of 30 to 50 percent of the total demand. Landscaping use is likely to be variable, depending highly on the area, plant types, climate, rainfall, water costs, maintenance practices (e.g., frequency of sidewalk cleaning), and the number of golf courses, swimming pools, and fountains. Fortunately, there are many available water-saving landscape options designed to promote water conservation.

The best time to water is in the predawn hours. Frequent shallow watering encourages shallow rooting, which makes grass (and other plants) more susceptible to drought and other stresses. After watering adequately, sprinklers should be turned off until the plants indicate further watering is necessary. Grass needs watering when the blades begin to curl (a sign of wilting), or when the grass becomes bluish-gray instead of bright green. Also grass does not spring back readily when walked upon when it needs water. If lawns are not overwatered, disease and fungus problems will be minimized.

Daytime watering bans and other measures. Daytime watering bans are useful simply because a significant percentage of water used for irrigation during the day is lost through evaporation and can cause plants to be burned.

Automatic controllers and valves. Weather- or sensor-based irrigation control technology uses local weather and landscape conditions to tailor irrigation schedules to actual conditions on the site or historical weather data. Instead of irrigating according to a preset schedule, advanced irrigation controllers allow irrigation to more closely match the water requirements of plants. These new control technologies offer significant potential to improve irrigation practices in homes, businesses, parks, and schools.

WaterSense plans to label weather-based irrigation controllers and soil moisture sensors. Automatic controllers and automatic valves replace manually operated valves in permanent irrigation systems to make accurate irrigation more convenient. Automatic controllers and valves use clock mechanisms to open and close solenoid-operated valves at predetermined times, permitting regular and unattended irrigation as programmed by the user.

Automatic/evapotranspiration controller. Automatic timer shutoffs for manual hose systems make it more convenient to irrigate accurately. The homeowner sets the timer for the desired length of time, rather than returning to turn off the sprinkler. The success of this measure also depends on proper irrigation according to the guidelines presented in a lawn watering guide (described in a following section).

Moisture sensors. Soil moisture sensors for automatic electronic systems permit irrigation only when the turf needs water; otherwise, the moisture sensor overrides the automatic controller. In conjunction with an automatic control system, moisture sensors can produce impressive water savings, thus permitting accurate and unattended irrigations. Proper installation and maintenance are required to ensure good performance. This measure overlaps with low-water-using plants, demonstration gardens, automatic controllers and valves, irrigation guides, and irrigation audits.

Irrigation efficiency. Any efficient irrigation alternative should include the provision of advice on improving irrigation efficiency based on a site visit. Utility personnel could distribute information on turf irrigation requirements and low-water-use landscaping. Water savings for this measure are estimated at 25 percent of outdoor water use. The saturation level of residential customers who are now watering below the recommended levels is estimated to be 5 percent. Efficiency in irrigation, however, extends beyond just equipment.

Lawn watering guides

A lawn watering guide illustrates an easy-to-use technique to determine the watering time required to meet irrigation demands of turf grass. The objective is to provide homeowners with information to permit application of accurate irrigation amounts throughout the year. Additional fliers describing the causes and cures of maintenance problems in residential irrigation systems may also be distributed at the same time.

An irrigation guide describes the causes and cures of scheduling and maintenance problems in large turf irrigation systems. The objective is to provide landscape managers of commercial and multifamily buildings having large turf landscapes with information to schedule application of accurate irrigation amounts throughout the year. In contrast, for the testing procedure in the landscape water management audit program, irrigation audits are carried out by water utility representatives (i.e., mobile irrigation lab), and a complete optimized schedule is provided to the site manager.

Among the issues to be addressed in any lawn watering guide are those listed below.

> ***Turf watering schedule.*** This measure would mandate which days of the week and during which hours of the day that watering for irrigation is permissible.
>
> ***Improved irrigation systems.*** This measure would educate the public on efficient irrigation systems, such as drip irrigation. A multipronged promotional campaign would target landscapers, sprinkler contractors and suppliers, nurseries, customers, and developers. The campaign will describe the types, costs, and advantages of efficient irrigation systems and tune-ups of existing irrigation systems. It will also describe proper retrofitting of inefficient irrigation systems with efficient system mechanisms, such as moisture detectors.
>
> ***Lawn watering public information.*** Because many homeowners overirrigate, this measure would provide the information, devices, and techniques for applying the correct amount of water for lawn irrigation, and the correct times to water. The measure would ensure preparation and distribution of watering guides based on the seasonal evapotranspiration (ET) rate as calculated from weather data. Evapotranspiration describes the total water used by a lawn. Water applied in excess of ET is not used by the lawn and thus is wasted. The two factors that govern the ET rate and, hence, the lawn watering requirements are grass species and climate. Workshops can be scheduled for sharing this information.
>
> ***Limited turf area.*** This measure would limit the square footage of turf for a new residence. Restrictions on turf will be similar to those identified in the codes for new developments.

Turf water management. This type of program would offer free workshops to teach landscape workers about progressive turf management practices and devices. It would encourage upgrading and retrofitting inadequate irrigation systems and installation of low-demand turf grasses.

Xeriscape

The term *xeriscape* originated with the Denver (Colo.) Water Department in 1981 in response to drought conditions occurring in Colorado and the western United States. A xeriscape program is an efficient-landscaping program that encourages homeowners to use water-saving techniques when they landscape. For example, most shade trees commonly used for landscaping require relatively large amounts of water, but there are many shrubs and flowering plants that can be used to reduce landscape water requirements. These are termed *xeric*. The fundamentals of xeriscape include the following concepts:

- Appropriate planning and design of landscapes
- Soil improvements
- Efficient irrigation systems
- Limited/reduced turf areas
- Use of native, drought-tolerant plant species
- Use of mulch to retain moisture and reduce evaporation
- Appropriate maintenance

Of these seven elements, limiting the size of turf areas is the most useful in reducing landscape water use. Water savings are estimated at 40 percent of outside water use.

Low-water-using plants—residential

Careful planning and site evaluation are necessary because of multiple climate zones, soil types, temperature ranges, and precipitation patterns over the course of a typical year. Furthermore, it is not uncommon for widely different conditions to exist within the same property. Local codes often dictate which species may be planted in certain municipalities. Therefore, the appropriate agencies should be consulted when developing a landscaping plan. Whenever possible, it is recommended to select drought-resistant plants that require less water.

Demonstration gardens

Many building owners and residential and commercial property landscape managers are not familiar with reduced turf-area landscapes using indigenous plants and other low-water-using plants. A demonstration garden could be used to show that xeriscaping can be attractive and functional for both shading and food production. Public buildings and medians in roadways are excellent places for demonstration gardens. This measure allows for overlapping use of low-water-using plants, xeriscaping, electronic controllers and valves, automatic timer shutoffs, soil moisture sensors, lawn watering guides, and irrigation audits.

Retrofit of existing landscape

The goal of this measure is to retrofit existing residential turf landscaping to reduce water use. This measure could be made mandatory when the property ownership is transferred. The utility could prepare standard designs and specifications through a knowledgeable landscape architect for retrofitting single-family and multifamily homes. These standards and specifications would cover plant lists, acceptable irrigation systems, turf types, square footage of allowable turf, and soil preparation. The utility will need supporting brochures and mailers as part of the promotional campaign. This measure should target high-water-use multifamily units as an initial priority. Recommended specifications include those listed below.

- Limiting turf to 50 percent of total irrigated landscape.
- Scarifying ground and adding a minimum of 6 in. (15 cm) of topsoil.
- Installing a surface layer of organic mulch around all shrubs and plant material.
- Consolidating turf areas into large, flat areas, thus eliminating small "islands."
- Using native plants in place of turf, wherever possible.
- Setting up automatic underground and drip irrigation systems with separate controls for unique water requirements.

Efficient landscaping techniques

There are a number of miscellaneous techniques that could be used to improve overall landscape efficiency. Some of these include the following:

Landscape water audits. A typical landscape audit consists of an irrigation inspection and a horticultural inspection. In the irrigation inspection, auditors (mobile irrigation labs) determine the existing system's efficiency and condition, test different sprinkler head types for field precipitation, and develop an irrigation schedule. In developing an irrigation schedule, the auditors inventory species, analyze the landscape's overall condition, and make recommendations for reducing water use. Water savings can be as high as 30 percent (Mayer and Deoreo, 1999).

Soil polymers and additives. Soil conditioners can be added to the soil to retain moisture. There are a number of soil additives on the market. These products maintain soil moisture, reduce salts (and chlorides), and increase growth, while using half the water. These additives inhibit evaporation by forming a subsurface membrane that water can pass through going down, but cannot pass through as evaporation. High sand content soils are prime candidates for such additives.

Porous piping. Porous pipes and similar products allow water to ooze through the pipe or through pores as if the pipe were sweating. Pipes are placed 24 in. (61 cm) apart, a few inches (centimeters) beneath the sod, and the water travels via capillary action.

Low-volume irrigation. Often referred to as *drip irrigation*, this measure comes in many forms, but basically consists of piping systems with small holes through which water drips into the soil. Usually this system is not used for lawns, but it can be with slight modification. Water is delivered slowly for long periods of time to the soil root zones.

Water consumption is expected to be half of current use due to direct application to the root zone and virtual elimination of evaporative losses. Drip systems work best where the water source is treated potable water, which prevents clogging in the system. Low-volume jet spray systems use more than drip irrigation systems, but far less than conventional systems. The system delivers tiny droplets through microjet sprinkler heads at the rate of 5 gph (19 L/hr) to confined areas. Conventional systems can easily be converted to low-volume systems by using pressure and flow regulators and new sprinkler heads.

Example—water conservation program

The city of Hollywood, Fla., has for years used a formal water conservation program. This program is composed of six elements: development/maintenance of an accurate database of Hollywood's water consumption, reduction of municipal water waste, a retrofit program, modification of relevant city codes (plumbing, irrigation, and landscaping), promotion of xeriscaping, and public information/education programs.

The city has launched four public education/information projects since the program's adoption, including the institution of a lending library for all Hollywood schools, annual involvement in AWWA's Drop Saver Contest (coupled with a fire hydrant decorating contest), and informative materials created especially for distribution. The public information/education projects focused on providing information to the public and on the role of children in future conservation efforts. Additional public information programs include a T-shirt decorating contest, celebration of Water Week, city commission presentations, and speaker engagements in the schools. A significant amount of material has been gathered by the staff on water conservation measures. Coordination with the South Florida Water Management District program is ongoing. Periodic newsletters may be initiated in the future to further the effort.

The ideas for Hollywood's public information/education programs originated from involvement with teachers, the public, municipal employees, and children. These water conservation efforts have proven to be highly successful in saving water and informing the public, as evidenced by static water usage in a growing community and by the city winning second place in the Florida Water and Pollution Control Operators' Association's annual contest and receiving an American City and County Award of Merit for its efforts.

REFERENCES

Bloetscher, F., and D.M. Meeroff. 2009. *Broward County Water Use Profile*. Boca Raton, Fla.: Florida Atlantic University.

Brown and Caldwell Consulting Engineers. 1984. *Residential Water Conservation Projects—Summary Report*. Washington, D.C.: US Department of Housing and Urban Development (HUD).

HUD Survey of Water Fixture Use. 1988. HUD.

Mayer, P.W., and W.B. Deoreo. 1999. *Residential End Uses of Water*. Denver, Colo.: AWWARF.

NCDENR. 1999. *Water Efficiency: Water Management Options—Kitchen and Food Preparation.* North Carolina Department of Environmental and Natural Resources, Division of Pollution Prevention and Environmental Assistance. DPPEA-FY99-36.

Tampa Bay Water. 2006.

USEPA. 1998. *USEPA Water Conservation Plan Guidelines*. Washington, D.C.: US Environmental Protection Agency.

———. 2008. *WaterSense*. http://www.epa.gov/watersense. Washington, D.C.: US Environmental Protection Agency.

Vickers, A. 2001. *Handbook of Water Use and Conservation*. Amherst, Mass.: WaterPlow Press.

West, W.W. 2006. *Hotel and Motel Water Conservation.* Florida Green Lodging Workshop, Best Western–The Westshore Hotel, May 2, 2006.

White, B.M. 2004. *Hotel and Motel Water Conservation: Saving Water by Implementing Conservation Measures*. Tampa Bay, Fla.: Tampa Water Department.

CHAPTER 16

Public Responsibility

The primary purpose of a public water and sewer system, as defined by law, is to provide adequate quantities of reliable service to its customers. How that is accomplished has both regulatory and policy implications. Beyond that, many utilities use mission statements to outline their goals to the public. The following are actual examples of mission statements provided by four utility systems. Each could be interpreted to say something slightly different about the outlook and priorities of the utility management staff and its governing board:

> "...water and sewer utility system is created to develop safe, reliable and financially self-sufficient potable water and sanitary wastewater systems which will meet the water and sewage needs of the residents of the service area."

> "...Utility will provide high quality water to our customers in amounts which meet their need and protect their health at a fair price."

> "...Department's goal is to provide a high level of services to our residents/ customers at the lowest possible cost."

> "...Authority will provide our customers with high quality water and wastewater services through responsible, sustainable, and creative stewardship of the resources and assets we manage."

In each case there is a statement about meeting the public health goals via a water supply that is safe and reliable. Providing this service confirms the health, safety, and welfare obligation of the owners, managers, and operators of the system to meet purposes to the best of their ability.

The first mission statement intends to provide service without subsidy from other sources, i.e., the customers of the system will be responsible for the costs to provide the service. This mission statement indicates that the utility may be very much involved with public opinion over quality concerns, and has a user base that may be willing to pay for higher-quality water than the minimums required by the regulations.

The second statement notes fairness in price. While "fairness" is not defined, the intention could be that the utility means to be competitive with its neighbors or that fairness will be exhibited among all users. The next mission statement includes "lowest possible cost." The indication is that the primary focus of policymakers is on finances, a common thread with elected officials, which may appear to contradict public health goals.

The final mission statement clearly looks at the long-term operation of the utility, using the phrase "responsible, sustainable, and creative stewardship of the resources and assets," which indicates that the utility needs to be mindful of limited water sources and the potential for contamination problems. Cost of service does not appear to be a factor except in the context of "responsible use of resources." However, it is clear that protection of water resources is a major focus, which is in keeping with the intent of the Safe Drinking Water Act.

Public perception of these mission statements may create unintended expectations. In each mission statement, it is obvious that some interaction with customers is ongoing. All four agencies that generated these mission statements are public entities with governing bodies that respond to an electorate. Each agency is attempting to demonstrate how it intends to meet the needs of its customers. Customers dissatisfied with the service of a water system operated by governments may exert pressure on elected officials to make changes in the operation of the utility. When the problem is being caused by incompetence or lack of leadership from the public officials responsible for the water system, then the public can act to replace them at the ballot box. Public agencies also will readily replace those in leadership positions to address real or perceived problems.

In contrast to the public-sector models, customers dissatisfied with the service provided by a private or investor-owned system usually cannot obtain improved service by appealing to the company, either as an individual or as a group, unless the situation is serious and stockholder attention can be raised. In serious cases, a regulatory agency such as a public service commission may become involved, but beyond serious health and safety issues, the bottom line for private or investor-owned utilities is profit. If not corrected, customer satisfaction issues can affect rate requests before a public utility commission. Because local official control over operations of private or investor-owned utilities is limited, local governments may acquire private systems where problems persist.

Water supply and delivery and sewer collection and treatment have a long history of provision by local governments in the United States. Several factors are at the root of this delivery format, e.g., the significant investments in infrastructure that must be made, the perception that water is a basic human service that should be available to everyone regardless of income or social status, and the use of water supplies and wastewater services as a means to attract, retain, or control growth in a given area. The purpose of this book was to introduce the reader to the basic goals of operating a water and/or wastewater utility, to understand its infrastructure, and to gain an understanding of the various aspects of management and supervision that must be undertaken to protect the significant investments that have been made in this infrastructure. All three areas provide utility managers with control of the future of their operations, and the ability to hold accountable those delivering the service.

Given public control of water and sewer systems, an informed electorate has begun to become interested in water and wastewater utility system issues. Customers often want to know if there are concerns that might threaten their water system and what system managers are doing to ensure that their drinking water is safe. Beyond the traditional quality concerns, the industry is now seeing issues raised about safety, security in light of the 9/11 incidents, fiscal control, and fairness among consumers.

Similarly, wastewater customers want to make sure that the wastewater service is available when they need it, and that the disposal of wastes is done in a manner that protects both the public health and the ecosystem. The latter will include not only the natural system, but also the ability of humans to use the ecosystem for fishing, swimming, boating, and other recreational activities. The potential for public health concerns, as seen in the Cuyahoga River in Cleveland in the 1930s to 1960s, is not acceptable to today's electorate.

RESPONSIBILITY FOR COMPLIANCE

The responsibilities for compliance with these regulations lie with water system owners, public officials, managers, and operators. Enforcement action for failure to meet regulations is usually directed against the responsible officials of a water system, water district, municipality, or company. These responsible officials must also answer to their constituents for these failures, especially if the cause is neglect, incompetence, or misconduct in operations. The most common legal liability results from failure to comply with specific regulations. Such violations are typically periodic violations of federal, state/provincial, and local water quality standards or reporting regulations for plant operations. There are external issues with water supplies that may affect water quality and cause failures in meeting regulatory requirements. The key to limiting this type of liability is training of operators, providing appropriate treatment and monitoring technology, reinvestment in infrastructure on a regular basis, and redundancy in treatment processes.

Direct, personal legal liability can result from negligence and misconduct in the management and/or operation of utility systems if the public health is affected or regulations are violated. *Negligence* is defined as the failure to exercise due care in the performance of the work, or something which an ordinarily prudent person would foresee as a risk of harm to others if not corrected. Negligence is differentiated from incompetence, which is the lack of ability or qualification to perform, and misconduct, which is a willful act that carries with it the potential for criminal penalties. Managers, officials, and employees of a water system may be exposed to a civil suit for damages if the negligent operation of the system results in injury or property damage or misconduct is noted. Although municipalities were once considered immune from suits (under the guise of sovereign immunity), the use of insurance by local governments has permitted the courts to be more sympathetic to claims where the cause can be shown to be the result of misconduct, negligence, or failure to employ generally accepted operating practices.

HOW SERIOUS IS THE RESPONSIBILITY?

Water and wastewater utility systems are oriented toward protecting public health. Even though the results of neglect or negligence occur infrequently, sickness or death could follow. As a result, the legal system takes these responsibilities seriously. A Canadian example, the Walkerton, Ont., outbreak, illustrates the extent for which responsibility may be assigned in both the United States and Canada.

Walkerton, Ont., Outbreak

The city of Walkerton, Ont., is a community of 4,800 people in a predominantly rural area. The city relies on a well field for its water supply. In the spring of 2000, nearly half the residents became ill with what was identified as *E. coli* O157:H7, and seven residents died of the bacteria. A formal inquiry into the matter was undertaken by Justice Dennis O'Connor over the ensuing two years. Ultimately the investigation focused on a particular well that appeared to have a surficial connection that allowed contamination from a nearby field, upon which manure had been spread, to contaminate the water system. Among many findings of the investigation, Justice O'Connor noted these deficiencies in operating the system (O'Connor, 2000).

- Insufficient well field protection (manure field in the cone of influence of the well)
- Failure of the utility to address well field protection efforts as required by regulatory agencies

- Poor operations by the utility staff (the water was insufficiently disinfected)
- Misrepresentation of water quality by operators, including concealing fecal coliforms in the water
- Failure of utility commissioners to respond to a 1998 report of deficiencies by the province
- City council did not appropriate adequate funding
- Province's legislative budget cuts reduced funding for routine lab services and enforcement despite knowledge of improper utility and private lab practices and warnings of health consequences
- Province's Office of Environment did not adequately inspect treatment facility

Much can be learned from this example, and most of the points covered in this document address those issues, including the need for regulatory response (wellhead protection efforts and water quality monitoring), staff oversight, planning for future development, acceptable maintenance practices, implementation of capital, the need for fiscal responsibility (failure to fund), and emerging issues (a new strain of *E. Coli*).

What the Walkerton example points out is that the responsibility for protecting the public health, safety, and welfare extends beyond the operator and utility director, whose priorities are compliance with regulations and ensuring an uninterrupted flow of water to customers. The Walkerton example shows that because most of the decision-making responsibility for the financial obligations and the overall management lies with the utility's elected and appointed officials, they also have a responsibility to the public to uphold.

In some areas of Australia, newspapers evaluate local official performance based on

- Increases in asset value
- Increases in fiduciary position
- Decrease or minimization of system failures or breakdowns
- Absence of catastrophic incidents

To meet these criteria, expenditures must be made on the principles of asset management to optimize repair and replacement of infrastructure, maintenance should be preventive in nature instead of reactive, and new or replacement facilities should be funded as a part of the regular budget process. This means a campaign promise to freeze rates would probably not be viewed as being responsible (nor the one proposing it deserving of reelection) in Australia.

Most utility systems in North America are underfunded, and most have deferred maintenance needs that are unmet. Work order systems are not consistently pursued, which makes assessment of infrastructure condition more difficult. Most have limited understanding of their asset values, life, and condition (asset management), and most struggle with maintenance because this is usually the first item cut in the budget since it is very difficult to quantify. Deferring replacement of infrastructure has perils and it increases the likelihood of failure of a component. Once monies are allocated for these items, there is a responsibility to ensure that these programs and projects are effectively implemented by managers and operators.

REFERENCES

O'Conner, D. 2000. Walkerton Commission Inquiry. http://www.attorneygeneral.jus.gov.on.ca/english/about/pubs/walkerton.

Index

NOTE: *f* indicates a figure; *t* indicates a table; A indicates an entry found in the Appendixes (on the CD).

Numerics

A

D

E

J

L

M

R

S

T

U

V

W

X